Akhyar · Syifaul Huzni ·
Mohd Iqbal
Editors

Proceedings of the 3rd International Conference on Experimental and Computational Mechanics in Engineering

ICECME 2021, Banda Aceh, October 11–12

Springer

Editors
Akhyar
Department of Mechanical Engineering
Syiah Kuala University
Darussalam, Indonesia

Syifaul Huzni
Department of Mechanical Engineering
Syiah Kuala University
Darussalam, Indonesia

Mohd Iqbal
Department of Mechanical Engineering
Syiah Kuala University
Darussalam, Indonesia

ISSN 2195-4356 ISSN 2195-4364 (electronic)
Lecture Notes in Mechanical Engineering
ISBN 978-981-19-3628-9 ISBN 978-981-19-3629-6 (eBook)
https://doi.org/10.1007/978-981-19-3629-6

This Springer imprint is published by the registered company Springer Nature Singapore Pte Ltd.
The registered company address is: 152 Beach Road, #21-01/04 Gateway East, Singapore 189721, Singapore

Lecture Notes in Mechanical Engineering

Lecture Notes in Mechanical Engineering (LNME) publishes the latest developments in Mechanical Engineering—quickly, informally and with high quality. Original research reported in proceedings and post-proceedings represents the core of LNME. Volumes published in LNME embrace all aspects, subfields and new challenges of mechanical engineering. Topics in the series include:

- Engineering Design
- Machinery and Machine Elements
- Mechanical Structures and Stress Analysis
- Automotive Engineering
- Engine Technology
- Aerospace Technology and Astronautics
- Nanotechnology and Microengineering
- Control, Robotics, Mechatronics
- MEMS
- Theoretical and Applied Mechanics
- Dynamical Systems, Control
- Fluid Mechanics
- Engineering Thermodynamics, Heat and Mass Transfer
- Manufacturing
- Precision Engineering, Instrumentation, Measurement
- Materials Engineering
- Tribology and Surface Technology

To submit a proposal or request further information, please contact the Springer Editor of your location:

China: Ms. Ella Zhang at ella.zhang@springer.com
India: Priya Vyas at priya.vyas@springer.com
Rest of Asia, Australia, New Zealand: Swati Meherishi at swati.meherishi@springer.com
All other countries: Dr. Leontina Di Cecco at Leontina.dicecco@springer.com

To submit a proposal for a monograph, please check our Springer Tracts in Mechanical Engineering at https://link.springer.com/bookseries/11693 or contact Leontina.dicecco@springer.com

Indexed by SCOPUS. All books published in the series are submitted for consideration in Web of Science.

More information about this series at https://link.springer.com/bookseries/11236

Contents

The Integration of Six Sigma and FMEA Methods in Determining Crude Palm Oil (CPO) Quality Improvement Factors at the Cot Girek Palm Oil Mill Unit of Pt. Perkebunan Nusantara-1

Muhammad Ikhsan[1](✉), Yuwaldi Away[1], and Irwansyah Irwansyah[2]

[1] Master of Industrial Engineering, Faculty of Engineering, Universitas Syiah Kuala, Banda Aceh, Indonesia
tgk.mhdikhsan92@gmail.com

[2] Department of Mechanical and Industrial Engineering, Universitas Syiah Kuala, Banda Aceh, Indonesia

Abstract. The Cot Girek Palm Oil Mill (POM) Unit of PT. Perkebunan Nusantara-1 adheres to high-quality standards in order to meet the company's objectives. According to the study's findings, there is still many data on the quality levels of Crude Palm Oil (CPO) that exceed the upper limit of the standards set, such as in 2019, the average free fatty acid (FFA) content of 3.59%, moisture content of 0.26%, average free fatty acid (FFA) content of 3.65%, and moisture content of 0.27% in 2020. Meanwhile, CPO impurity levels remain within the company's control limits. The purpose of this study is to prioritize actions to increase CPO quality. According to the findings of this study, the raw material of fresh fruit bunches (FFB) containing free fatty acids and high moisture content with an RPN value of 504 is one of the elements that reduce CPO quality levels. Furthermore, the equipment at the continuous clarifier station, pure oil tank, oil purifier, and vacuum dryer do not operate efficiently, preventing them from effectively separating water, impurity, and CPO with an RPN value of 448.

Keywords: CPO · *Six Sigma* · FMEA (*Failure Mode and Effect Analysis)*

1 Introduction

The international market competitiveness for palm oil exports has resulted in Indonesian Crude Palm Oil (CPO) goods with quality that meets international market demand. The Cot Girek Palm Oil Mill (POM) Unit of PT. Perkebunan Nusantara-1 is a company in the Crude Palm Oil (CPO) industry founded in 1996 and is located in Cot Girek Lama Village, Cot Girek subdistrict of North Aceh Regency. In order to meet market demand, the Cot Girek POM Unit of PT. Perkebunan Nusantara-1 employs CPO quality standards during the manufacturing process. The CPO quality criteria for PT. Perkebunan Nusantara-1's Cot Girek POM Unit include free fatty acid (FFA) maximum of 3.5%,

Akhyar et al. (Eds.): ICECME 2021, LNME, pp. 1–13, 2023.
https://doi.org/10.1007/978-981-19-3629-6_1

maximum moisture content 0.15%, and impurity content 0.02%. While the goal of PT. Perkebunan Nusantara is to improve quality in order to compete in the international market and increase profits; the company uses CPO quality standards with a maximum FFA category of 3.5%, a maximum moisture content of 0.15%, and an impurities content of 0.02%.

Meanwhile, based on the data obtained, the average achievement of CPO quality levels of Cot Girek POM Unit of PT. Perkebunan Nusantara-1 in 2019 is an average FFA content of 3.59%, a moisture content of 0.26%, and an average FFA content of 3.65%, a moisture content of 0.27% in 2020. To increase the CPO quality of the Cot Girek POM Unit of PT. Perkebunan Nusantara-1 control efforts must be carried out, as well as an analysis of the causes of the drop in CPO quality and modifications to the work system. Meanwhile, regulating and supervising the work system in the company is one method that can be utilized to improve the quality of CPO so that it can meet the criteria that have been set. The method used in quality control is the Six Sigma method through the Define, Measure, Analyze, Improve, Control and Failure Mode and Effect Analysis (FMEA) methods.

2 Research Methodology

The research was carried out at the Cot Girek POM Unit of PT. Perkebunan Nusantara-1 in Cot Girek Lama Village, Cot Girek subdistrict, North Aceh Regency. This study collected data on FFA levels, moisture content, and CPO impurity levels.

The following are the stages of data analysis performed using the Six Sigma and FMEA (Failure Mode and Effect Analysis) methods:

1. Define
   Defining Six Sigma projects and creating SIPOC diagrams at this stage determines the proportion of defects that are the most significant cause of damage and the source of production failure.
2. Measure
   The measurement phase is carried out through several stages by taking data on CPO production at the Cot Girek POM Unit of PT. Perkebunan Nusantara-1. In the meantime, this stage is carried out by establishing the quality features of essential workstations, developing a data collection plan, and testing the performance baseline for data output levels.
3. Analyze
   The analysis is carried out at the analyze stage to determine the sources of quality problems on the capabilities of the CPO production process by using Setting performance targets from quality characteristics (CTQ), Cause-effect diagrams (Fishbone Diagram).
4. FMEA
   Based on the highest RPN value, the FMEA determines the importance of the causes causing the quality deterioration that must be remedied promptly.

5. Improve
   The 5W+1H approach will be employed in the analysis that will be used at the stage of building this action plan; with this method, it is hoped that this plan may be used to implement quality improvement.
6. Control
   At this stage, the objectives are to determine process capability or sigma level, document and disseminate the results of improving CPO quality, standardize successful best practices in improving CPO quality, document procedures and serve as standard work guidelines.

The steps to be carried out in this study can be seen in the description of the diagram below:

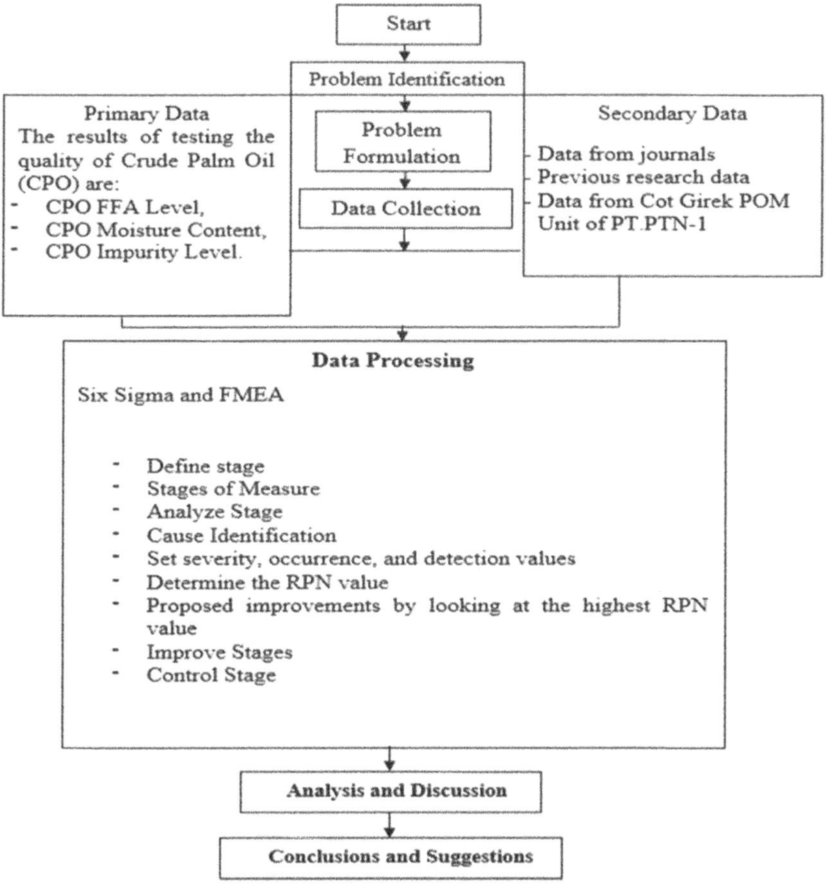

Fig. 1. Research flow chart

3 Research Result

3.1 Data Collection

The data collected in this study is from the Quality Control part, which includes free fatty acid content (FFA), moisture content, impurities content, and reference data for enhancing CPO quality using PTPN-1 standards and SNI 7182:2015 biodiesel and SNI 01-2901-2006 CPO foodstuffs. In the meanwhile, the information is presented in the table below.

Table 1. Achievements and quality standards of CPO according to SNI and PTPN-1

Quality Criteria	CPO Standard			average PTPN-1 achievement	
	SNI 7182:2015 biodiesel	SNI 01-2901-2006 CPO foodstuffs	PTPN-1 Standard	2019	2020
FFA	Max 0.5%	Max 0.5%	2.5%–3.5%	3.59%	3.65%
water	Max 0.05	Max 0.5%	Max 0.15%	0.26%	0.27%
Impurities	Max 0.05%	Max 0.5%	Max 0.02%	0.14%	0.15%

3.2 Data Processing

The Six Sigma and Failure and Mode Effect Analysis (FMEA) method were used to process the data in this study. The following are the stages of data processing in this study:

1. Define

In this study, the first stage is referred to as 'define.' Several levels of analysis are also used at this step, namely, by deciding on and defining the Six Sigma project. This research's Six Sigma project is in the CPO Quality control section. The Cot Girek POM Unit of PT. Perkebunan Nusantara-1 tested the quality of CPO on CPO Quality Control using three criteria. The three criteria are free fatty acid (FFA) levels, the moisture content in CPO, and impurity levels in CPO. Quality standards are set by the Cot Girek POM Unit PT Perkebunan Nusantara-1 is FFA content of 2.5–3.5%, maximum moisture content of 0.15% and impurities content of 0.02.

2. Measure

The 'measure' stage attempts to establish performance requirements for the Cot Girek POM Unit of PT. Perkebunan Nusantara-1. At the Cot Girek POM Unit, the performance standard that serves as the measurement target is the CPO quality standard. The quality requirements that will be followed are free fatty acids (FFA), moisture content, and impurity levels.

2.1 An X Control map and R content (FFA), Moisture content and Impurity Content

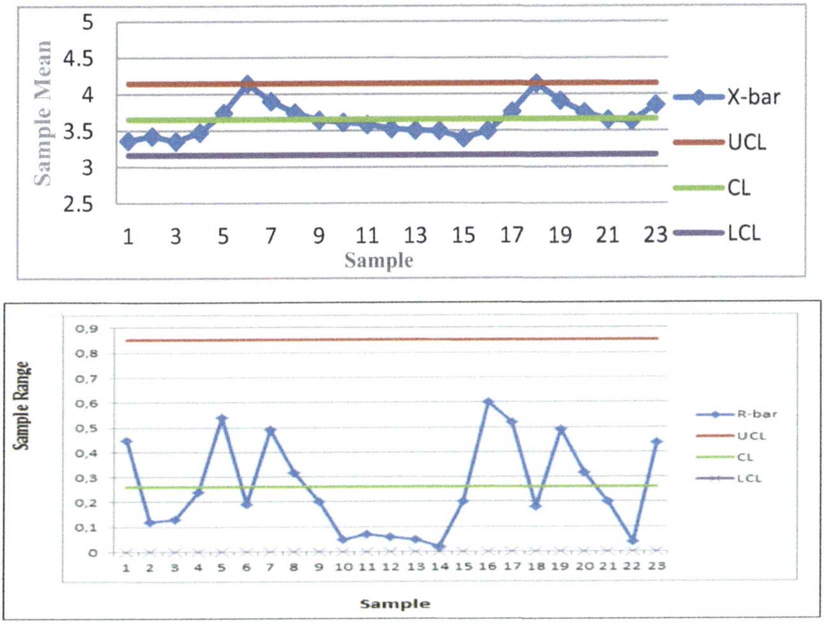

Fig. 2. X-bar Map and R-bar control for FFA levels

According to Fig. 2, all data are within the control limits that have been established, and there is no need to make adjustments or enhancements before proceeding to the next testing stage.

Figure 3 demonstrates that all data on CPO moisture content is within the control limits that have been established, both the upper and lower control limits, indicating that no adjustments or enhancements are required and that the next test phase can begin immediately.

According to Fig. 4, the CPO impurity content data is still beyond the upper control limit. Therefore analysis and improvement are required to determine the source of the data exceeding the upper control limit of the X control chart determination. There is no need to alter the R-bar map because all data is within control limits. The X-bar map can be revised by removing data that is outside the control limits. Meanwhile, Fig. 5 shows the outcomes of the improved X-bar control map.

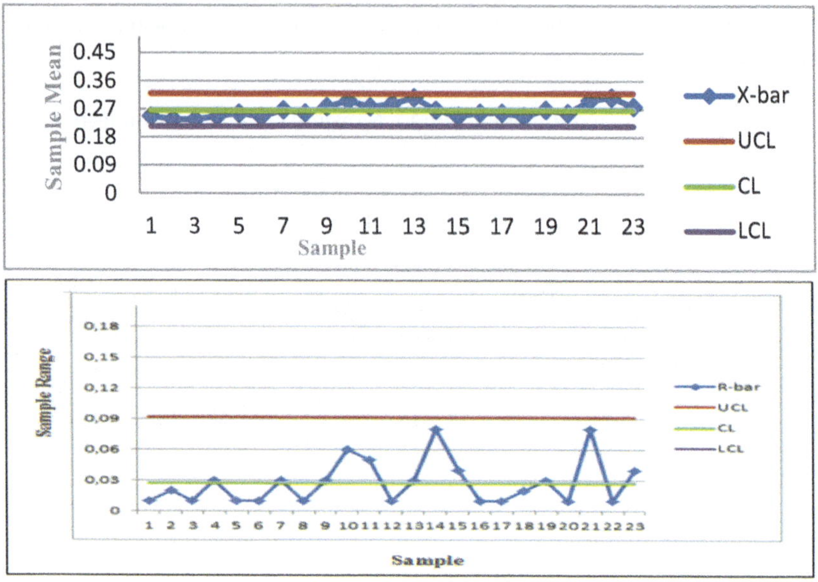

Fig. 3. X-bar map and R-bar control of moisture content

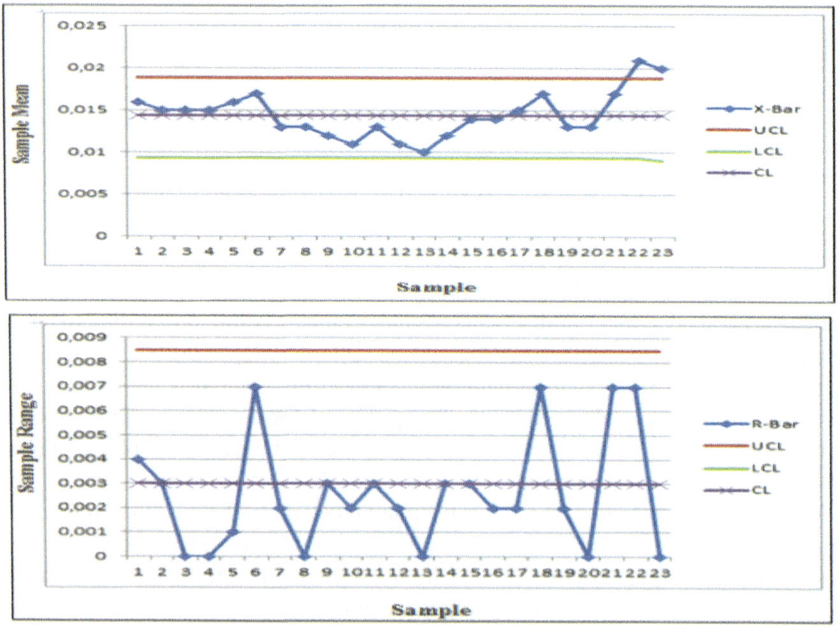

Fig. 4. X-bar map and R-bar control of impurities

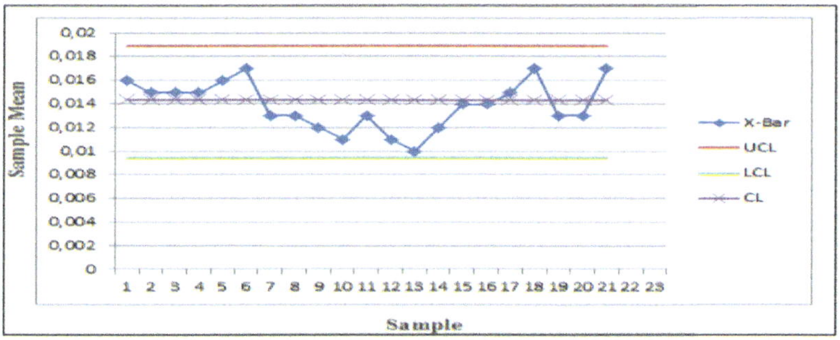

Fig. 5. X-bar control map after revision

2.2 Capability Process, DPMO dan Six Sigma

a. Capability Process, DPMO and Six Sigma for FFA content

$$Capability\,Process(CP) = \frac{USL - LSL}{6\sigma} = \frac{3,77 - 3,52}{1.38} = 0.18$$

The probability of a failure rate that is above the USL per million opportunities (DPMO) is:

$$\text{DPOM} = \text{P}\{\text{Z} \geq \frac{USL - \underline{\underline{X}}}{S}\} \times 1,000,000 = \text{Z} \geq$$
$$\frac{3,77 - 3,52}{0,230}\frac{3.77 - 3,65}{0.230}\frac{3,77 - 3,66}{0,230}\frac{USL - \underline{\underline{X}}}{S}\} \times 1,000,000$$
$$= \text{P}\{\text{Z} \geq 0.52\frac{USL - \underline{\underline{X}}}{S}\} \times 1,000,000$$
$$= 1 - \{\text{Z} \geq 0.67847\frac{USL - \underline{\underline{X}}}{S}\} \times 1,000,000$$
$$= 301530$$

Possible failure rates that are above LSL per million opportunities (DPMO) are:

$$\text{DPOM} = \text{P}\{\text{Z} \leq \frac{\underline{\underline{X}} - LSL}{S}\} \times 1,000,000 = \text{Z} \geq \frac{3.65 - 3.52}{0,230}\frac{USL - \underline{\underline{X}}}{S}\} \times 1,000,000$$
$$= \text{P}\{\text{Z} \geq 0.57\frac{USL - \underline{\underline{X}}}{S}\} \times 1,000,000$$
$$= 1 - \{\text{Z} \geq 0.71566\frac{USL - \underline{\underline{X}}}{S}\} \times 1,000,000$$
$$= 284340$$

$$\text{Value of Sigma}(\sigma) = \text{Normsinv} = \text{DPMO USL} + \text{DPMO LSL}$$
$$= 301530 + 284340.$$
$$= 585870.$$

Value of Sigma $(\sigma) = \text{Normsinv}\,\{\frac{1.000.000 - DPMO}{1.000.000}\frac{1.000.000 - DPMO}{1.000.000}\frac{10^6 - DPOM}{10^6}\} +$ 1,5

Value of Sigma (σ) = Normsinv $\{\frac{1.000.000-137860}{1.000.000}\frac{1.000.000-137860}{1.000.000}\frac{1.000.000-422200}{1.000.000}\frac{1.000.000-422200}{1.000.000}\frac{1.000.000-585870}{1.000.000}\}$ + 1,5

Value of Sigma (σ) = 2,1

The process capability value is 0.18, which is still less than the maximum value of the process capability, 1.33, as shown by the calculations above. While the sigma value is 2.1, this indicates that the Cot Girek POM Unit of PT. Perkebunan Nusantara-1's level of achievement in meeting the predetermined FFA CPO quality standards is in the range of 70% to 75%, and the Cot Girek POM Unit of PT. Perkebunan Nusantara-1 must improve its performance system in order to meet the standards that have been established.

b. Capability Process, DPMO and Six Sigma for moisture content

$$CapabilityProcess(CP) = \frac{USL - LSL}{6\sigma} = \frac{0.28 - 0.25}{0.015} = 0.2$$

The probability of a failure rate that is above the USL per million opportunities (DPMO) is:

$$\text{DPOM} = \text{P}\{\text{Z} \geq \frac{USL - \underline{X}}{S}\} \times 1,000,000 = \text{Z} \geq$$
$$\frac{0,15 - 0,27}{0,025}\frac{0,28 - 0,27}{0,025}\frac{0.28 - 0.25}{0.025}\frac{3,77 - 3,52}{0,230}\frac{3,77 - 3,66}{0,230}\frac{USL - \underline{X}}{S}\} \times 1,000,000$$
$$= \text{P}\{\text{Z} \geq 0.56\frac{USL - \underline{X}}{S}\} \times 1,000,000$$
$$= 1 - \{\text{Z} \geq 0.71226\frac{USL - \underline{X}}{S}\} \times 1,000,000$$
$$= 287740$$

The probability of failure rate above LSL per million opportunities (DPMO) is:

$$\text{DPOM} = \text{P}\{\text{Z} \leq \frac{\underline{X} - LSL}{S}\} \times 1,000,000 = \text{Z} \geq \frac{0,27 - 0,25}{0,027}\frac{0.27 - 0.25}{0.025}\frac{USL - \underline{X}}{S}\} \times 1,000,000$$
$$= \text{P}\{\text{Z} \geq 0.8\frac{USL - \underline{X}}{S}\} \times 1,000,000$$
$$= 1 - \{\text{Z} \geq 0.78814\frac{USL - \underline{X}}{S}\} \times 1,000,000$$
$$= 211860$$

$$\begin{aligned}\text{Value of Sigma}(\sigma) &= \text{Normsinv} = \text{DPMOUSL} + \text{DPMO LSL}.\\ &= 287740 + 211860.\\ &= 499600.\end{aligned}$$

Value of Sigma (σ) = Normsinv $\{\frac{1.000.000-DPMO}{1.000.000}\frac{1,000,000-DPMO}{1,000,000}\frac{10^6-DPOM}{10^6}\}$ + 1.5

Value of Sigma (σ) = Normsinv $\{\frac{1.000.000 - 137860}{1.000.000}\frac{1.000.000 - 137860}{1.000.000}\frac{1.000.000 - 422200}{1.000.000}\frac{1.000.000 - 422200}{1.000.000}\frac{1.000.000 - 585340}{1.000.000}\frac{1.000.000 - 585340}{1.000.000}\frac{1,000,000 - 499600}{1,000,000}$

+ 1.5

Value of Sigma (σ) = 2.0

The process capability value is 0.2, which is still far below the maximum process capability value of 1.33, according to the results of the foregoing computations. While the sigma value is 2.0, this indicates that the Cot Girek POM Unit of PT. Perkebunan Nusantara-1 has achieved 60-70% of the established CPO water quality standards, and that the Cot Girek POM Unit of PT. Perkebunan Nusantara-1 needs to improve its performance system in order to meet the standards that have been set.

c. Capability Process, DPMO and Six Sigma for impurity levels

$$CapabilityProcess(CP) = \frac{USL - LSL}{6\sigma} = \frac{0.02 - 0}{0.0133} = 1.49$$

The probability of a failure rate that is above the USL per million opportunities (DPMO) is:

$$\text{DPOM} = \text{P}\{\text{Z} \geq \frac{USL - \underline{X}}{S}\} \times 1,000,000 = \text{Z} \geq$$
$$\frac{0,028 - 0,27}{0,027}\frac{0,28 - 0,27}{0,027}\frac{0,0150 - 0,0137}{0,0022}\frac{0,02 - 0,016}{0,0022}\frac{0.02 - 0.016}{0.0022}\frac{3,77 - 3,52}{0,230}$$
$$\frac{3,77 - 3,66}{0,230}\frac{USL - \underline{X}}{S}\} \times 1,000,000$$
$$= \text{P}\{\text{Z} \geq 1.95\frac{USL - \underline{X}}{S}\} \times 1,000,000$$
$$= 1 - \{\text{Z} \geq 0.9744\frac{USL - \underline{X}}{S}\} \times 1,000,000$$
$$= 25600$$

The probability of failure rate above LSL per million opportunities (DPMO) is:

$$\text{DPOM} = \text{P}\{\text{Z} \leq \frac{\underline{X} - LSL}{S}\} \times 1,000,000 = \text{Z} \geq$$
$$\frac{0,27 - 0,25}{0,027}\frac{0.016 - 0}{0.0022}\frac{USL - \underline{X}}{S}\} \times 1.000.000$$
$$= \text{P}\{\text{Z} \geq 7.14\frac{USL - \underline{X}}{S}\} \times 1,000,000$$
$$= \{\text{Z} \geq 0\frac{USL - \underline{X}}{S}\} \times 1,000,000$$
$$= 0$$

$$\text{Value of Sigma}(\sigma) = \text{Normsinv} = \text{DPMO USL} + \text{DPMO LSL}.$$
$$= 25600 + 0.$$
$$= 25600.$$

Vaue of Sigma (σ) = Normsinv $\{\frac{1.000.000 - DPMO}{1.000.000}\frac{1,000,000 - DPMO}{1,000,000}\frac{10^6 - DPOM}{10^6}\} +$ 1.5

$$\text{Value of Sigma }(\sigma) = \text{Normsinv}\left\{\frac{1.000.000 - 137860}{1.000.000}\ \frac{1.000.000 - 137860}{1.000.000}\ \frac{1.000.000 - 422200}{1.000.000}\ \frac{1.000.000 - 422200}{1.000.000}\ \frac{1.000.000 - 585340}{1.000.000}\ \frac{1.000.000 - 585340}{1.000.000}\ \frac{1.000.000 - 568680}{1.000.000}\ \frac{1,000,000 - 25600}{1,000,000}\right\} + 1.5$$

Value of Sigma (σ) = 2.5

Based on the calculation results above, it can be seen that the process capability value is 1.49 and has exceeded the maximum process capability value, which is 1.33. In contrast, the sigma value is 2.5, which means that the level of achievement of the Cot Girek POM Unit of PT Perkebunan Nusantara-1, in meeting the established CPO moisture quality standards, is in the range of 70%–90% and the Cot Girek POM unit of PT. Perkebunan Nusantara-1 must improve its performance system in order to improve quality in order to meet the standards that have been set.

3. Analyze

At this stage, to examine the problem of CPO quality at the Cot Girek POM Unit of PT. Perkebunan Nusantara-1, a cause and effect diagram or a Fishbone diagram, as well as an analysis of Cause and Effect Diagrams, must be used.

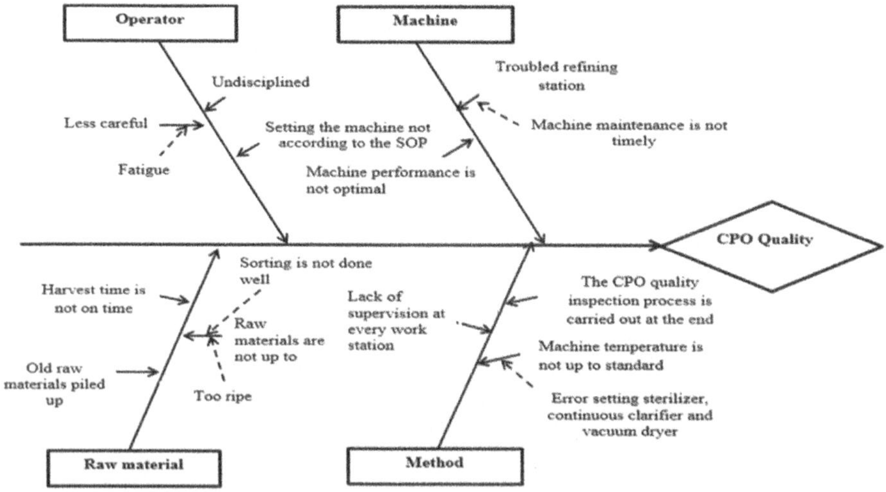

Fig. 6. Cause-and-effect diagram of FFA levels, water and impurities

1. Failure Mode and Effect Analysis (FMEA)
   The steps for implementing FMEA are (1) Identifying processes and products, (2) Listing potential problems that can arise, (3) Assessing problems based on the severity of the failure (severity), the probability of failure (occurrence), and the ability to detect failures (detection), (4) Calculating the Risk Priority Number or RPN and

prioritizing actions starting with the problem that has the most significant RPN value and (5) Taking action to reduce risk in the form of problem-solving. As for how to calculate the value of RPN (Risk Priority Number) for the failure mode that causes a decrease in the quality of CPO at the Cot Girek POM Unit of PT. Perkebunan Nusantara-1 is as follows:

Table 2. Calculation of RPN Value

	S	O	D	RPN	How to overcome
Average free fatty acid content > 3.5%	9	8	7	504	Guide the plantation coordinator and FFB providing agents on good palm oil harvesting practices, as well as the effect of time and location on oil palm quality
Average moisture content > 0.15%	8	7	7	392	Assist the field coordinator and supplier agents in understanding how to harvest high-quality palm oil
The level of impurities is still present > 0.02%	8	7	5	280	Always supervise and educate the operator of the clarifying station to be aware of the machine's status at the station

2. Improve
   The next step is to design improvement strategies and set goals once the dominating cause has been identified. Improvement plans are created by using brainstorming techniques to come up with a variety of alternate problem-solving ideas. Several employees from the company contributed to the creation of this improvement plan. The improvement strategy is laid out in a matrix format based on the 5 W (why, what, where, when, and who) and 1 H (how) principles, spelt out in great detail. Based on the findings of interviews utilizing the 5 W and 1 H approaches, the following are the results of enhancing the quality of CPO at the Cot Girek POM Unit of PT. Perkebunan Nusantara-1's:
3. Control
   Control and oversight of implemented changes are carried out at this stage to ensure that there are no more process failures that will cause the quality of CPO to deteriorate. Documentation, records, and references to corrective actions are required in the control analysis so that failures that have resulted in low CPO quality will not occur again.

Table 3. Proposed improvements

Criteria	Proposed improvements
Average free fatty acid (FFA) content > 3.5%	– Provide insight into how palm oil raw materials adhere to industry standards. – Explain the significance of machine upkeep. – Improve control of digester/blending stations.
Average Moisture Content > 0.15%	– Provide an understanding of the palm piles that generate increasing moisture content. – Monitor and maintain machines regularly. – Explain how to set the machine following the SOP.
CPO impurities are still present > 0.02%	– Provide understanding to the production process operator in order to run the machine according to the SOP. – Before beginning the manufacturing process, teach the operator of the clarity station how to use the repair and cleaning system.

Table 4. Control and improvement plan

Improvement plan	Control mechanism
Conduct coaching and training to improve HR	Conduct training and coaching to impart a grasp of theory and practice to develop the skills and expertise of operators at work
Selection of people's FFB purchases and sorting with a maximum of FFB from company plantations	Provide understanding and warning to palm oil suppliers from the people to enter good FFB
SOP for using the machine	Provide the operator with an understanding of the method for utilizing the machine in line with the SOP and not being careless in using the machine
Machine maintenance system	Implement a Total Productive Maintenance (TPM) system to ensure that the condition of machines and equipment is always maintained and not easily damaged

4 Conclusions

Based on the research that has been done related to quality engineering, some things that can be concluded are as follows:

1. According to the data analysis results, the process capability value for free fatty acid (FFA) is 0.18, and moisture content is 0.2, indicating that the process capability value

for FFA and moisture content is deficient. The impurities content, which is 1.49, is also extremely good for process capability.

2. Meanwhile, based on the analysis using the Six Sigma method, it is known that the sigma value of FFA content is 2.1, the moisture content is 2.0, and the impurities content is 2.5, indicating that the level of sigma value achieved by the Cot Girek POM unit of PT. Perkebunan Nusantara-1 is still low and still has a high failure rate in improving the quality of CPO until it reaches the standard that has been set.
3. According to the analysis utilizing a causal diagram, the reason for the reduction in CPO quality at the Cot Girek POM Unit of PT. Perkebunan Nusantara-1 is that the FFB sorting process does not adhere to the method; as a result, workers are not performing correctly, and the operator is not disciplined and lacks human resources at work. While the proposed improvements to improve the quality of CPO so that it reaches the standards that have been set are by increasing supervision and control of the FFB sorting process, providing counselling to plantations and oil palm farmers about the correct oil palm harvesting process, emphasizing the discipline of FFB sorting operators; Clarification station operators carry out routine repairs and maintenance of machines at the clarification station.

References

1. Kurniawan, W., dan Ricky Saputera, D.S.: Usulan penerapan metode six sigma untuk meningkatkan mutu Crude palm oil (cpo) di pt. X. [The proposed application of the six sigma method to improve the quality of Crude Palm Oil (CPO) at pt. X]. Jurnal ilmiah teknik industri **5**(2), 85–91 (2017). Teknik industri fti universitas trisaktil
2. Ishak, A., Naibaho, H.: *Quality Control* of impurity content on crude palm oil using define, Measure, analyze, improve, control method and grey failure mode and effect analysis. Industrial engineering department, universitas sumatera utara Seminar nasional vii manajemen & rekayasa kualitas (2018)
3. Christian, W.: Pengendalian kualitas Dengan metode statistik pada xyz [Quality control by statistical method on xyz]. Calyptra jurnal ilmiah mahasiswa universitas surabaya **2**(2) (2013). Jurusan manajemen fakultas bisnis dan ekonomika universitas surabaya
4. Fachruddin, A., Rahayu, L.: Evaluasi Prasyarat Keberhasilan Sistem Resi Gudang di Kabupaten Bantul [Evaluation of the prerequisites for the success of the warehouse receipt system in Bantul Regency]. Penerbit AGRARIS-J. Agribusiness Rural Dev. Res. **3**(2) (2017)
5. Siregar, K., Ishak, A., Sinaga, H.A.: Quality control of crude palm oil (CPO) using define, measure, analyze, improve, control (DMAIC) and fuzzy failure mode and effect analysis. Industrial Engineering Department, Faculty of Engineering, Universitas Sumatera Utara, Medan, Indonesia (2018)
6. Hasibuan, E.P.: Bagan Alir Proses Pengolahan Kelapa Sawit. [Palm Oil Process Flow Chart] Tugas Teknologi Pengolah Hasil Pertanian Universitas Sumatra Utara Medan (2018)

Simulation of Early Virtual Manufacturing for Polytechnic State of Bandung

Fachri Koeshardono(✉), Muhammad Indra Fikri, Muhammad Idham, Bagus Sasmito Haryo Yudhanto, Haidar Nashir Aunurrahman, and Aril Gunawan

Polban, Jalan Gegerkalong Hilir, Ciwaruga, Bandung Barat, West Java, Indonesia
{fachri.koeshardono,muhammad.indra.pmf17,muhammad.idham.pmf17, haidar.nashir.pmf17,aril.gunawan.pmf17}@polban.ac.id

Abstract. Virtual manufacturing gained momentum from the emerging technology of information. With the current infrastructure, it is possible to start the concept for Polytechnic both as education tools and productions. In this study, the current resources in 2 workshops are simulated by Tecnomatix in manufacturing 4 products simultaneously. Those products (Plant Stander, Coffee Mixture Machine, Cassava Peeler and Analog Rice Machine) differ by the number of parts. The result is that each day these 2 workshops could produced 78 Plant Standers, 9 Coffee Mixture Machines, 6 Cassava Peelers and 9 Analog Rice Machines. It is also planned that the data could be gathered in real time by the installation of cameras from the layouts of workshops.

Keywords: Virtual manufacturing · Tecnomatix · Mechanical product · Real time monitoring

1 Introduction

Virtual Manufacturing is developed to improve the whole industrial process. The concept itself has been developed for several decades. It could gain the momentum from the emerging of Information Technology recently. The concept is explained in its scope, socio-economic aspects also the future by Chablat [1]. There also three main aspects in manufacturing: Logistics, Production and Assembly. Souza [2] developed this concept by divide the product data into three categories: tolerance and specifications, geometric and manufacturing information. Hence exchange those data between process planning, resources management, CN programming and validations, neutral type file is utilized. Those neutral files have its own advantages and disadvantages. The data then simulated in virtual reality developed from C++ and the library from PC and Java. Virtual Reality Environment provide the immersive experience of the floor shop. The similar concept with more advanced tool developed by Peng [3]. It is called Networked Virtual Manufacturing (net-VM). It consists of System Engine, Input/Output Control and User Interface, Virtual Manufacturing database and data Network. Tolio [4] developed it also by integrating Computer Aided Design (CAD), Computer Aided Manufacturing

Akhyar et al. (Eds.): ICECME 2021, LNME, pp. 14–23, 2023.
https://doi.org/10.1007/978-981-19-3629-6_2

(CAM) and Computer Aided Process Planning (CAPP). It verified by using NC Simulation Software VERICUT. Horvárth [5] attempted to integrate manufacturing process modeling by virtual manufacturing. The model itself came from the tool for evaluation of manufacturability of mechanical products. Virtual Manufacturing not only as the tool in industry. Li Wang [6] use the virtual manufacturing concept to innovate in teaching manufacturing technology course in Jilin University.

2 Materials

The students are given the task to design products using CAD. CATIA is used as the designing tool. There were some surveys about the demand of product in the society especially from the farmers. In conclusion, 4 products are planned to be manufactured: Plant-Stander (a), Coffee Mixing Machine (b), Cassava Peeler Machine (c) and Analog Rice Machine (d) as seen in Fig. 1. The products are sorted based on its complexities in terms of number of parts and the manufacturing processes needed.

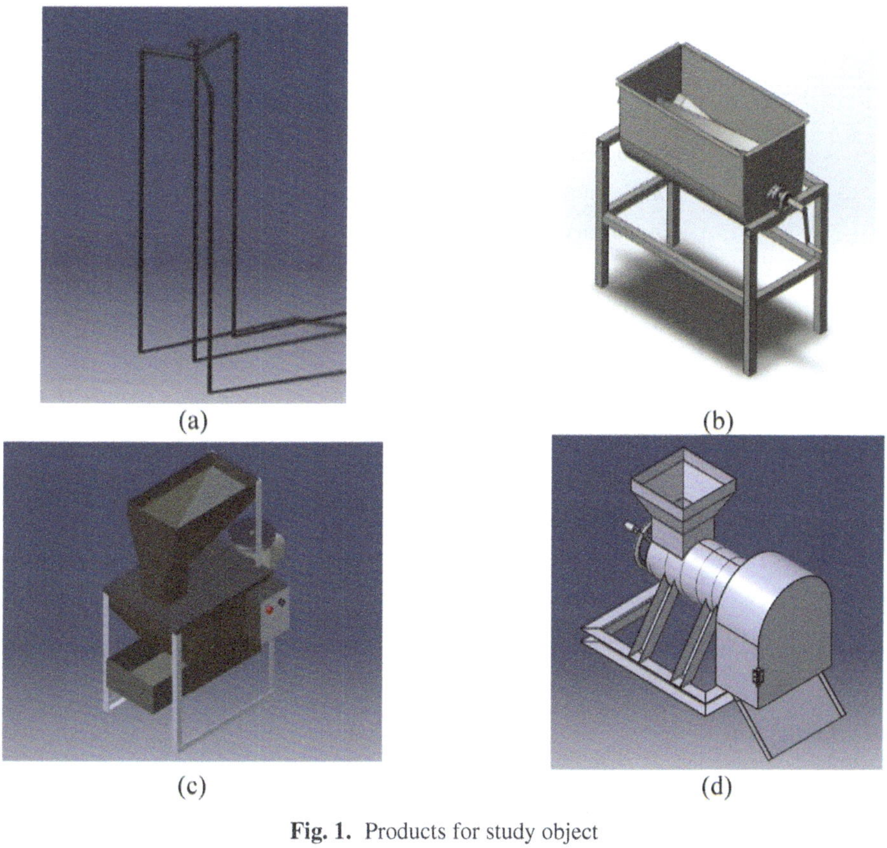

(a) (b)

(c) (d)

Fig. 1. Products for study object

The estimation time for each process of making each product and the time needed is presented in Table 1 through Table 4. Plant Stander has the lowest processing time while Cassava Peeler the highest one. Some processes in the 3 products are similar, like drilling and turning. Hence, there is an opportunity to use the same machine to manufacture different products and almost at the same time.

Table 1. Plant-stander manufacturing processing time

No	Manufacturing tools	Processing time (minute)
1	Measuring equipment	13
2	Hand grinding	13
3	Assembly equipment	4

Table 2. Coffee mixing machine manufacturing processing time

No	Manufacturing tools	Processing time (minute)
1	Hand grinding	73
2	Bending machine	5
3	Rolling machine	6
4	Drilling machine	36
5	Turning machine	28
6	Welding machine	80
7	Assembly equipment	20

Table 3. Cassava peeler machine manufacturing processing time

No	Manufacturing tools	Processing time (minute)
1	Hand grinding	15
2	Bending machine	57
3	Drilling machine	8
4	Shearing machine	63
5	Pressing tool	15
6	Turning machine	8
7	Welding machine	47
8	Assembly equipment	21

Table 4. Analog Rice Machine manufacturing processing time

No	Manufacturing tools	Processing time (minute)
1	Hand grinding	51
2	Turning machine	10
3	Drilling machine	10
4	Bending machine	49
5	Shearing machine	18
6	Welding machine	65
7	Assembly equipment	19

From those tables, the requirement for the machines is mapped. All of them are provided in Polytechnic State of Bandung in 2 shops: Machining and Fabrication. The layouts of those labs are as in Fig. 2 and Fig. 3.

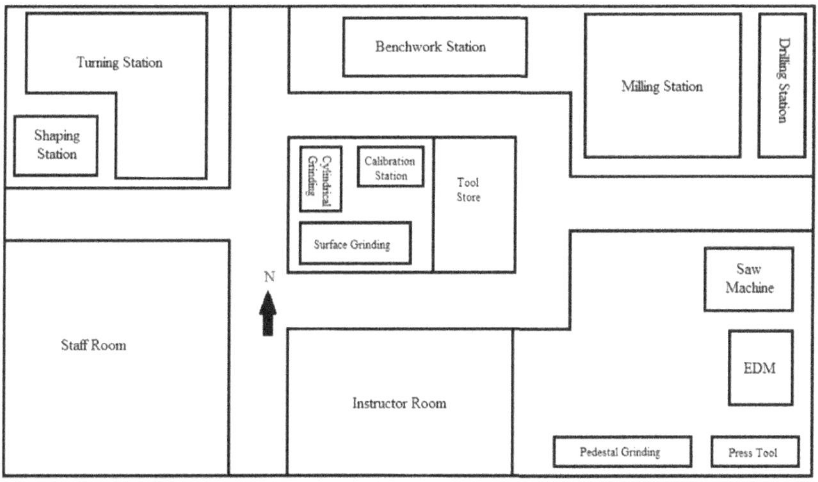

Fig. 2. Machining shop layout

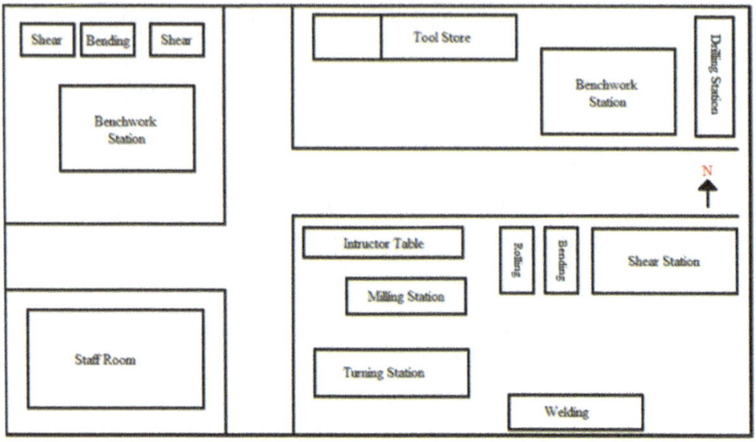

Fig. 3. Fabrication workshop layout

Machining Shop working stations are as followed: Turning Station, Shaping Station, Benchwork Station, Milling Station, Saw Machine, EDM, Grinding and Press Tool. It is to be noted that the tools are supplied by station in the middle. Monitoring center is in that area. The second lab is Fabrication Shop. The working stations are as followed: Benchwork Station, Shear, bending, Milling, Turning, Rolling Drilling and Welding. The machining lab is the center for parts that is machined while fabrication for plate working and welding.

3 Methods

It is assumed that the production is embedded with the studying process in Polytechnic. It is continuous process for Plant Stander. On the other hand, Coffee Mixing, Cassava Peeler and Analog Rice are planned for 2, 4 and 2 shifts respectively. The software utilized is Tecnomatix from Siemens. The data from Table 1 through Table 4 then inputted in module provided with the assumptions from each shift (Fig. 4).

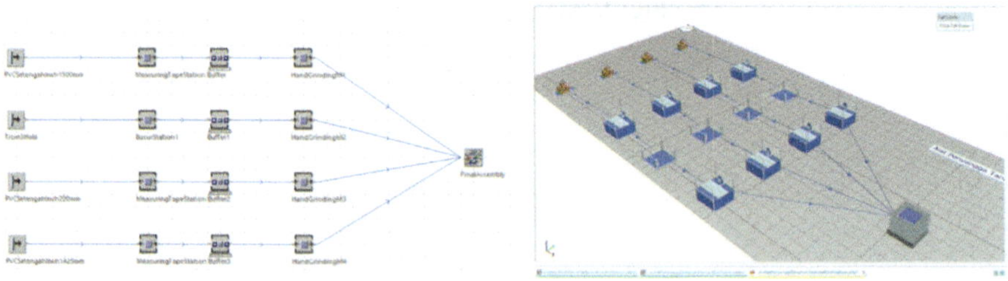

Fig. 4. Plant stander process input

In Plant Stander, there are 4 parts as input. This product could be used as the practice for the student in the first year to familiarized with production process in the shop since it only need hand grinding and measuring tools. The process getting more complex with Coffee Mixing Machine with three shifts. First shift is from 07.00–12.00, second shift 13.00–15.00. In the first shift, Sub Assembly 1 and 2 could be produced parallel and the assembly process at second shift. Same arrangement almost identical for Cassava Peeler with different time management. The first shift is divided into two parts, from 07.30–09.30 and 09.30–12.00. The first one for Sub Assembly 1 and second one for Sub Assembly 2. The difference also, it takes longer to produce the Cassava Peeler so Sub Assembly 3 at 13.00–14.00 with the assembly process from 14.00–16.00. The process for Analog Rice Machine is not as complex in previous machine so it only needs 2 shifts as Coffee Mixing Machine. The sequence for each shift I resented in Fig. 5, Fig. 6 and Fig. 7.

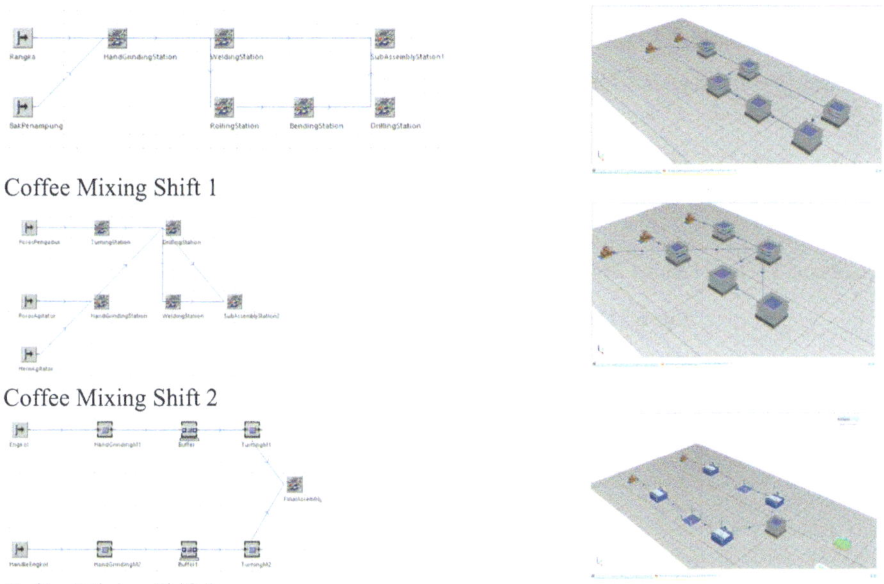

Fig. 5. Data input in Tecnomatix for Coffee Mixing

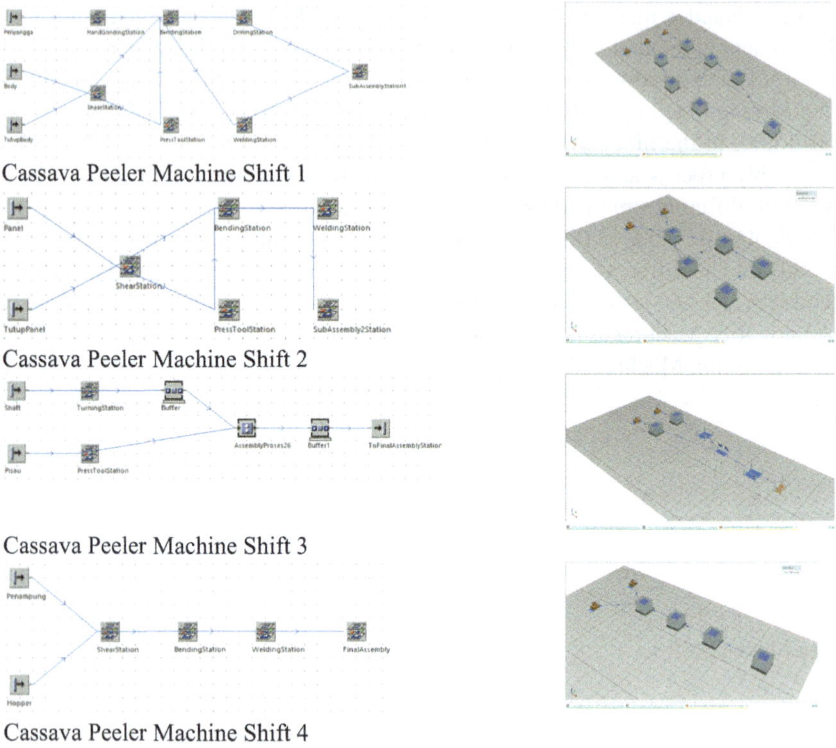

Fig. 6. Data input in Tecnomatix for Cassava Peeler

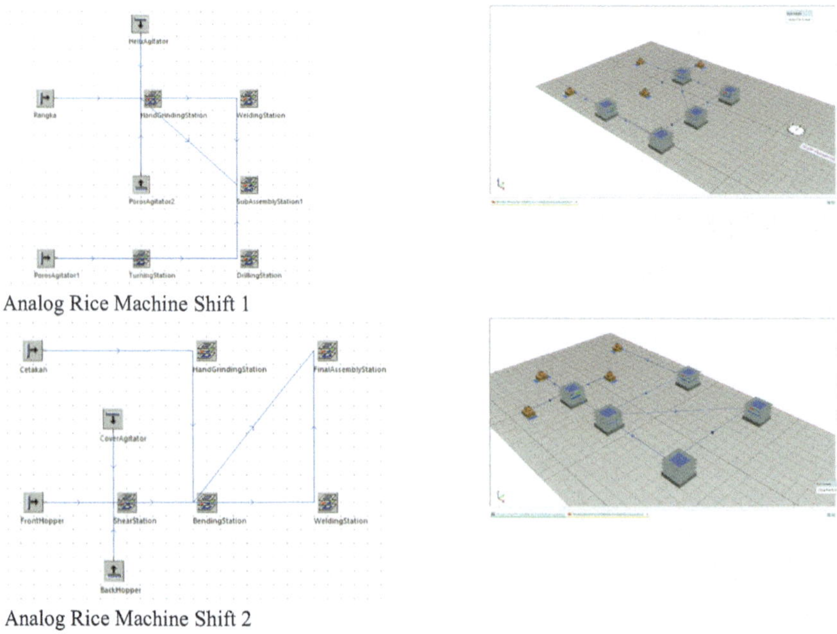

Fig. 7. Data input in Tecnomatix for Analog Rice

4 Results

Simulation is run to clarify the maximum capacities of the shops in producing those products in one day. It is needed as early feasibility study for the shops in Polytechnic has the required capability. The goal is the product could be used as study case in several subjects in the lecturing process. It could be concluded that the quantity per day for each product is: 78 Plant Standers, 9 Coffee Mixture Machines, 6 Cassava Peelers and 9 Analog Rice Machines. However, it needs careful consideration since the assumption that the buffer is quite small. The detail of quantity for each subassembly and its assembly is presented in Table 5, 6, 7 and 8.

Table 5. Plant-stander quantity

Shift	Time schedule	Product	Unit
1	07.00–16.00	Plant stander assembly	78

Table 6. Coffee mixing quantity

Shift	Time schedule	Product	Unit
1	07.00–12.00	Coffee Mixing Machine Sub Assembly 1	9
1	07.00–12.00	Coffee Mixing Machine Sub Assembly 2	9
2	13.00–15.00	Coffee Mixing Machine Assembly	9

Table 7. Cassava peeler quantity

Shift	Time schedule	Product	Unit
1	07.30–09.30	Cassava Peeler Sub Assembly 1	8
2	09.30–12.00	Cassava Peeler Sub Assembly 2	9
3	13.00–14.00	Cassava Peeler Sub Assembly 3	7
4	14.00–16.00	Cassava Peeler Assembly	6

Table 8. Analog rice quantity

Shift	Time schedule	Product	Unit
1	07.30–09.30	Analog Rice Machine Sub Assembly 1	9
2	09.30–12.00	Analog Rice Machine Assembly	10

There is a difference in Assembly and its Subassembly in Cassava Peeler and Analog Rice, so the result needs to be confirmed further. There is an urgency further study by examining in the simulation result. Another step for this study is by gathering data from the floor by installing camera in the workshops. It is needed to give more virtually environment in shops. The planned points of the camera are as Fig. 5 and Fig. 9 (Fig. 8):

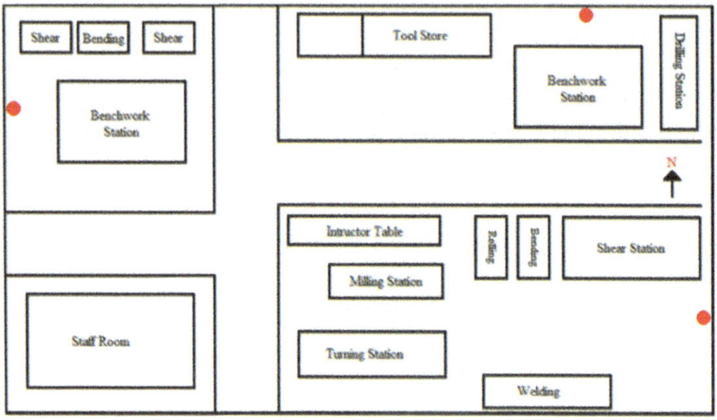

Fig. 8. Camera position in machining shop layout

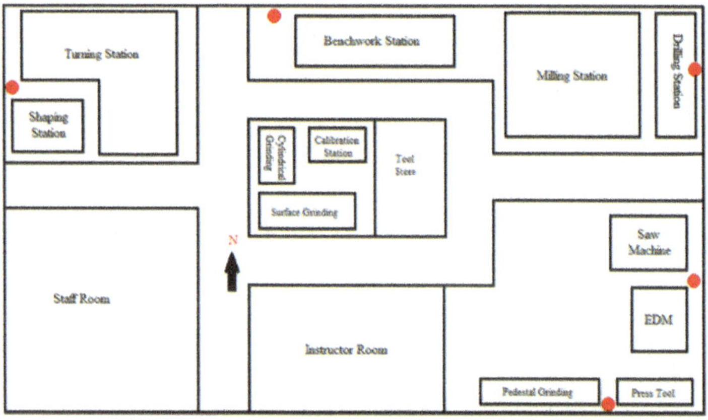

Fig. 9. Camera position in machining shop layout

The final aim of the research is to achieve real-time ability in production monitoring, production control and logistics. The challenge is in developing the intelligent processing and analyzing the data in the production process as explained by Pielmeier [7]. State of the art in analyzing the data stream in manufacturing process are as followed: Real-Time System and Analytics, Event, Domain Experts and Data Mining.

5 Conclusion

The number of products from the simulation is quite satisfactory with 78 Plant Standers, 9 Coffee Mixture Machines, 6 Cassava Peelers and 9 Analog Rice Machines in one day.

However, the data structure and analyzing the simulation result for implementing virtual manufacturing in Polytechnic State of Bandung need to be studied further.

References

1. Chablat, D., Dépincé, P., Noël, E., Woelk, P.O.: The virtual manufacturing concept: scope, socio-economic aspect and future trends. In: Proceedings of DETC, pp. 1–6 (2004)
2. Souza, M.C.F., Sacco, M., Porto, A.J.V.: Virtual manufacturing as a way for the factory of future. In: IFAC, pp. 467–472 (2004)
3. Peng, Q., Chung, C., Yu, C., Luan, T.: A networked virtual manufacturing system for SMEs. Int. J. Comput. Integr. Manuf. **20**, 71–79 (2007)
4. Tolio, T., Sacco, M., Terkaj, W., Urgo, M.: Virtual factory: an integrated framework for manufacturing systems design and analysis. Procedia CIRP **7**, 25–30 (2013)
5. Horváth, L., Rudas, I.J., Machado, J.A.T., Hancke, G.P.: Application of part manufacturing process model in virtual manufacturing. In: ISIE, pp. 1367–1372 (1999)
6. Wang, L., Zhang, E.-G.: The application of virtual manufacturing technology in the course reform of manufacturing technology. Adv. Soc. Sci. Educ. Humanit. **83**, 191–194 (2017)
7. Pielmeier, J., Braunreuther, S., Reinhart, G.: Approach for defining rules in the context of complex event processing. In: 11th CIRP Conference on Intelligent Computation in Manufacturing Engineering, pp. 8–12 (2018)

Comparative Study of Dry and MQL Condition on Hole Quality and Tool Vibration in Drilling Titanium Ti-6Al-7Nb Alloy

Zulfadli[1(✉)], Muhammad Rizal[2], and Muhammad Dirhamsyah[2]

[1] Department of Mechatronic, Aceh Polytechnic, 23119 Banda Aceh, Indonesia
zulfadli@politeknikaceh.ac.id

[2] Department of Mechanical Engineering, Faculty of Engineering, Universitas Syiah Kuala, 23111 Banda Aceh, Indonesia
{muh.rizal,mdirhamsyah}@unsyiah.ac.id

Abstract. Titanium Ti-Al-Nb is titanium-aluminum-niobium alloy which is commonly used in biomedical applications. However, because this is a difficult-to-cut material, cutting conditions and parameters must be thoroughly investigated in order to ensure component quality. Because minimum quantity lubrication (MQL) has been identified as a technologically possible strategy for improving the machinability of particular titanium alloys, the objective of this study is to compare the effects of MQL on the hole quality and tool vibration of titanium Ti-6Al-7Nb with the dry cutting condition. Drilling experiments were conducted to assess hole quality and vibration when drilling Titanium bar using carbide coated drills. The obtained findings show that MQL improves hole quality, resulting in better surface holes and reduced titanium burr formation under the MQL condition. Furthermore, tool vibration may be decreased under all tool wear and cutting parameter conditions.

Keywords: Minimum Quantity Lubrication (MQL) · Drilling · Tool vibration · Hole quality · Titanium alloy

1 Introduction

Titanium alloys are a biomaterial of choice for orthopedic and dental implants, screws, pacemakers, and cardiovascular stents because of their great biocompatibility, corrosion and fatigue resistance, and low density [1]. However, titanium alloys are difficult to manufacture due to their limited heat conductivity, high hardness, adherent particle creation near the tool's cutting edge, and tendency to harden in the machined zone [2].

Recently, more environmentally friendly machining processes, such as the minimum quantity lubrication (MQL) approach, are being used in machining proceses to increase the machining quality and machinability of titanium alloys in the previous study. Sadeghi et al. [3] investigated the MQL for grinding Ti6Al4V with synthetic ester and vegetable oils. They reported that the use of synthetic ester oil resulted in enhanced surface quality. Rahim and Sasahara [4] evaluated different types of lubricants affected the machining

Akhyar et al. (Eds.): ICECME 2021, LNME, pp. 24–31, 2023.
https://doi.org/10.1007/978-981-19-3629-6_3

quality of Ti6Al4V alloys. It was discovered that using the MQL from palm oil resulted in a significant improvement in the cut titanium surface integrity. Perçin et al. [5] investigated the torque during when drilling titanium alloy under a number of machining conditions. They claimed that the MQL condition produced the least torque and the better surfaces, which they ascribed to the MQL lubricants lowering friction at the tool-chip contact. The machinability of Ti6Al4V under the nanofluid MQL condition was investigated by Nam and Lee [6]. The findings reveal that the nano-diamond particles significantly reduced the frictional at the cutting zone. The latest report is an investigation of titanium alloys drilling to compare the surface hardness performance between two types of titanium alloys, Ti-6Al-4V and Ti-6AL-7Nb [1]. For certain cutting conditions, the alloy Ti-6Al-7Nb exhibits clearly lower roughness values, indicating that it is more machinable than Ti-6Al-4V.

However, it is still rare for researchers to study the machinability and machining performance of this type of titanium alloy, even though it is Ti-6Al-7Nb alloy that was developed to minimize the proven toxicity problems related to vanadium, so that it is more biocompatible than other types of alloys. The purpose of this study was to analyze and compare the effect of dry cutting and MQL using patchouli oil MQL in drilling Ti-6Al-7Nb. The effect of cutting parameters and cutting conditions are also investigated to the hole quality and tool vibration during processes.

2 Experimental Method

2.1 Material

Titanium Ti-6Al-7Nb was chosen in this experimental study. The benefit of this material is that it is biocompatible and reduces the known toxicity issues associated with vanadium. According to Challa et al. [7] revealed that the titanium alloy containing niobium had a lower toxicity, demonstrating its greater biocompatibility. The dimensions of the material were 100 mm in length and 20 mm in diameter. The ultimate tensile strength was in the range 900–1021 MPa, modulus elasticity was 105 GPa with an elongation in the range 10–15% [8]. Four types of 3-mm-diameter coated drills with different wear conditions were utilized as depicted in Fig. 1. Tool 1 (T1) was new tool (no wear), tool 2 (T2) was tool with $VB = 0.10$ mm, tool 3 (T3) was tool with $VB = 0.15$ mm, and tool 4 (T4) was tool with $VB = 0.25$ mm.

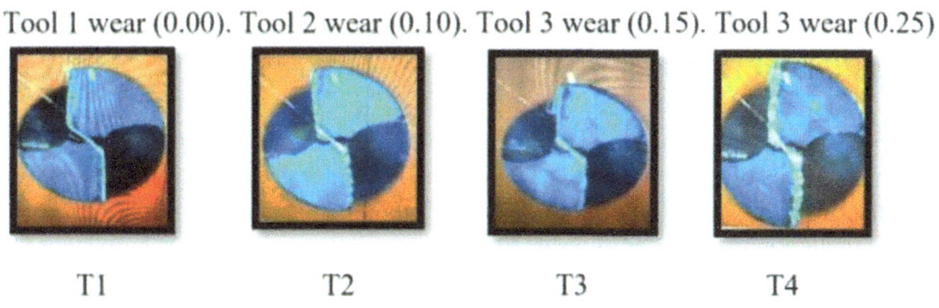

Fig. 1. Four wear states of the tool used.

2.2 Experimental Setup

A four-axis machining center (Feeler VMP-45 (A)) was used to conduct drilling test in both dry and MQL conditions. The MQL condition was achieved using a minimal quantity lubrication system coupled to the machining center's spindle system, using patchouli oil as coolants. The flow rate of MQL was 15 mL/h and a compressed air pressure of 1.2 MPa.

To measure the vibration signals, the accelerometer was attached in the spindle tool housing, and the signal was acquired using a c-Daq NI-9250 data acquisition and recorded using signal express software at a sampling rate of 10 kHz. Figure 2 depicts the drilling test experimental setup. Drilling experiments were carried out under dry and MQL conditions using a full factorial trial design with cutting speeds (*Vc*) of 20, 30, and 40 m/min and feed rates (*f*) of 0.01, 0.03, and 0.05 mm/rev.

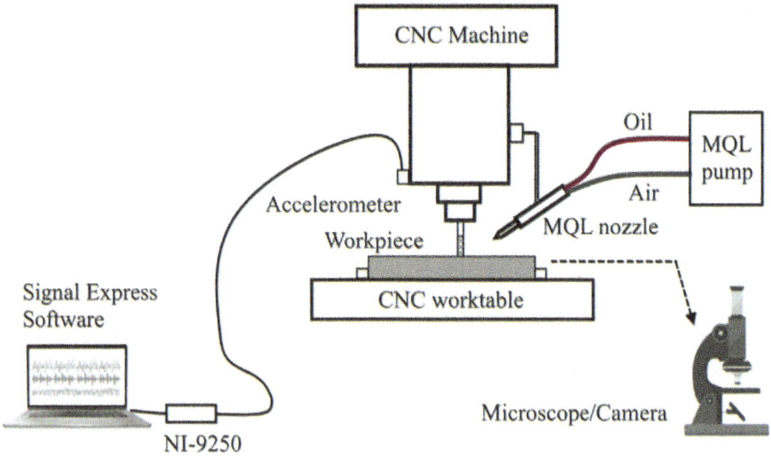

Fig. 2. Experimental setup.

After the drilling experiments were completed, the hole surface of the machined part under dry and MQL conditions were examined with a microscope camera with magnification 1000x. This digital microscope was used to observe the burr and hole quality of drilled parts.

3 Results and Discussion

Figure 3 shows observations result from drill holes with dry machining taken as samples from several holes of when using T1, T2, T3, and T4 under dry condition. It is clear that the results of drilling holes in dry conditions, visible formation of burrs at the edges of the holes in all conditions of tool wear. But only a little on the tool in a sharp state (T1).

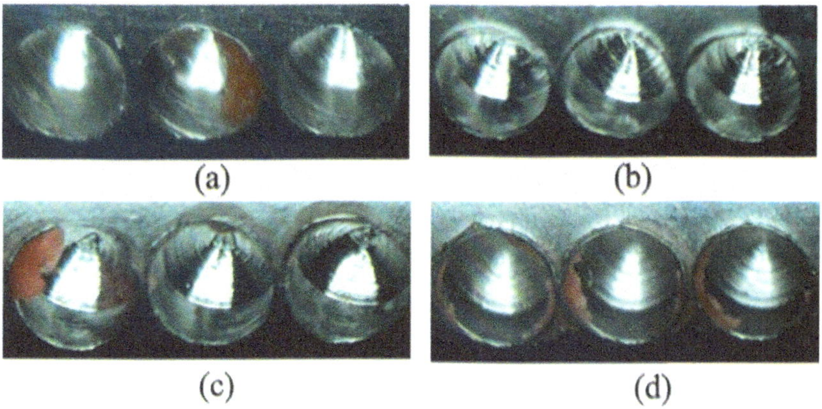

Fig. 3. Hole quality from dry condition.

Figure 4 shows observations result from drill holes with MQL condition taken as samples from several holes of when using T1, T2, T3, and T4. It is obvious that the results of drilling holes under MQL circumstances, just little creation of burrs at the borders of the holes, especially when tool wear exceeded 0.15 mm. When compared to the dry condition, the hole quality under MQL is better.

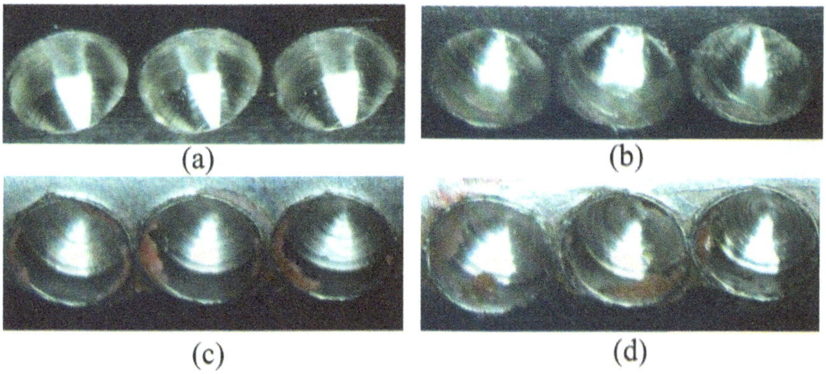

Fig. 4. Hole quality under MQL condition.

Figure 5 shows the vibration signal in Experiment 1 (Vc = 40 m/min, and f = 0.05 mm/rev), was carried out with dry air drilling, from the initial drilling process T1 to the final stage of T4 with conditioned tool wear conditions. From the time domain graph, it can be seen that the vibration changes during the machining experiment are very significant. At the beginning of drilling T1 the drill tool is in new condition, VB = 0.00 mm, the vibration signal shows 1.0 (m/s^2). Then due to the increased wear of the drill bit, the measured vibration signal shows at T2 VB = 0.10 mm, the cutting vibration signal shows 0.8 (m/s^2). at T3 VB = 0.15 mm the vibration signal shows 0.6 (m/s2). at T4 VB = 0.25 mm vibration signal shows 0.6 (m/s^2) The increase in vibration due to tool wear from T1 to T4 is 80%.

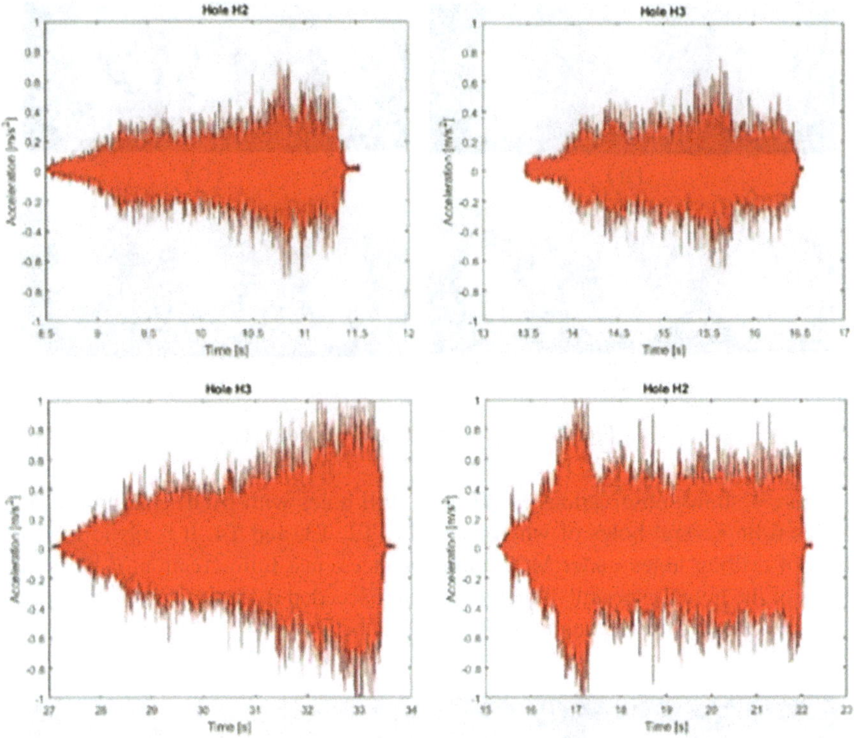

Fig. 5. Vibration signal of T1, T2, T3 and T4 under dry condition.

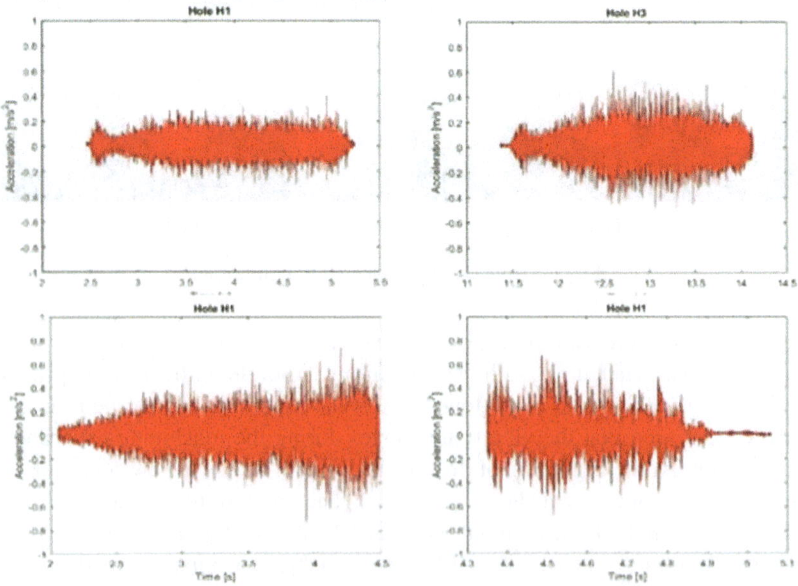

Fig. 6. Vibration signal of T1, T2, T3 and T4 under MQL condition.

Figure 6 shows vibration signal during drilling process under MQL condition using patchouli oil at experimental set 1 (Vc = 40 m/min and f = 0.05 mm/rev), from the initial drilling process T1 to the final stage of T4 with the tool wear conditions that have been conditioned. From the time domain graph, it can be seen that the vibration changes during the machining experiment are very significant. At the beginning of the T1 drilling, the drill tool is in new condition, VB = 0.00 mm, the vibration signal shows 1.0 (m/s^2) Then, due to the increase in drill bit wear, the measured vibration signal shows at T2 VB = 0.10 mm the cutting vibration signal shows 0.8 (m/s^2). at T3 VB = 0.15 mm the vibration signal shows 0.8 (m/s^2), at T4 VB = 0.25 mm the vibration signal shows 0.8 (m/s^2) The increase in vibration force due to tool wear is very visible from T1 to T4.

The variation of feed rate on the RMS signal in the dry cutting displays a graph where it can be seen in Fig 7, which the RMS value is related to the feed value (mm/rev). Vibration signal graph where it can be seen that the greater the feed speed the RMS value decreases, it can be seen at T1, T2, T3. While there are differences in T4. This difference occurs with the increase in the RMS value at T4 at f = 0.05 mm/rev caused by the trapping of the burrs on the drill chisel causing excessive vibration as shown in the image above at T4.

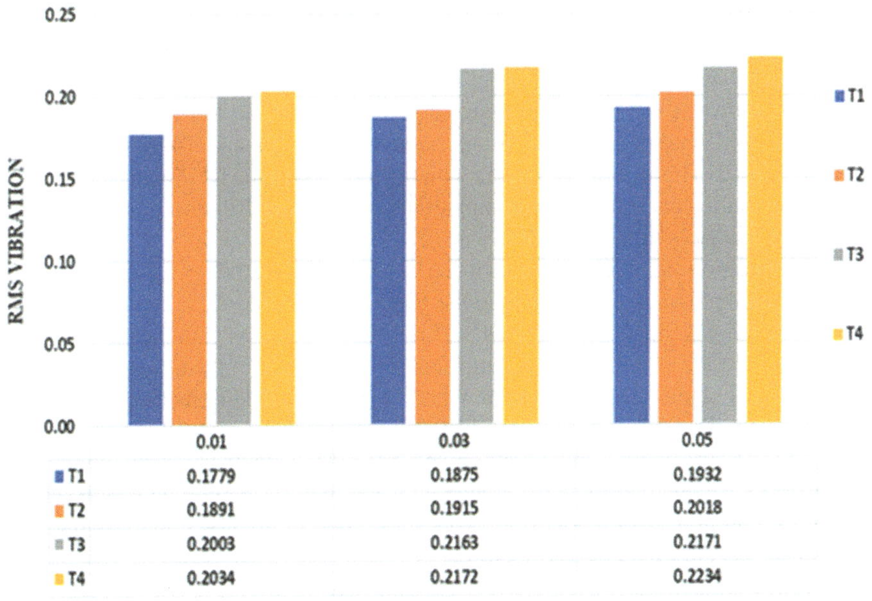

Fig. 7. RMS vibration vs feed rate in dry condition under various of tool wear

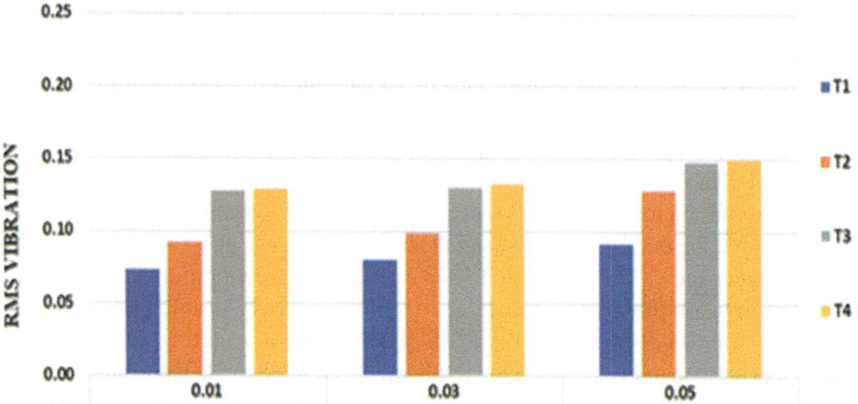

	0.01	0.03	0.05
T1	0.0733	0.0803	0.0915
T2	0.0922	0.0988	0.1286
T3	0.1275	0.1299	0.1477
T4	0.1291	0.1332	0.1506

Fig. 8. RMS vibration vs feed rate in MQL condition under various of tool wear

The variation of the constant feeding speed in the MQL process displays a graph where it can be seen that the RMS values of T1, T2, T3, T4 are related to the feed value (mm/rev) is shown in Fig. 8. Vibration signal graph where it can be seen that the greater the feed the RMS value decreases, it can be seen at T2, T1 and T4. While there are differences in T2. This difference occurs with the increase in the RMS value at T2 at $f = 0.05$ mm/revolution caused by the trapping of the burrs on the drill bit, causing excessive vibration as shown in the graphic above at T2.

In this vibration signal graph where it can be seen T1 L7 in the dry sampling method, it can be seen that the greater the intake the RMS value decreases, it can be seen at T1 MQL, while there is a difference in T1 dry This difference occurs with the increase in the RMS value at T1 at $f = 0.05$ mm/revolution caused by the heat of the tool surface and the wear of the tool blade surface so that the vibration signal in the dry method is very large in difference in the RMS value on the drill tool, causing excessive vibration as shown in the graph above on T1 L7 dry method and T1L7 MQL method. In the comparison chart of RMS values, there is a difference between the dry method and MQL at 0.05 mm/rev feeding in average about 0.07. This shows that in the RMS analysis the MQL performance is better than the dry method.

4 Summary

The hole quality and tool vibration under drilling Ti-6Al-7Nb were investigated under dry and MQL conditions. The comparison results with dry cutting and under MQL using patchouli oil were presented. The quality of the holes is neater and smoother using MQL patchouli oil and so is the vibration of the drill when the machining process shows a significant decrease. The difference between the vibration RMS values in dry machining is greater about 0.05 to 0.07 than with the vibration RMS at MQL conditions. These results show that MQL using patchouli oil is very effective in reducing vibration and improving hole quality in titanium Ti-6Al-7Nb alloys.

References

1. Festas, A.J., Pereira, R.B., Ramos, A., Davim, J.P.: A study of the effect of conventional drilling and helical milling in surface quality in titanium Ti-6Al-4V and Ti-6AL-7Nb alloys for medical applications. Arab. J. Sci. Eng. **46**(3), 2361–2369 (2020). https://doi.org/10.1007/s13369-020-05047-8
2. Barman, A., Adhikari, R., Bolar, G.: Evaluation of conventional drilling and helical milling for processing of holes in titanium alloy Ti6Al4V. Mater. Today Proc. **28**, 4–9 (2020)
3. Sadeghi, M.H., Haddad, M.J., Tawakoli, T., Emami, M.: Minimal quantity lubrication-MQL in grinding of Ti–6Al–4V titaniumalloy. Int. J. Adv. Manuf. Technol. **44**(5–6), 487–500 (2009)
4. Rahim, E.A., Sasahara, H.: Effect of machining parameters and MQL liquids on surface integrity of high speed drilling Ti-6Al-4V. Key Eng. Mater. **447–448**, 816–820 (2010)
5. Perçin, M., Aslantas, K., Ucun, İ, Kaynak, Y., Çicek, A.: Micro-drilling of Ti–6Al–4V alloy: the effects of cooling/lubricating. Precis Eng. **45**, 450–462 (2016)
6. Nam, J., Lee, S.W.: Machinability of titaniumalloy (Ti-6Al-4V) in environmentally-friendly micro-drilling process with nanofluid minimum quantity lubrication using nano-diamond particles. Int. J. Precis. Eng. Manuf.-Green Technol. **5**(1), 29–35 (2018). https://doi.org/10.1007/s40684-018-0003-z
7. Challa, V.S.A., Mali, S., Misra, R.D.K.: Reduced toxicity and superior cellular response of preosteoblasts to Ti-6Al-7Nb alloy and comparison with Ti-6Al-4V. J. Biomed. Mater. Res. – Part A **101**(7), 2083–2089 (2013)
8. Kunčická, L., Kocich, R., Lowe, T.C.: Advances in metals and alloys for joint replacement. Prog. Mater Sci. **88**, 232–280 (2017)

Supply Chain Risk Mitigation Strategy of Gayo Arabica Coffee Using House of Risk Method in Bener Meriah Regency

Putera Muzakkir(✉), Hasan Yudie Sastra, and Husni

Magister of Industrial Engineering, Faculty of Engineering,
Universitas Syiah Kuala, Banda Aceh, Indonesia
muzakkirputera@gmail.com

Abstract. The decline in the selling value of Gayo Arabica coffee produced by farmers in Bener Meriah Regency indicates that there has been a phenomenon of price stability disturbance. One of the reasons is related to the quality standards of the coffee produced. elements in handling in every process of activity starting from the cultivation process to post-harvest carried out will cause risk events that can worsen the quality of gayo arabica coffee produced, it is necessary to map the existing system so that it can identify risk events that arise. The method used in this study is the House of Risk (HOR). This HOR method serves to analyze and prioritize the sources of risk causes (risk agents) and establish priority risk mitigation strategies that can reduce risk agents. By mitigating risks in the Gayo Arabica coffee supply chain, it is expected to reduce existing risks and the supply chain can run optimally. Based on this study, analyzing the supply chain using the HOR method, there were 32 risk events and 46 risk agents grouped into 31 risk agents. Using HOR I method, it measures the severity of risk events, occurrences of risk agents, and produces an Aggregate Risk Potential (ARP) value where 28 risk agents are prioritized to be resolved where the three risk agents whose priority is to be reduced are the mistakes of collectors in the technical quality control process (A12), poor planning from farmers (A31), and poor quality green bean packaging (A18).HOR II produces 14 mitigation strategies and 8 mitigation strategies are prioritized based on the value of effectiveness and level of difficulty. The three priority mitigation strategies are compiling and establishing SOPs for Gayo Arabica coffee cultivation which is integrated with current coffee plantation technology (PA11), collaborating with fertilizer and packaging companies (PA09), and providing quality control training to conduct quality assessments in accordance with SNI standard (PA06).

Keywords: Supply chain · Gayo Arabic coffee · House of Risk method

1 Background

Coffee is one of Indonesia's main export commodities, which plays an important role in the Indonesian economy. Aceh province is the area that has the highest production

Akhyar et al. (Eds.): ICECME 2021, LNME, pp. 32–42, 2023.
https://doi.org/10.1007/978-981-19-3629-6_4

arabica coffee (Indonesian Coffee Statistics 2020). The largest coffee producing areas in Aceh province are Aceh Tengah and Bener Meriah Regency. The area of coffee plantations in these two districts reaches 80% or 96,000 hectares of the total coffee land area in Aceh Province, which is 121,000 hectares (BPS Aceh province 2018). Coffee plantations are the main livelihood of the community, most of whom work as coffee farmers. The development of Gayo arabica coffee cultivation has increased from 2014 to 2020 both in terms of the area planted and the level of production. In 2018, the productivity of Arabica coffee in Bener Meriah Regency experienced a large growth compared to previous years. However, after 2018 arabica coffee experienced a decrease in production from 30,408 to 29,101 tons as well as in terms of the average selling price of arabica coffee, the price decreased very drastically, from 65,824 to 29,287 rupiah per can. The Table 1 below explains that the productivity level of Arabica coffee from 2014 to 2018 has increased Arabica coffee production, but in 2019 and 2020 Gayo Arabica coffee productivity has decreased. The decline in productivity in 2019 and 2020 was also followed by a decline in the selling price.

Activities in the supply chain, from plan, source, make, delivery and return, can pose risks that can have a negative impact in the form of losses for the three stakeholders, for example in the planning process of harvest time that is not appropriate it will reduce the quality of coffee beans.

Table 1. Production value and average selling price Gayo arabica coffee in Bener Meriah Regency

Year	Area (ha)	Production (ton)	Average selling price
2014	47.370	11.026	41.162
2015	47.370	11.526	41.431
2016	46.283	26.357	36.349
2017	46.273	28.358	–
2018	46.273	30.408	65.284
2019	46.273	29.101	29.287
2020	48.950	29.000	39.370

Source: Researcher 2021.

Supply chain risk management serves to understand and overcome the chain effect when a large or small risk occurs in a supply network. By designing a risk mitigation in the supply chain, it is expected to reduce the risks that arise. The risk mitigation effort that will be carried out at the Bener Meriah Regency Coffee Plantation is an effort to create reliable supply chain conditions. By designing a risk mitigation in the supply chain, it is expected to reduce the risks that arise. The risk mitigation effort that will be carried out at the Bener Meriah Regency coffee Plantation is an effort to create reliable supply chain conditions. Therefore, the research on supply chain risk mitigation strategies at the coffee plantation of Bener Meriah Regency, the method that will be used is the House of Risk (HOR) method. The House of Risk method aims to eliminate risks and their sources of risk and provide an analysis of the level of difficulty and effectiveness of the planned risk mitigation actions. The House of Risk (HOR) method has two phases, the first phase functions to identify risks and has an output in the form of a risk agent priority rating and the second phase is risk management which has an output in the form of priority risk mitigation actions in preventing the occurrence of risk agents.

2 Methodology

Supply chain (supply chain) is an organizational system that distributes its production goods and services to customers. Supply chain is also a relationship about how raw materials are obtained, then processed, to become a product that is distributed to consumers. Supply Chain Management (SCM) is a unified process and production activity starting from raw materials obtained from suppliers, the value-adding process that converts raw materials into finished goods, the storage process, inventory of goods to the delivery of finished goods to retailers and consumers. The House of Risk HOR method is used to analyze risk causes, risk events, and determine risk mitigation (Mustaniroh et al. 2018; Pamungkassari et al. 2018; Purwaditya et al. 2018). The HOR method is more suitable to be used to mitigate risk, because this HOR method identifies all risks starting from risk agents, risk events so that they can determine the most appropriate mitigation for the risk. Risk management is an effort to find out, analyze and control risk in every company activity with the aim of obtaining higher effectiveness and efficiency (Ulfah, dkk, 2017). Risk management can be done through 3 processes which are risk identification, risk evaluation and measurement, and risk management. Risk identification is carried out to identify what risks are faced by an organization.

The purpose of risk management is as a tool for system to achieve their goals through the allocation of resources to plan, make decisions, and carry out productive activities. The House of Risk (HOR) is a modification of the FMEA (Failure Modes and Effects of Analysis) and the House of Quality (HOQ) Approach. In HOR phase 1, it connects one or a set of needs (what) and a response (how) that indicates one or more needs/needs (Suparjo 2018). HOR phase 1 aims to identify risk events and risk agents that are the source of risk events. In HOR phase 2 it is used to set risk mitigation strategies in handling risk agents that have been previously identified and in accordance with the order of levels of the priority scale that has been determined in HOR phase 1. It is necessary to choose an action that is not difficult to implement but can effectively reduce possible sources of risk.

3 Result and Discussion

Identification is conducted based on literature study, which is then is made suitable with interview process with the regency related; in order to find out the risks and risks causes that may potentially have happened in the regency. After the risks are identified then they will be classified according to the activities. Table 2 shows that are 32 risk events where they are 11 risks from farmer, 13 risks from collector, and 8 risks from supplier.

Table 2. Risk events identification

	Risk code	Risk events
Farmer	E1	Risk of delay in planting seeds
	E2	Harvest time planning scheduling error
	E3	The risk of the number of seeds not on demand
	E4	Limited amount of fertilizer
	E5	Risk of seeds can't grow well even die
	E6	The risk of being late in giving fertilizer
	E7	Risk of late pruning shoots
	E8	Risk of crop failure
	E9	The risk of growth parasite plants on coffee trees
	E10	Risk of harvest delay
	E11	Risk of late requested collector to take the harvested coffee
Collector	E12	Risk of delay in picking up coffee from farmers
	E13	Risk of worse planning to check the coffee quality
	E14	Risk of lack of coffee raw material supply
	E15	Risk of limitation pulper machine
	E16	Risk of limitation huller machine
	E17	Risk of limitation of good quality drying materials
	E18	Risk of too many coffee fruits that are too young
	E19	Risk of not using dry house
	E20	Risk of machine damage during the production process
	E21	Risk of disease and dirt contamination during production
	E22	Risk of leakage packaging in storage
	E23	Risk of unreachable collectors in taking harvest
	E24	Risk of return green bean product
Supplier	E25	Risk of limited green bean required
	E26	Risk of unappropriated time to supply coffee
	E27	Risk of lack packaging quality
	E28	Risk percentage of water level received from collectors not the same
	E29	Risk of defected bean too high
	E30	Risk of high humidity and low temperature in storage warehouse
	E31	Risks to the vehicles used when shipping are not sterile
	E31	Risk of coffee products returned

Table 3 shows that are 31 risk agents identifies, where one of the risk causes can start one or more risk events. These risks are then given ratings to see how far they can affect the risk activities of gayo arabica coffee chain system. The identified risks are then given risk ratings to mark severity, occurrence and correlation. The result of each criterion is then cumulated with the ARP values calculation.

Table 3. Risk agent identification

Aj	Risk agents
A1	Extreme weather
A2	Farmers fail to harvest
A3	Low green bean selling price
A4	Low coffee fruit selling price
A5	Inappropriate of harvesting technique
A6	Farmer lack of plantation technique
A7	Inappropriate in quality control technique
A8	Supplier errors in supply planning
A9	Supplier lack of supervision in warehouse conditions
A10	Supplier error on final product quality control
A11	Collector inappropriate in post-harvest process technical
A12	Collector lack in technical quality control process
A13	Farmer misplace the coffee harvest
A14	Inclement communication between farmer and collector
A15	Poor warehouse quality
A16	Lack of enthusiasm of farmers in seedling
A17	Collector lack of capital
A18	Poor quality green bean packaging
A19	Quality product packaging is bad
A20	Improper coffee plantation process
A21	Improper coffee plant treatment
A22	Limited stock of seeds
A23	Limited stock huller machine from local market
A24	Limited stock pulper machine from local market
A25	Limited stock seeding fertilizer from local market
A26	Limited stock growth fertilizer from local market
A27	Plants affected by pests, viruses and diseases

(*continued*)

Table 3. (*continued*)

Aj	Risk agents
A28	Limited transportation for take the coffee harvest
A29	No maintenance on production machines
A30	Limited special transportation for delivery product coffee
A31	Farmers don't have a good plan in coffee cultivation

Table 4 shows the recapitulation of ARP value from all risks that have happened. The next step is to do a depiction using pareto diagram to see what risks that must be mitigated based on the pareto diagram principles.

Table 4. ARP value recapitulated

Rank	Code	ARP	Cum	%Cum
1	A12	600	600	10,91%
2	A31	567	1167	21,22%
3	A18	472	1639	29,81%
4	A15	375	2014	36,62%
5	A19	360	2374	43,17%
6	A17	301	2675	48,65%
7	A27	297	2972	54,05%
8	A7	252	3224	58,63%
9	A11	252	3476	63,21%
10	A20	252	3728	67,79%
11	A14	204	3932	71,50%
12	A10	189	4121	74,94%
13	A29	189	4310	78,38%
14	A30	144	4454	81,00%
15	A6	135	4589	83,45%
16	A2	99	4688	85,25%
17	A26	90	4778	86,89%
18	A3	84	4862	88,42%
19	A5	84	4946	89,94%
20	A9	84	5030	91,47%
21	A8	75	5105	92,84%

(*continued*)

Table 4. (*continued*)

Rank	Code	ARP	Cum	%Cum
22	A22	63	5168	93,98%
23	A23	63	5231	95,13%
24	A4	60	5291	96,22%
25	A21	60	5351	97,31%
26	A25	54	5405	98,29%
27	A16	28	5433	98,80%
28	A24	21	5454	99,18%
29	A1	18	5472	99,51%
30	A28	18	5490	99,84%
31	A13	9	5499	100,00%

The principles of the pareto diagram are that 20/80, where 80% of the risk condition came from 20% of the main cause of the risk condition. The pareto diagram for ARP values can be seen in Fig. 1. Based on Fig. 1, it can be suggested that the risk because chosen was Collector's Error in Technical Quality Control Process (A12) and poor planning from farmers (A31).

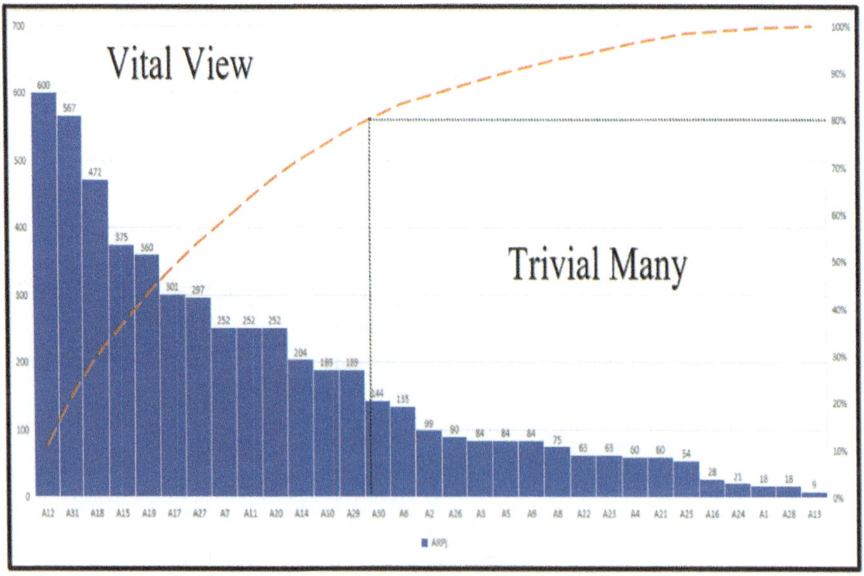

Fig. 1. Pareto diagram for ARP value

A12 and A31 became the first and the second highest risk causes because of having the higher ARP value. After the sorted from highest to lowest, there were high percentages, they are: 10,91% and 10,31%, so that when it was done cumulatively it reached 21,22%.

The Pareto chart above shows that the intersection point formed by the x-axis and the 80% line separates the important risk agents on the left and the less important risk agents on the right. Furthermore, a mapping of risk agents is carried out which is a vital view which is a top priority for intensive handling. Mitigation of risk agents is an effort to suppress the emergence of risk agents that will cause risk events. one risk agent that appears can have an impact on the occurrence of at least one risk event. Mitigation actions of risk agents are mapped based on the results of primary and secondary data collection so that complete information is obtained both fundamentally and data obtained in the field.

One risk agent can be eliminated by one or more mitigation actions, similarly one mitigation action can impact one or more risk agents. The sequence of identification code mapping (PAj) of mitigation actions from the table above is presented in the following Table 5 below.

Table 5. Action mitigation risk identification

PAj	Action mitigation risk
PA1	Evaluate planning of Gayo arabica coffee cultivation
PA2	Removing trees infected by diseases/viruses
PA3	Routine cultivating of coffee plantation
PA4	Maintenance production machine intensevile
PA5	Completing warehouse with ventilation dan exhaust machine proper to Promote good air circulation
PA6	Provide quality control training to assess quality in accordance with SNI
PA7	Provide organic pesticides as necessity
PA8	Establish cooperative relationships with shipping service providers
PA9	Establish cooperative relationships with Packaging companies
PA10	Provide facility of post-harvest process like machines and dry house
PA11	Develop and establish an SOP for the post-harvest implementation of Gayo Arabica coffee which is integrated with current coffee plantation technology
PA12	Gayo Arabica Coffee Harvesting Integrated with Current Coffee Plantation Technology
PA13	Develop and establish standards for the implementation of green bean quality control
PA14	Develop and establish inventory standard procedure

Determining effective mitigation actions, it is also necessary to identify the relationship level of a mitigation action taken against a risk agent. Before sorting priority scale for mitigation actions (Dk). Although it is possible that one risk action taken will have an impact on several risk agents, determination of this correlation is needed to assess each action is an appropriate and effective action in overcoming the targeted risk agent (Table 6).

Table 6. Pareto ratio of effectiveness to difficulty

Pj	PAj	ETDk	Kum	%Kum
1	PA11	1335	1335	14,10%
2	PA09	1265,4	2600,4	27,47%
3	PA06	1153,8	3754,2	39,65%
4	PA12	929,25	4683,45	49,47%
5	PA03	756	5439,45	57,46%
6	PA01	744	6183,45	65,31%
7	PA05	736,2	6919,65	73,09%
8	PA04	567	7486,65	79,08%
9	PA14	486	7972,65	84,21%
10	PA10	450	8422,65	88,97%
11	PA13	331,8	8754,45	92,47%
12	PA02	297	9051,45	95,61%
13	PA07	297	9348,45	98,75%
14	PA08	118,8	9467,25	100,00%

First, a calculation is made for the Total Effectiveness (TEk) of each mitigation action for each impact generated on each risk agent based on the ARP value. From the results of The Total Effectiveness (TEk) of each mitigation action that has been obtained, then calculated the Total Effectiveness to Difficulty Ratio (ETDk). The results of these calculations are based on consideration and determination of the level of difficulty (Dk) for the implementation of each mitigation action. The Degree of difficulty (Dk) scale is used to determine the level of difficulty encountered in implementing mitigation actions. Consideration of costs, facilities and infrastructure, as well as other resources is an inhibiting factor in implementing mitigation actions (Fig. 2).

PA11 mitigation actions impacted seven risk agents or 17.9% of the identified risk agents. While the mitigation actions that have an impact on one risk agent are found in five mitigation actions, namely PA03, PA04, PA13, PA02 and PA07. To filtering the potential for Total Effectiveness (TEk) against Difficulty (Dk), Pareto Analysis is used again, so that the results obtained are mitigation actions that must be prioritized to be realized as priority actions in the initial handling of risk agents (Fig. 3).

From the Pareto analysis diagram that has been presented above, it is known that the risk mitigation actions that become the vital view or top priority located between the x-axis and the 80% line are eight mitigation actions, while the remaining six other mitigation actions are included in the category trivial many or less priority actions. The

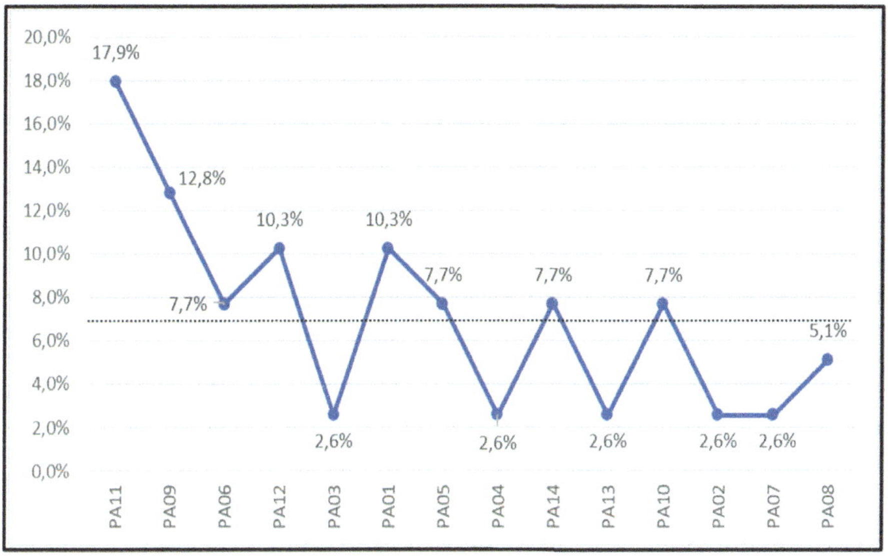

Fig. 2. Percentage impact of mitigation action on risk agent

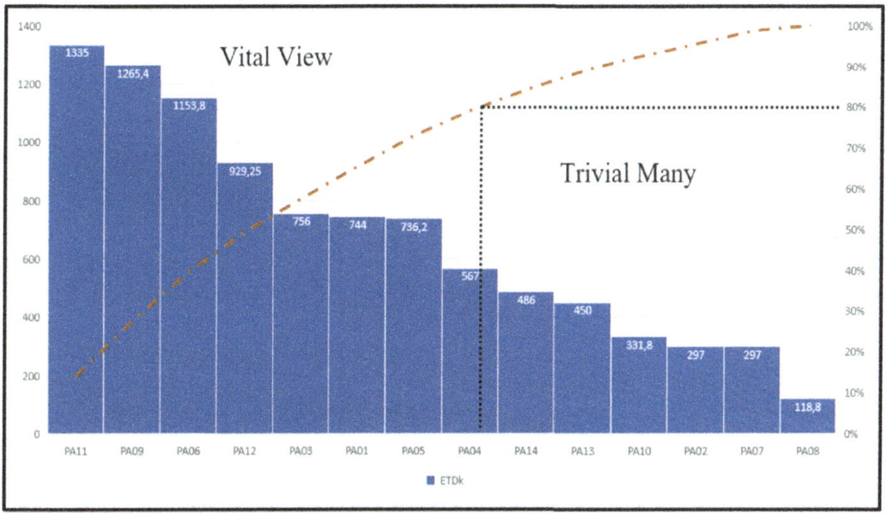

Fig. 3. Pareto diagram for ETDk value

risk mitigation strategy in the data above contributes as much as 79.08% of the total value of the Effectiveness to Difficulty Ratio (ETDk). By taking these mitigation actions, the emergence of risk agents can be reduced. Priority Risk mitigation action are develop and establish an SOP for the post-harvest implementation of Gayo Arabica coffee which is integrated with current coffee plantation technology (PA11), Establish cooperative relationships with Packaging companies (PA09), Provide quality control training to assess quality in accordance with SNI (PA06), Gayo Arabica Coffee Harvesting Integrated with Current Coffee Plantation Technology (PA12), Routine cultivating of coffee plantation (PA03), Evaluate planning of Gayo arabica coffee cultivation (PA01), Completing warehouse with ventilation dan exhaust machine proper to Promote good air circulation (PA05), and the last is maintenance production machine intensevile (PA04).

4 Conclusion

There are 32 risk events and 313 risk agents that were identified on the supply chain system of gayo arabica coffee process in Aceh. A12 variable (Collector's Error in Technical Quality Control Process) and A31 (poor planning from farmers) were 2 risk causes with the highest ARP value. The processing result with House of risk Method Phase II The total risk mitigation actions are 14 actions which will then be calculated and analyzed to determine the priority mitigation actions. A sequence of risk mitigation strategies that can be applied to the supply chain of gayo arabica coffee plantations in Bener Meriah regency, making and forming Standard Operational Procedure (SOP) for gayo arabica coffee cultivation which is integrated with current coffee plantation technology (PA11), collaborating with fertilizer and packaging companies (PA09) were 2 mitigation risk strategy with the highest ETDk values which are prioritize to implimanting in Bener Meriah Regency.

References

Statistik, B.P.: Statistik Kopi Indonesia 2020. Penerbit: BPS-Statistics Indonesia (2020)

Statistik, B.P.: Provinsi Aceh Dalam Angka 2018. BPS Provinsi Aceh, Penerbit (2018)

Statistik, B.P.: Kabupaten Bener Meriah Dalam Angka 2021. BPS Kabupaten Bener Meriah, Bener Meriah, Penerbit (2021)

Ulfah, M., Siti, M., Nindy, C.S., Muhamad, G.M.S., Fitri, A.: Analisa Perbaikan Manajemen Risiko Rantai Pasok dengan Pendekatan House of Risk. Penerbit Journal Industrial Services (2017)

Suparjo, A.R.: Manajemen Risiko Operasional pada PT ABC dengan Menggunakan Metode FMEA. LPPM Untag, Penerbit (2018)

Mustaniroh, S.A., Dewi, A.N., dan Dhita, M.I.: Mitigation Strategies for Supply Chain SME using House of Risk Method (A case Study in Langgeng Jaya Abadi SME, Malang Regency), Malang (2018)

Maximum Application of Power Point Tracker on Microcontroller Based Dual Energy Solar Powered Dryer

Saifuddin Muhammad Jalil[1,4], Husaini[1,3(✉)], Rizal Munadi[1,2], and Ira Devi Sara[1,2]

[1] Doctoral Program, School of Engineering, Post Graduate Program, Universitas Syiah Kuala, Banda Aceh 23111, Indonesia
saifuddin@unimal.ac.id, {husainiftm,rizal.munadi, ira.sara}@unsyiah.ac.id

[2] Department of Electrical and Computer Engineering, Universitas Syiah Kuala Darussalam, Banda Aceh 23111, Indonesia

[3] Department of Mechanical Engineering, Universitas Syiah Kuala Darussalam, Banda Aceh 23111, Indonesia

[4] Department of Electrical, Universitas Malikussaleh, Lhokseumawe 24351, Indonesia

Abstract. Solar-powered dryers are still expensive and difficult to obtain for farmers with medium to low production capacity. For this reason, it is necessary to develop technology on a solar-powered dryer with a solar panel capacity of 200WP is equipped with a maximum power point tracker as an integrated solar collector with a photovoltaic system. The system in this solar-powered dryer consists of a solar panel tracker, a rotating aluminum cylinder net, a photovoltaic tracker, a battery bank, a microcontroller, and a DC/AC inverter. This system improves the performance of solar panels in the drying process of Moringa leaves. This design is expect to be the right system to ensure maximum results by providing solar radiation energy in the drying process of agricultural products. The aims goal to minimize problems in the drying of agricultural products experienced by farmers using an automatic dual-energy solar-powered dryer as a development innovation for the drying process. This study uses an experimental methodology and field data collection using a solar-powered dryer that has been analyzed and measured to obtain data comparisons. shows the maximum data of voltage and power response in positioning photovoltaic to produce the increased output power of photovoltaic array. Maximum application of powerpoint tracker on microcontroller based dual energy solar powered dryer on solar panels for drying moringa leaves produces the advantage of maximum energy absorption in solar panels and shortens the time in the drying process of moringa leaves.

Keywords: Maximum power point tracker · Solar powered dryer · Dual energy · Moringa leaves · Microcontroller

1 Introduction

The drying process of agricultural products, especially moringa leaves is an important process in the traditional agricultural sector and the agribusiness industry, such as the

Akhyar et al. (Eds.): ICECME 2021, LNME, pp. 43–52, 2023.
https://doi.org/10.1007/978-981-19-3629-6_5

production of medicines, health foods and beverages [1]. This drying process is related to the applied technology which is economically valuable and easy to operate by farmers. This environmentally friendly drying process takes into account the depletion of natural fuel resources which quickly run out due to the increasing cost of fossil fuels, solar drying is expected to be the main solution in the future in the process of drying moringa leaves using solar panels [2]. In addition, environmental considerations and the damage caused by humans due to increasing consumption of fossil fuels encourage farmers and industry to use renewable energy as a clean and sustainable resource, so that the use of solar energy for drying will continue and become a top priority [3–5].

From some literature, there are many dryers that have been designed and developed using renewable energy applications [6]. From some of these tools, the drawbacks are solar panels that do not apply a maximum power point tracker on a solar-powered dryer which results in no maximum efficiency in reducing the conversion of the resulting output voltage, to overcome the energy conversion efficiency of the solar panel, a maximum power point tracker is needed on the solar panel. solar panels so that the solar panels are right at the maximum point of energy efficiency, which is an angle of 90° [5, 7–11].

Drying using the sun is classified into several main categories, one of which is the agricultural sector and the moringa leaf processing industry, many of which can be utilized from solar energy as the main ingredient for the drying process, one of which is the drying process of moringa leaves by using solar panels as a medium to absorb heat rays. solar energy and then converted in watts which is generated and sent to the battery and several other supporting components to maximize heating so that the time needed will be faster in the drying process of moringa leaves compared to what is done by traditional moringa leaf farmers or the moringa industry as it is today. do the drying of moringa leaves in the traditional way, namely drying in the sun [6, 10–18].

The solar tracking panel is one of the solar drying sub-systems used in the design of agricultural dryers. This device produces an integrated solar collector, the principle of solar tracking works based on a sensor-based electronic design sub-system that is used to collect solar radiation. The use of a microcontroller in several studies conducted is as a logic control, with an automatic system, fuzzy logic control is also chosen as an integrated control of the maximum power point tracker to improve the function of the solar system as one of the selection methods to get better solar tracking system performance [5, 19–22].

Maximum power point tracking has been widely developed by previous research studies using different control systems to improve the performance of tracking the direction of motion of solar radiation in generating solar energy supply to photovoltaics, by generating electric power and generating heat radiation to the target object to be dried in an integrated manner in the collecting system. solar power, which utilizes agricultural products as the main object. In this study the microcontroller was chosen because it is easy to program as the main logic of the processing unit sequentially using research to produce better performance of solar panel drying units for agricultural products, especially Moringa leaves as Fig. 1 below is a design of the maximum block diagram system for tracking power point. which is applied to the drying of Moringa leaves (Fig. 2).

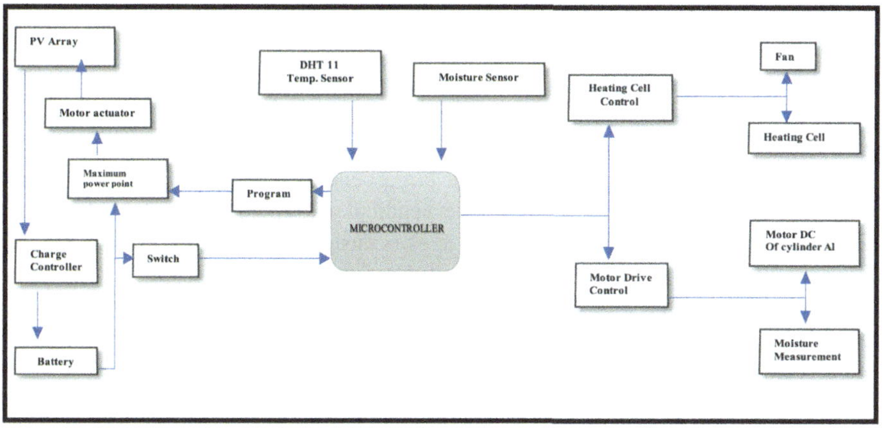

Fig. 1. Power point tracker maximum block diagram system design

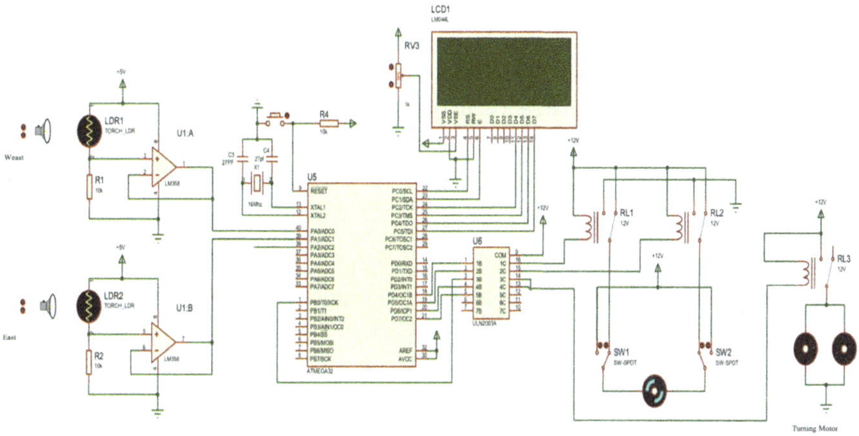

Fig. 2. Maximum power point tracker and prototype block diagram system using proteus software

The method of incremental conductance occurs based on the principle of maximum power yielding the formula as presented:

$$\frac{dp}{dv} = -\frac{I}{V} \tag{1}$$

The photovoltaic output power is P, V and I, respectively, voltage, current and photovoltaic output power. According to Chiang et al. [19, 20] the dP/dV power slope was used to calculate digitally by sampling the output current and voltage of the photofollaic array at the following time intervals (n-1) as shown below:

$$\text{where;}\quad P(\eta) = V(\eta)I(\eta) \tag{2}$$

2 Stepper Motor as a Driver for Solar Panels and Collectors

Solar panels to get maximum light require a solar tracking device [7, 5, 19, 23]. An active tracker is a tracker that works based on coordination or a combination of a microprocessor with an electro-optical or time-based sensor. Using a solar tracker system can increase the effectiveness of solar panels, because the largest energy is received by the solar panel is the direction of solar radiation that is perpendicular to the plane of the solar panel. The solar tracking system with 2 rotating axes is designed to increase effectiveness in energy reception.

Sun trackers basically have the same working principle. Both of these sun trackers use a one-axis tracking system. This system can only follow the movement of the sun from east to west caused by the rotation of the earth. The sensor used is a photo resistor. efficiency in the sun tracker, the two-axis system can be applied to the sun tracker. The two-axis system allows the sun tracker to control the azimuth and latitude position of the sun's position [9, 10].

The stepper motor in the tracking system is used as the drive of the panel. In electric motors, there are two kinds of driving motors, namely DC/AC motors and stepper motors. DC/AC motors consist of 2 coils as ststor and a magnetic core that can rotate on its axis as a rotor..Explanation of the stepper motor in the stepper motor there are 4 stator coils arranged in a certain position so that it cannot produce continuous rotation. The magnetic core will stop rotating if the stator provides a magnet that is opposite to the magnetic core, because both attract each other. If an electric current is applied to coil 1, the magnetic core will be facing coil 1, then if the electric current is shifted to coil 2, the magnetic core will be facing coil 2, and so on. The magnetic core can also stop rotating in a position between the two stator coils with how to give electric current to two coils at the same time. This can be used to produce a smaller turning angle, and a smoother motor rotation (Danang Susilo 2010).

3 Microcontroller and Microcontroller Working Principle

Microcontroller if interpreted literally means a micro-sized controller. In its use, the microcontroller is usually embedded in the device to be controlled. At first glance, the microcontroller is the same as the microprocessor on a computer. However, the microcontroller has many components integrated with it such as a timer/counter, while the microprocessor does not. Microcontroller is a microprocessor chip equipped with CPU, memory (RAM and ROM), and Input-Output [10].

Microcontroller is a computer on a chip used to control electronic equipment, which emphasizes efficiency and cost effectiveness [7]. It can literally be referred to as a "small controller" where an electronic system that previously required many supporting components such as TTL and CMOS ICs can be reduced/minimized and finally centralized and controlled by this microcontroller. With the use of this microcontroller then:

1. Electronic systems will become more compact.
2. Electronic system design will be faster because most systems are software that is easy to modify.

3. Troubleshooting is easier to track because of the compact system.

However, the microcontroller is not fully capable of reducing the TTL and CMOS IC components which are often still needed for high-speed applications or simply increasing the number of input and output (I/O) channels. In other words, a microcontroller is a mini or micro version of a computer because the microcontroller already has several peripherals that can be directly utilized, such as parallel ports, serial ports, comparators, digital to analog conversion (DAC), analog to digital. conversions and so on using only the minimum system that is not complicated or complex.

Technically, there are only 2 types of microcontrollers, namely RISC and CISC and each has its own pedigree/family. RISC stands for Reduced Instruction Set Computer: limited instructions but has more facilities CISC stands for Complex Instruction Set Computer: instructions can be said to be more complete but with adequate facilities. Microcontroller is a microprocessor system in which there is already a CPU, ROM, RAM, I/O, Clock, and other internal equipment that are interconnected and well organized (observed) by the manufacturer and packaged in one package. the chip is ready to use, so we only need to program the contents of the ROM according to the preparation rules made by the manufacturer.

Understanding Microcontroller is an electronic component that contains an interconnection system between the Microprocessor, RAM, ROM, I/O interface, and several peripherals. Microcontroller is also called On-chip-Peripheral. with naked eye control is controller, hence controller is controller of something. Microcontroller is a micro-sized controller.

An easy way to operate internal microcontroller devices such as (timer/counter, ext, user, etc.) is to study the registers that control the equipment. Microcontrollers are also integrated chips that are often part of embedded systems. In the application of the microcontroller there are advantages of system performance using a microcontroller as follows:

1. The controller on the microcontroller uses assembly programming language based on digital basic rules so that the operation of the system becomes very easy to do (this assembly language is easy to understand because it uses assembly language applications where input and output parameters can be directly accessed without using many commands). Assembly language design does not use much programming language so that the writing requirements such as uppercase and lowercase letters for assembly language are still reasonable.
2. The microcontroller is arranged in a single chip where the processor, memory, and I/O are integrated in a single control system so that the microcontroller can be regarded as a mini computer that can work innovatively according to system requirements.
3. The running system is independent without relying on the computer, while the computer parameters are only used to download instructions or programs. The steps to download a computer with a microcontroller are very easy to use because it does not use many commands.
4. The microcontroller provides additional facilities for the development of memory and I/O tailored to the needs of the system.
5. The price to obtain this tool is cheaper and easier to obtain.

4 Sensor Light Dependent Resistor

LDR (Light Dependent Resistor), is a type of resistor whose resistance changes due to the influence of light. When it's dark, the resistance value is getting bigger, while when it's bright, the value is getting smaller. LDR is a type of resistor that is commonly used as a light detector or meter. amount of light conversion [9, 19, 23, 24]. Light dependent resistor consists of a semiconductor disk that has two electrodes on its surface. The LDR resistance changes with changes in the intensity of the light hitting it. In the dark, the LDR resistance is about 10M, and in the light, it's 1k or less. LDRs are made of semiconductor materials such as cadmium sulfide. With this material the energy from the falling light causes more charge to be released or the electric current to increase. This means that the resistance of the material has decreased.

Light dependent resistors are used to convert light energy into electrical energy. However, because the response to light is quite slow, LDR is not used in situations where the light intensity changes drastically. This sensor will change its resistance value when there is a change in the brightness of the light.

5 Load Resistance

Battery voltage is the operating voltage of the solar cell panel module if the battery is connected directly to the solar cell panel module [22, 25, 26]. For example, the battery is generally 12 V, the battery voltage is usually between 11.5 to 15 V. To be able to charge the battery, the solar cell panel must operate at a higher voltage than the battery bank voltage. The highest efficiency is when the solar panel cells operate close to maximum powerpoint. In the example above, the battery voltage should be close to Vmp. If the battery voltage decreases below Vmp or increases above Vmp, then its efficiency decreases.

6 Results and Discussion

The prototype equipped with a maximum power point tracker has been designed to achieve better solar drying performance by dual energy drying based on an integrated maximum power point tracking system, overcoming the drawbacks of a single photovoltaic system using a microcontroller as the main logic control in maintaining performance whole solar collector drying system. The results show in Fig. 3 and Fig. 4, the maximum power point tracker is a system used to increase reception and voltage at different temperature conditions. This tool is efficient in drying if the LDR1 and LDR2 voltages are right at an angle of 25° to 5° during the day and an angle of 0° to −30° in the afternoon with an outside weather temperature of 31 °C.

Changes in the tilt angle of the solar panel position to the input voltage LDR 1 and LDR 2 to the microcontroller can be seen in Table 1 and Table 2 below.

Table 1. Changes in the tilt angle of the solar panel from 35° panel to 0° with respect to LDR1 and LDR2. Voltage

Solar panel tilt angle	Voltage	
	LDR1 (VDC)	LDR2 (VDC)
35°	4,48	4,87
30°	4,75	4,90
25°	4,85	4,9
20°	4,86	4,85
15°	4,86	4,76
10°	4,86	4,71
5°	4,86	4,68
0°	4,66	4,57

Table 2. Changes in the angle of the solar panel from 0° to −40° to the LDR1 and LDR2. voltages

Solar panel tilt angle	Voltage	
	LDR1 (VDC)	LDR2 (VDC)
0°	4,66	4,57
−5°	4,86	4,6
−10°	4,85	4,56
−15°	4,84	4,54
−20°	4,84	4,52
−25°	4,83	4,47
−30°	4,82	4,44
−35°	4,78	4,41
−40°	4,70	4,37

Information:

Data were obtained from noon to evening,
Coordinate 5°09′56.5″N 97°07′51.2″E
5.165682, 97.130877
Outdoor Temperature 31 °C.

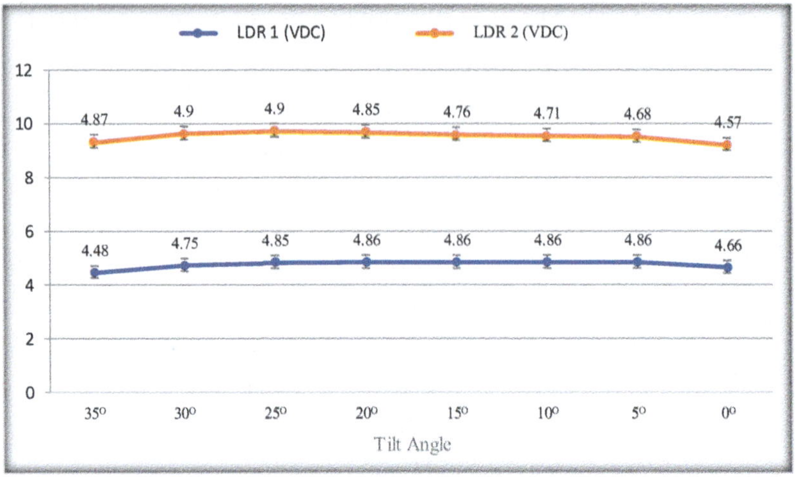

Fig. 3. Graph of voltage against tilt angle of solar panel position from angle 35° to 0°

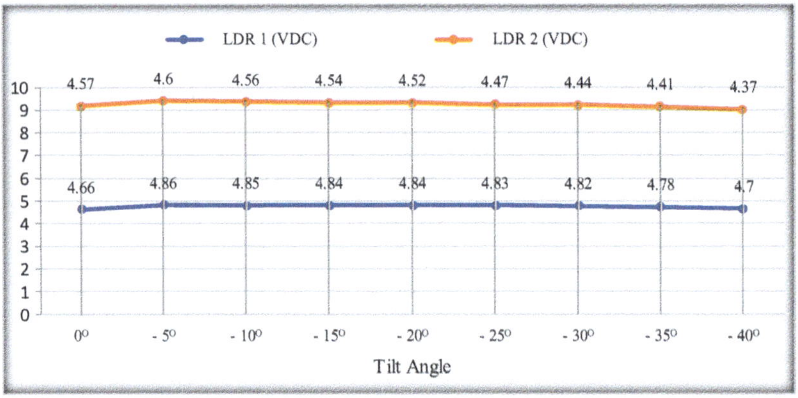

Fig. 4. Graph of voltage against tilt angle of solar panel position from angle 0° to −40°

7 Conclusion

The model prepared with a most extreme control point tracker has been planned to realize way better sun oriented drying execution by double vitality drying based on an coordinates greatest control point following framework, overcoming the downsides of a single photovoltaic system employing a microcontroller as the most rationale control in keeping up execution. entirety sun oriented collector drying framework. Greatest application of powerpoint tracker on microcontroller based double vitality sun based fueled dryer on sun based boards for drying moringa takes off produces the advantage of greatest vitality retention in sun powered boards and abbreviates the time within the drying prepare of moringa takes off.

Acknowledgements. The authors would like to thank Universitas Syiah Kuala for the laboratory equipment and the Ministry of Research Technology and the Higher Education Republic of Indonesia for the research grant program under the grant of Penelitian Disertasi Doktor No: 33/UN11.2.1/PT.01.03/DRPM/2021.

References

1. Iwansyah, A.C., et al.: Effects of various drying methods on selected physical and antioxidant properties of extracts from moringa Oliefera leaf waste. Sustainability **12**(20), 1–12 (2020). https://doi.org/10.3390/su12208586
2. Lingayat, A., Chandramohan, V.P., Raju, V.R.K.: Design, development and performance of indirect type solar dryer for banana drying. Energy Procedia **109**, 409–416 (2017). https://doi.org/10.1016/j.egypro.2017.03.041
3. Costa, S.C.S., Diniz, A.S.A.C., Kazmerski, L.L.: Solar energy dust and soiling R&D progress: literature review update for 2016. Renew. Sustain. Energy Rev. **82**, 2504–2536 (2018). https://doi.org/10.1016/j.rser.2017.09.015
4. Fudholi, A., et al.: Solar drying technology in Indonesia: an overview. Int. J. Power Electron. Drive Syst. **9**(4), 1804 (2018). https://doi.org/10.11591/ijpeds.v9.i4.pp1804-1813
5. Jalil, S.M., Abdurrahman, F., Meliala, S., Rosdiana, R.: Design of maximum power point tracking for solar collector drying system: an experimental study. Int. J. Power Electron. Drive Syst. **9**, 1799–1803 (2018). https://doi.org/10.11591/ijpeds.v9n4.pp1799-1803
6. Tiwari, A.: A review on solar drying of agricultural produce. J. Food Process. Technol. **7**(9) (2016). https://doi.org/10.4172/2157-7110.1000623
7. Hasanah, R.N., Setyawan, A.B., Maulana, E., Nurwati, T., Taufik, T.: Computer-based solar tracking system for PV energy yield improvement. Int. J. Power Electron. Drive Syst. **11**(2), 743–751 (2020). https://doi.org/10.11591/ijpeds.v11.i2.pp743-751
8. Genobiagon, C.P., Alagao, F.B.: Performance of low-cost dual circuit solar assisted cabinet dryer for green banana. J. Mech. Eng. Res. Dev. **42**(1), 42–45 (2019). https://doi.org/10.26480/jmerd.01.2019.42.45
9. Meliala, S., Saifuddin, Rosdiana: Drying design model for dried-anchovy using solar collector and solar cell panel position control. Int. J. Recent Technol. Eng. **7**(6S3), 1430–1435 (2019). http://www.scopus.com/inward/record.url?eid=2-s2.0-85067196861&partnerID=MN8TOARS
10. Saifuddin, Meliala, S., Albra, R.W., Ikramuddin: Solar cell position control building design using microcontrollerf optimizing power sunway in the coffee drying application. Int. J. Recent Technol. Eng. **7**(6S5), 1490–1497 (2019). http://www.scopus.com/inward/record.url?eid=2-s2.0-85066619390&partnerID=MN8TOARS
11. Rabha, D.K., Muthukumar, P.: Feasibility study of the application of a latent heat storage in a solar dryer for drying Green Chili. In: 2nd International Conference on Energy, Power Environment Towards Smart Technology ICEPE 2018, pp. 1–6 (2019). https://doi.org/10.1109/EPETSG.2018.8658770
12. ELkhadraoui, A., Kooli, S., Hamdi, I., Farhat, A.: Experimental investigation and economic evaluation of a new mixed-mode solar greenhouse dryer for drying of red pepper and grape. Renew. Energy **77**, 1–8 (2015). https://doi.org/10.1016/j.renene.2014.11.090
13. Njt, E.: Effect of drying techniques of moringa leaf on the quality of Chin-Chin enriched with Moringa leaf powder. IOSR J. Environ. Sci. **10**(4), 65–70 (2016). https://doi.org/10.9790/2402-1004016570

14. Layek, A., Muthukumar, P.: Evaluation of convective heat transfer coefficient of herbs dried in a mixed mode solar dryer. In: ICUE 2018 on Green Energy for Sustainable Development, pp. 24–26 (2018)
15. Vaghela, D., Bhautik, G., Sengar, S.H.: Comparative study of solar tunnel and open sun drying for moringa oleifera leaves. Int. J. Sci. Environ. Technol. **7**(2), 472–476 (2018). www.ijset.net
16. Lakshmi, D.V.N., Layek, A., Muthukumar, P.: Evaluation of convective heat transfer coefficient of herbs dried in a mixed mode solar dryer. In: Proceedings of the Conference on Industrial and Commercial Use of Energy, ICUE, October 2018 (2019). https://doi.org/10.23919/ICUE-GESD.2018.8635684
17. Çiftçioğlu, G.A., Kadırgan, F., Kadırgan, M.A.N., Kaynak, G.: Smart agriculture through using cost-effective and high-efficiency solar drying. Heliyon **6**(2), 1–7 (2020). https://doi.org/10.1016/j.heliyon.2020.e03357
18. Singh, P., Shrivastava, V., Kumar, A.: Recent developments in greenhouse solar drying: a review. Renew. Sustain. Energy Rev. **82**, 3250–3262 (2018). https://doi.org/10.1016/j.rser.2017.10.020
19. Hafez, A.Z., Yousef, A.M., Harag, N.M.: Solar tracking systems: technologies and trackers drive types – a review. Renew. Sustain. Energy Rev. **91**, 754–782 (2018). https://doi.org/10.1016/j.rser.2018.03.094
20. Suriadi, S., Fajri, I.N., Munadi, R., Gapy, M.: Reduksi Osilasi Daya Pada MPPT Panel Surya Dengan Metode Kombinasi PNO dan Fuzzy. J. Rekayasa Elektr. **15**(2), 110–117 (2019). https://doi.org/10.17529/jre.v15i2.13682
21. Dronachari, M.: Application of different types solar dryers in agriculture crops- a review. Int. J. Pure Appl. Biosci. **7**(2), 303–326 (2019). https://doi.org/10.18782/2320-7051.5586
22. Senthil Nayagam, V., Premalatha, L.: Green energy based coupled inductor interleaved converter with MPPT technique for BLDC application. Int. J. Power Electron. Drive Syst. **9**(4), 1725 (2018). https://doi.org/10.11591/ijpeds.v9.i4.pp1725-1732
23. Fadil, S., Capar, A.C., Caglar, K.: Two axis solar tracker design and implementation. In: ELECO 2013 - 8th International Conference on Electrical and Electronics Engineering, pp. 554–557 (2013). https://doi.org/10.1109/eleco.2013.6713907
24. Indriani, A., Mt, S.T., Hendra, E., Witanto, Y.: Utilization of Drying System Using Genuino Uno Microcontroller and Solar Cell for Processing of Agricultural Products of Pasar Pedati and Harapan Village People (2017)
25. Chandel, M., Agrawal, G.D., Mathur, S., Mathur, A.: Techno-economic analysis of solar photovoltaic power plant for garment zone of Jaipur city. Case Stud. Therm. Eng. **2**, 1–7 (2014). https://doi.org/10.1016/j.csite.2013.10.002
26. Hashim, N., Salam, Z., Johari, D., Nik Ismail, N.F.: DC-DC boost converter design for fast and accurate MPPT algorithms in stand-alone photovoltaic system. Int. J. Power Electron. Drive Syst. **9**(3), 1038 (2018). https://doi.org/10.11591/ijpeds.v9.i3.pp1038-1050

Enhancing Solar Power Harvest By Using Absorber Plates on Thermoelectric Generator Modules

Abdul Halim, Zuryati Djafar(✉), and Wahyu H. Piarah

Mechanical Engineering Department, Faculty of Engineering, Universitas Hasanuddin, Sungguminasa 92171, South Sulawesi, Indonesia
{zuryatidjafar,wahyupiarah}@unhas.ac.id

Asbtract. Research on thermoelectric as a power generator known as a thermoelectric generator (TEG) shows a positive trend. The current constraint is the relatively low efficiency, especially if it is used to absorb direct sunlight energy. To increase the output power of TEG, it is endeavored to increase the temperature of the hot side by placing the absorber plate on the hot side so that the temperature difference between the hot and cold side (ΔT) will also increase. This study was conducted to determine the magnitude of the increase in electrical energy generated from the TEG module using several variations of absorber plates such as aluminum, iron, and copper compared to without absorber plates. The results showed that the intensity of solar radiation greatly affects the temperature difference obtained from both sides of the TEG module so that it can produce an increase in the difference in output power voltage and efficiency. The power and efficiency of Copper, Iron, Aluminum and without plates were obtained: 0.0023 W and 1.831 %, 0.0019 W and 1.301 %, 0.0004 W and 0.628 %, and 7.14×10^{-5} W and 0.1014 %. It can be seen that there is an increase in output power using Al, Fe, and Cu plates by 5.6, 26.6, and 32.2 times, respectively, and an increase in efficiency by 6.2, 12.8, and 18.1 times.

Keywords: Absorber plate · Efficiency · Solar power · Thermoelectric generator · Voltage difference

1 Introduction

In everyday life, we cannot be detached from the use of energy, both on a small and large scale. The need for energy will continue to increase along with technological developments. Most of the energy sources currently used are non-renewable, such as oil and coal. They are classified as fossil energy and will run out due to their use in everyday life [1]. Therefore, it is necessary to find and utilize new renewable energy sources. Currently, the use of renewable energy is still in a relatively small percentage but is expected to be able to supply energy needs in the future. Alternative energy sources such as wind, sunlight, temperature differences in the sea, geothermal, river flows, and other sources, can be used to produce electrical energy [2].

Akhyar et al. (Eds.): ICECME 2021, LNME, pp. 53–59, 2023.
https://doi.org/10.1007/978-981-19-3629-6_6

The development of the use of new and renewable energy (EBT) has received support from the Government of Indonesia through the regulation of the minister of energy and mineral resources number 39 of 2017 concerning the implementation of physical activities using new and renewable energy and energy conservation. Through these official regulations, it is expected that the wider community will be willing and able to utilize EBT in order to ensure the availability of electrical energy [3].

In Indonesia, one of the renewable energy sources that can be utilized is sunlight. Based on solar radiation data collected from various locations in Indonesia, it shows that the potential in the western region of Indonesia is 4.5 kWh/m^2.day with a monthly variation of about 10%, and the eastern region of Indonesia is 5.1 kWh/m^2.day with a monthly variation. around 9% with an average Indonesian value of 4.8 kWh/m^2.day with a monthly variation of around 9%. This energy potential is equivalent to 112 thousand GWp. However, currently, only about 49MWp of solar energy has been utilized. This means that the potential for solar energy that has been utilized is still far from 1% [4].

As a form of support in the development of alternative energy sources, we offer the use of a thermoelectric generator (TEG). In this study, the source of heat energy is obtained from sunlight which is then absorbed by an absorbent plate that functions to increase the absorption of solar energy which is then converted into electrical energy

2 Literature Review

In 1821 a German researcher Thomas Johann Seebeck discovered a device that could convert heat energy into electrical energy [5–9]. Thermoelectric Generator (TEG) is a power generating device that uses the principle of the Seebeck effect [5–8]. The principle of the Seebeck effect itself is: when two metal rods or wires which are generally semiconductors are connected where one end is heated and the other end is kept cold, it will produce a voltage difference or electromotive force [10–13].

In Fig. 1, the structure of the thermoelectric generator (TEG) consists of an arrangement of elements of n-type (electron-excess material) and p-type (electron-deficient material). Heat enters on one side and is discharged on the other. The difference in temperature between the two sides causes diffusion of the material to receive the hot side which is actively moving and has a higher flow velocity to the cold side [11]. As a result of this diffusion creates an electric field. The number of electrons on the hot side of the

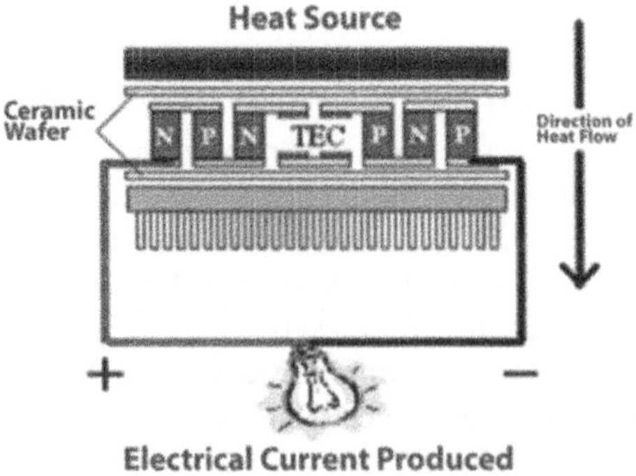

Fig. 1. Thermoelectric generator structure [13]

metal is more tenuous when compared to the cold side which is denser [12]. The amount of voltage generated is proportional to the temperature difference [13].

Thermoelectric technology is a solid-state technology that is unmoving and environmentally friendly [12, 14]. In addition, this technology is one of the alternative energy sources in overcoming the energy crisis from year to year [15]. The advantages of this thermoelectric technology are that it is safe for the environment, small in size, silent, does not require complicated and light maintenance [13, 16].

In addition to having advantages, thermoelectric generators also have disadvantages, namely, the efficiency value is still low. This is something that always gets attention to be improved, among others, by optimizing the performance of the components in the thermoelectric generator, or developing thermoelectric materials and improving the manufacturing quality of the thermoelectric generator itself [13].

The efficiency value of the thermoelectric module can be increased by lowering the temperature on the cold side by using a heatsink, fan, water jacket, or by providing a low ambient temperature above the cold side of the thermoelectric to keep the temperature difference high [17].

Another way to increase the efficiency offered here is to attach the absorber plate to the hot side surface of the thermoelectric.

In analyzing the performance of the thermoelectric module, the Seebeck coefficient which describes the voltage (electromotive force) arising due to temperature differences becomes very important. Seebeck coefficient can be expressed by the following equation [5, 12]:

$$\propto = \frac{\Delta V}{(T_h - T_c)} \tag{1}$$

where ΔV is Voltage difference (Volt), α is Seebeck Coefficient (V/°C), T_h is Hot side temperature (°C) and T_c is Cold side temperature (°C).

While the electric current generated is given by the following equation [5]:

$$I = \frac{\propto \Delta T}{R_i + R_L} = \frac{\propto (T_h - T_c)}{R_i + R_L} \tag{2}$$

In this equation, I is Current (Ampere), R_i is Internal Resistance (Ω), R_L is External resistance and ΔT is Temperature difference ($T_h - T_{c}$).

The heat transfer rate (Q_h) from the heat source on the hot side surface to the cold side surface is as follows [12]:

$$Q_h = (\propto I\, T_h) + k(T_h - T_c) \tag{3}$$

where k is the thermal conductivity of the module. The output power (P) generated to resist external loads [R_L] and efficiency (η) thermoelectric generator is as follows [5] :

$$P = I^2 R_L \tag{4}$$

$$\eta = \frac{\mathbf{P}}{\mathbf{Q_h}} \tag{5}$$

3 Research Method

The design of the equipment used in this study can be seen in Figure 2 below. The absorber plate has a thickness of 2 mm in a rectangular shape, placed on top of the thermoelectric. The thermoelectric bottom surface uses a heatsink-fan cooling system. Three variations of absorber materials are used, namely Aluminum (Al), Iron (Fe), and Copper (Cu) plates. Measurements were carried out simultaneously for the 3 variations of the absorber plate and without the absorber plate.

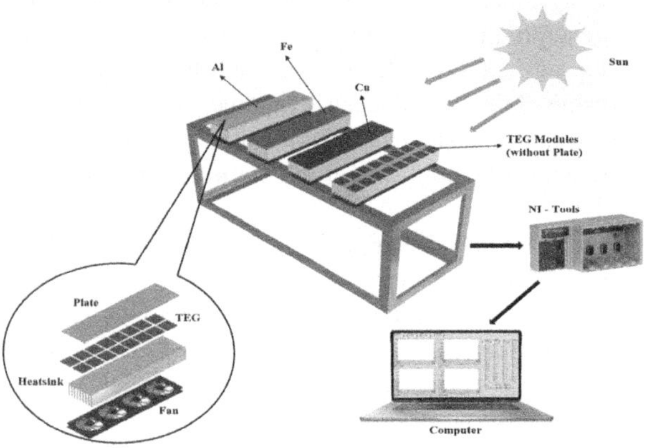

Fig. 2. Testing installation schematic

The measurements of the temperature of the hot side (Th), the cold side (Tc), and the voltage (ΔV) using the National Instrument and carried out with an interval of 15 min for 8 h, starting from 9.00 a.m to 17.00 p.m. Measurement data recording using LabVIEW software.

4 Results and Discussion

In this paper the testing parameters are focused on the temperature difference (ΔT), voltage difference (ΔV), heat absorbed (Q_h) output power (P) and efficiency (η). The results of measurements and calculations are displayed in graphical form.

Figure 3 shows the characteristics of the temperature difference between the hot and cold sides of the thermoelectric T and the voltage V generated during data collection. It appears that the pattern of temperature differences is identical to the resulting stress pattern. This is in accordance with the principle of the Seebeck effect. At first, there was an increase in temperature and voltage differences along with the increasing intensity of sunlight and was at a high condition between 10.30 a.m and 13.00 p.m. After that, there was a decrease as the intensity of the sun decreased in the afternoon. The difference in temperature and the highest voltage for each absorber plate is as follows: for copper (Cu) plates, it produces T and V of 4.767 °C and 0.871 V; iron plate (Fe), T and V of 6.295 °C and 0.805 V; aluminum plate (Al), T and V of 2,996 °C and 0.369 V;

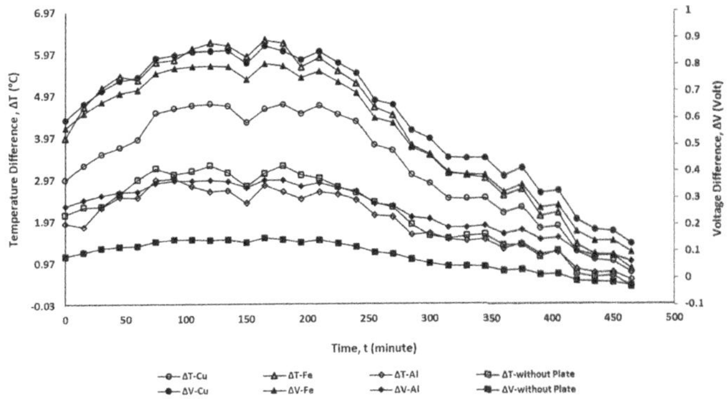

Fig. 3. Characteristics of temperature (ΔT) and voltage differences (V) to time (t)

Figure 4 shows the characteristics of the temperature difference with the heat absorbed by the thermoelectric module. The characteristic patterns that occur are very identical to each other where the increase and decrease in heat absorbed also depends on the intensity of sunlight. The measurement values obtained are: for copper (Cu) plates, T and Qh are 4.767 °C and 0.127 W; iron plate (Fe), T and Qh of 6.295 °C and 0.155 W; aluminum plate (Al), T and Qh of 2,996 °C and 0.0722 W and for non-plate produce T and Qh of 3.295 °C and 0.0744 W. It also appears that the heat absorbed is highly dependent on the absorber material used. For aluminum material, the rate of heat absorption is not too influential compared to iron and copper materials.

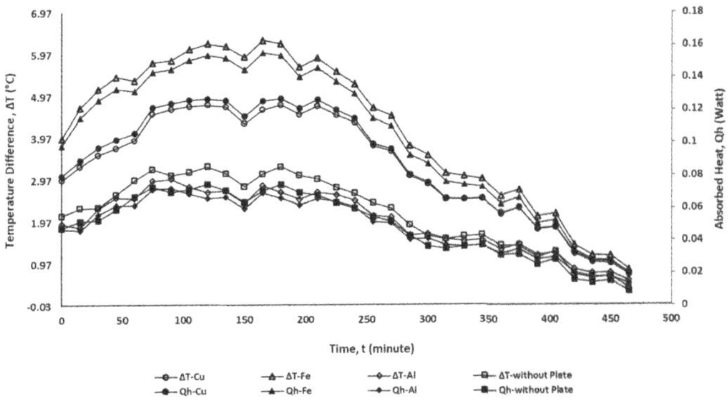

Fig. 4. Characteristics of temperature difference (ΔT) and absorbed heat (Q_h) to time (t)

Figure 5 shows the characteristics of the output power and the heat absorbed. The ability to convert the output power from the absorbed heat is relatively low for aluminum which is shown by the graph gap which is relatively large compared to iron and copper. In addition, the use of absorber plates shows an increase in the output power produced. The results are shown for copper (Cu), accounted for 0.0023 W for power and heat absorbed 0.1271 W; iron (Fe), produces a power of 0.0019 W and absorbed heat of 0.1550 W; aluminum (Al), produces a power of 0.0004 W and absorbed heat of 0.0722 W and for non-plate produces a power of 0.00007 W and absorbed heat of 0.0744 W. The intensity of the sun also shows a great influence on the output power produced.

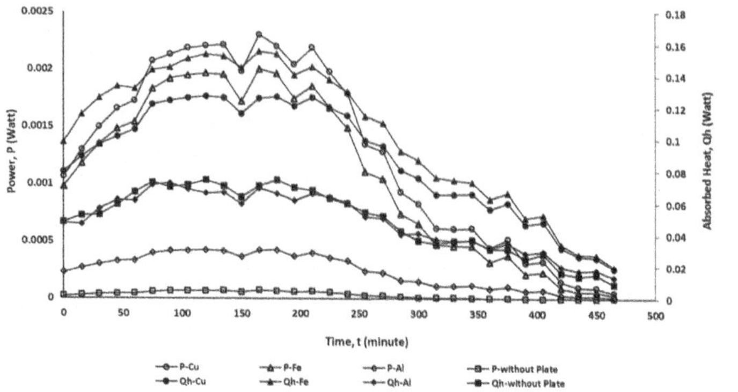

Fig. 5. Relationship between power (P) and absorbed heat (Q_h) to time

Figure 6 shows output power characteristics and thermoelectric efficiency which is resulted from measurements on the absorber plate. It can be seen that the output power and efficiency will increase with the use of absorber plates where the absorber plates that provide the best output power and efficiency are descending from copper, iron, and aluminum. The power and efficiency values produced are copper plate (Cu), producing P and η of 0.0023 W and 1.831 %; iron plate (Fe), producing P and η of 0.0019 W and 1.3014 %; aluminum plate (Al), producing P and η of 0.0004 W and 0.6287 %; as well as for without-plate produces P and η of 0.00007 W and 0.1014 %.

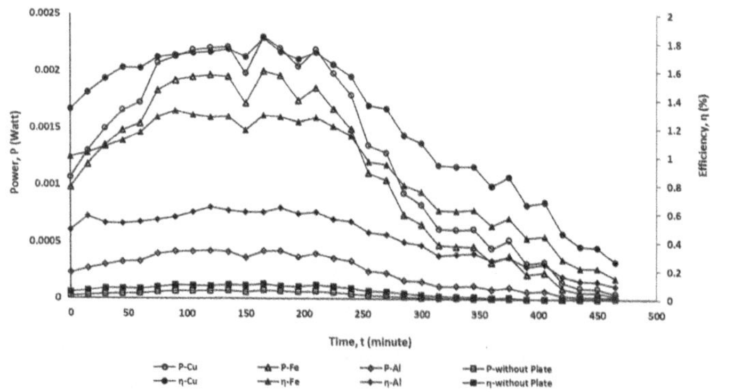

Fig. 6. Relationship of power (P) and efficiency (η) to time

5 Summary

The characteristics of the output power and thermoelectric efficiency indicate that the use of absorbent plates will have a positive impact on the thermoelectric performance.

The power and efficiency of Copper, Iron, Aluminum and without plates were obtained: 0.0023 W and 1.831%, 0.0019 W and 1.301%, 0.0004 W and 0.628%, and 7.14×10^{-5} W and 0.1014%. It can be seen that there is an increase in output power using Al, Fe, and Cu plates by 5.6, 26.6, and 32.2 times, respectively, and an increase in efficiency by 6.2, 12.8, and 18.1 times.

References

1. Kholiq, I.: Utilization of alternative energy as renewable energy to support fuel substitution. J. Sci. Technol. **19**(2) (2015)
2. Ramadhan, A.I., Diniardi, E., Mukti, S.H.: Analysis Design System Solar Power Generation Capacity 50 WP, Engineering (2016)
3. Budiman, A., Das, K., Mohammad, A., Tan, K.T., Tonby, O.: Ten ideas for reinvigorating Indonesia's energy sector. McKinsey & Company, Technical report (2014)
4. Directorate General of New, Renewable Energy and Energy Conversion, Strategic Plan of the Ministry of Energy and Mineral Resources 2015-2019, Ministry of Energy and Mineral Resources, Technical report (2015)
5. Djafar, Z., Putra, N., Koestoer, R.A.: The utilization of heat pipe on cold surface of thermoelectric with low-temperature waste heat. Appl. Mech. Mater. **302**, 410–415 (2013)
6. Mustofa, M., Djafar, Z., Syafaruddin, Piarah, W.H.: A new hybrid of photovolatic-thermoelectric generator with hot mirror as spectrum splitter. J. Phys. Sci. **29**, 63–75 (2018)
7. Usman, S., Djafar, Z., Tjandinegara, T., Djafar, Z., Piarah, W.H.: Heat utilization of incinerator chimneys as mini power generation based on thermoelectric. EPI Int. J. Eng. **1**(1), 1–2 (2018)
8. Cai, Y., Mei, S.J., Liu, D., Zhao, F.Y., Wang, H.Q.: Thermoelectric heat recovery units applied in the energy harvest built ventilation: parametric investigation and performance optimization. Energy Convers. Manag. **171**, 1163–1176 (2018)
9. Shittu, S., Li, G., Akhlagi, Y.G., Ma, X., Zhao, X., Ayodele, E.: Advancements in thermoelectric generator for enhanced hybrid photovolatic system performance. Renew. Sustain. Energy Rev. **109**, 24–54 (2019)
10. Min, G., Roe, D.M.: Peltier devices as generator. In: Handbook of Thermoelectrics, p. 479. CRC Press LLC, Florida (1994)
11. Willfahrt, A.: Screen printed thermoelectric devices. Linköping Studies in Science and Technology, no. 1663 (2014). 49
12. Goldsmid, H.J.: Introduction to Thermoelectricity, Springer Series in Material Science 121. Springer-Verlag, Heidelberg (2010). https://doi.org/10.1007/978-3-642-00716-3
13. Putra, N., Koestoer, R.A., Adhitrya, M., Roekettino, A., Trianto, B.: The potential of thermoelectric power generation for hybrid vehicles. J. Makara Teknologi **13**(2), 53–58 (2009). (in Bahasa)
14. Prasetyo, Y., et al.: Characteristic of thermoelectric TEC varies type with variation load resistor. J. Energi dan Teknol. Manufaktur **2**, 36–41 (2019). (in Bahasa)
15. Nugrah, S., Aziz, A., Mainil, R.I.: Testing a thermoelectric generator (TEG) with a 60 Volt electric heater using cooling water at environmental temperatures. Energy Conversion Laboratory, Department of Mechanical Engineering, Faculty of Engineering, University of Riau. J. TEKNIK **4**(2) (2017). (in Bahasa)
16. Riffat, S.B., Ma, X.: Thermoelectrics: a review of present and potential applications. Appl. Therm. Eng. **23**, 913–935 (2003)
17. Ismail, B.I., Ahmed, W.H.: Thermoelectric power generation using waste-heat energy as an alternative green technology. Recent Patents Electr. Eng. **2**, 27–39 (2009)

Fuzzy Logic Control System For Fuel-Saving Using Steering Behavior

Suroto Munahar[1,2](✉), M. Munadi[1], Aris Triwiyatno[3], and Joga Dharma Setiawan[1]

[1] Department of Mechanical Engineering, Universitas Diponegoro, Semarang, Indonesia
munahar@unimma.ac.id, munadi@lecturer.undip.ac.id,
joga.setiawan@ft.undip.ac.id

[2] Department of Automotive Engineering,
Universitas Muhammadiyah Magelang, Magelang, Indonesia

[3] Department of Electronics Engineering, Universitas Diponegoro, Semarang, Indonesia
aristriwiyatno@live.undip.ac.id

Abstract. Currently, the development of spark ignition engine technology leads to energy-efficient and low-emission vehicle policies, which is indicated by programs, such as the Low Cost Green Car (LCGC) program, Electric Vehicle (EV), Hybrid Vehicle (HV), and vehicles with alternative fuels. The direction of these policies is due to the declining availability of fossil energy and the issue of global climate change. Furthermore, various efforts have been taken to resolve this problem. To realize energy-efficient vehicles, the development of spark ignition engine currently leads to the achievement of AFR stoichiometry (14.67). This allows the vehicle to operate with optimal power and good fuel economy. However, further improvement can be made by carefully analyzing the power requirements and fuel control. Therefore, this study offers a new method to improve fuel economy based on driver behavior aspects, which was applied by integrating steering operations (the driver's behavior while operating the steering) and fuel control. The steering operation was designed with 3 clusters, namely the low, medium, and high. Similarly, vehicle speed was designed with 3 clusters, namely low, medium, and high., The fuel mode was designed based on the input steering operation, vehicle speed, and fuzzy logic. The simulation results were in the form of fuel modes, such as economizer and stoichiometry. The economizer regulates the AFR fuel in the lean mixture (15.4), while the stoichiometry adjusts the AFR at the ideal mixture (14.67), respectively. Therefore, the fuel economy was realized when the steering operation and vehicle speed were both in the low cluster. Based on the simulation, it was concluded that the technology application designed, has a very high potential of being applied in real vehicles to assist in realizing energy-efficient vehicles.

Keywords: Steering operation · Vehicle speed · Engine speed · Steering speed · Fuel control system

1 Introduction

Environmental friendly and energy-efficient vehicles are one of the policies for the development of spark ignition engine [1, 2]. Therefore, the availability of global energy is

Akhyar et al. (Eds.): ICECME 2021, LNME, pp. 60–73, 2023.
https://doi.org/10.1007/978-981-19-3629-6_7

a major factor [3]. However, there is a continuous rise in energy demand in the transportation sector along with an increase in the number of vehicles. The development of injection technology is one of the efforts to realize energy-efficient vehicles [4]. Lee [4] developed dual port injection technology, which led to a fair good fuel economy. Furthermore, Hydraulic Variable Valve Actuation (HVVA) was introduced to increase fuel efficiency [5]. The element of flexibility and application of genetic algorithms were also considered in this study. The results were quite interesting with about 13% achievement of fuel economy, but both studies did not include steering operation as a control element.

Electronic Control Unit (ECU) is a fuel system control module in vehicles, which is continuously being developed for the improvement of internal combustion engines. Meanwhile, air to Fuel Ratio (AFR) is the main element to increase fuel economy, which is a driver's need [6–8]. It has three ranges of values, including lean (AFR above (14.67), stoichiometry (14.67), and rich below (14.67) [9]. Currently, the achievement of stoichiometry values is the focus of AFR control. In general, it is maintained at the ideal value or stoichiometry (14.67).

Recently, artificial intelligence (AI) is being applied as a controller in the fuel control system [10, 11], an example is a fuzzy logic [12, 13]. Furthermore, this study used fuzzy logic to control AFR but is yet to consider steering operation. Therefore, there is a need for further studies using the intake manifold pressure variable as an AFR clustering with neuro-fuzzy. The output would be in the form of a fuel injection control system. Another study used oxygen sensors to control AFR [6]. Besides fuzzy logic, the neural network is an AI that can be used for control systems [14–16]. Therefore, there is a continuous increase in studies on neural networks as a fuel control system continue [17]. Arsei [17] utilized the Recurrent Neural Network (RNN) to achieve the AFR value in the stoichiometry range but has also not considered steering operation.

The behaviour of drivers in steering operation varies. This greatly affects the engine performance which needs an adjustment mechanism between the driver behavior and engine fuel requirements [18, 19] Recently, the safety system is a growing study theme in the field of driver behavior [20–22]. Grove [20] developed recognition of driver behavior in toll roads. Kohl [21] carried out an observation of vehicle graphics technology on driver behavior. Furthermore, Martinelli [22] developed a control system to prevent vehicle theft based on driver behavior. These studies observed driver behavior but did not consider the fuel control system. Besides security systems, driver behavior also extends to other scopes [23–25]. Hong [23] assessed driver behavior based on the level of aggressiveness. Sharma [24] developed the following car technique on driver behavior. Vaezipour [25] used a simulator tool to recognize driver behavior. However, these studies only discussed the recognition/assessment of driver behavior without assessing the fuel control system. Furthermore, no study has been carried out on the development of driver behavior in steering operations for fuel control. Therefore, this study developed this theme to improve fuel economy based on the driver behavior in the steering operation.

2 Methods

The steering operation control system model for the fuel system using a fuzzy logic controller has 4 sub-models, which include a fuel control system, engine gas dynamics, drivetrain, and steering control system as presented in Fig. 1. The fuel control system and engine gas dynamics models are designed based on the previous study [26].

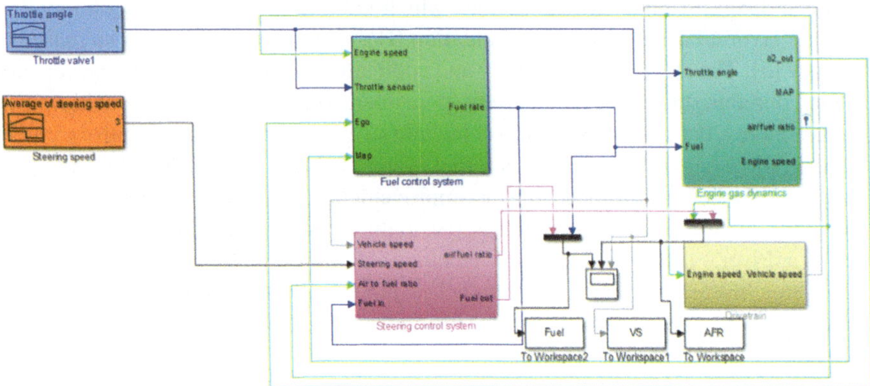

Fig. 1. The design of steering operation control system model on the fuel control system.

2.1 Fuel Control System

The fuel control system describes the dynamics of fuel control and has 4 inputs, namely engine speed, throttle sensor, EGO, and map. Engine speed is used for the dynamics of the crankshaft rotation on the engine in rpm. The throttle sensor acts from the throttle valve opening angle when the driver increases speed. As an input, EGO comes from the oxygen sensor monitoring oxygen gas in the exhaust manifold channel. Meanwhile, as an output, it corrects AFR settings on the engine. Furthermore, a map, known as manifold absolute pressure, is the dynamics of pressure in the intake manifold during engine operation. The fuel control system model is shown Fig. 2.

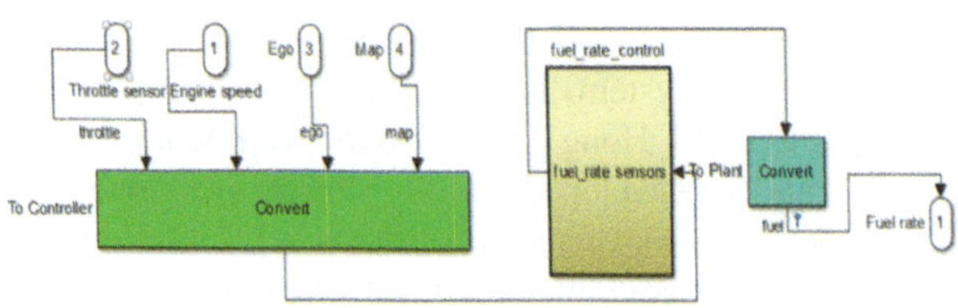

Fig. 2. Fuel control system model.

a. **Fuel Rate Control System**
The fuel rate control system has 3 components namely, control logic, airflow calculator, and fuel calculator. Control logic operates based on the state flow method with the output mode of fuel, normal oxygen, and sensor performance. The fuel rate control system model is shown in Fig. 3.

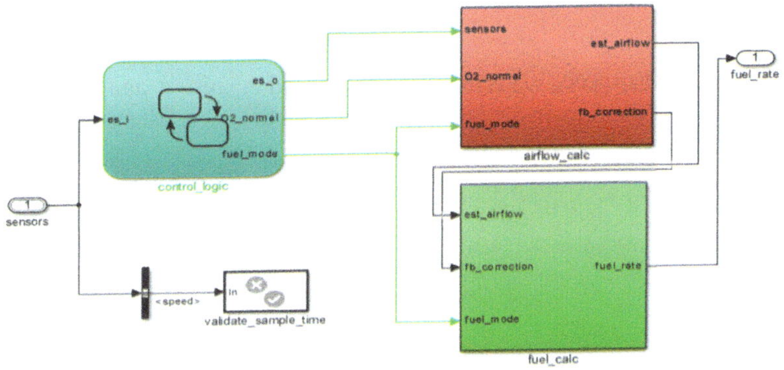

Fig. 3. Fuel rate control system model.

b. **Airflow Intake Manifold And Closed Loop Correction**

The throttle valve of opening angle, engine speed, and intake manifold pressure was used to calculate the estimated airflow in the intake manifold, while the engine was in operation. Meanwhile, the integration of intake manifold pressure and engine speed was calculated based on the characteristic database.

The operation of the closed loop correction is based on the integration between mode fuel, oxygen content in the exhaust manifold, oxygen switching, and intake manifold pressure. The intake manifold and closed loop correction airflow models are shown in Fig. 4.

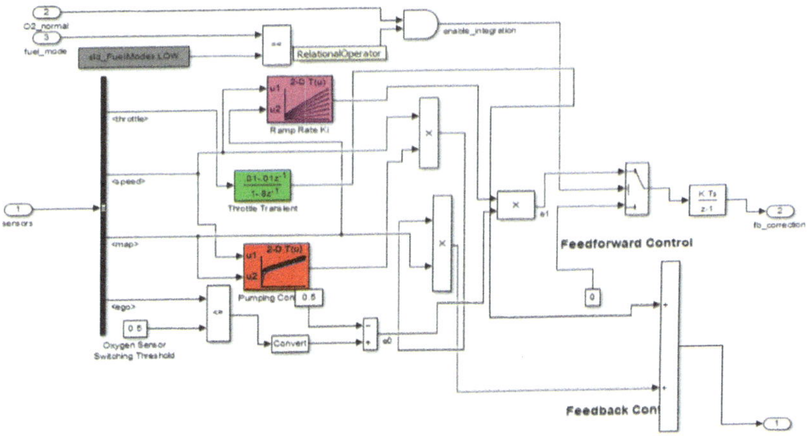

Fig. 4. Intake manifold and closed loop correction airflow models.

Estimated intake manifold airflow was calculated based on Eq. 1.

$$q = \frac{N}{4\pi}.V_{cd}.v\frac{P_m}{R.T}$$

$$= C_{pm}(N, P_m)NP \tag{1}$$

where N is the engine speed, V_{cd} is the stroke volume of the piston, and v is volume efficiency. P_m is the air pressure in the intake manifold and the gas constant and temperature specifications. Meanwhile, q is an intake mass flow.

The closed loop correction logic in the fuel control system refers to Eqs. 2, 3, 4, 5 and 6.

$$e_0 = 0.5 \text{ for EGO} \le 0.5 \tag{2}$$

$$e_0 = -0.5 \text{ for EGO} > 0.5 \tag{3}$$

$$e_1 = K_i(N, P_m)e_0 \text{ for EGO} \le 0.5 \tag{4}$$

$$\dot{e_2} = \dot{e_1} \text{ for LOW for EGO signal} \tag{5}$$

$$\dot{e_2} = 0 \text{ for RICH, disable/signal EG} \tag{6}$$

$$e_0, e_1, e_2 = \text{intermediate signal error}$$

2.2 Steering Control System

The steering control system has 2 components, namely fuzzy logic and behavior recognition of steering operation. Furthermore, it has 2 inputs and outputs, each. Fuzzy logic works based on vehicle speed and steering. Vehicle and steering speed is calculated in units of km/h and rotation/seconds, respectively. The vehicle speed, steering speed, and fuzzy logic work together to control the mode fuel and is achievable with fuzzy logic. Behavior recognition becomes a control system that works in tandem with the fuel control system. The steering control system model is shown in Fig. 5.

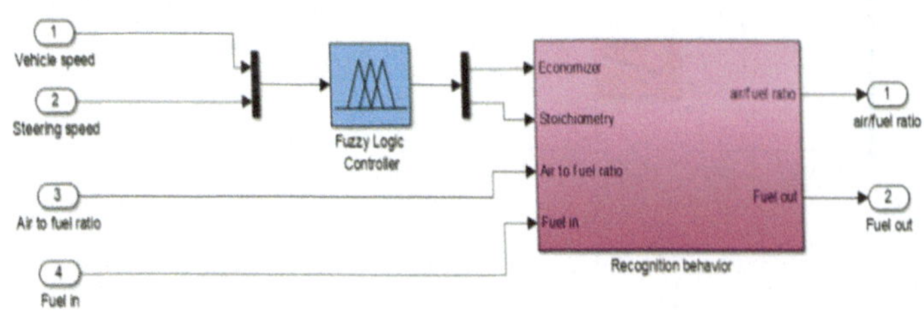

Fig. 5. Steering control system model.

The fuzzy logic used in this model has 2 membership functions, which include steering and vehicle speed. Steering speed has 3 variables, namely low, medium, and high with a value of 0–0.05, 0.04–0.11, and 0.07–0.12 rotation/seconds, respectively.

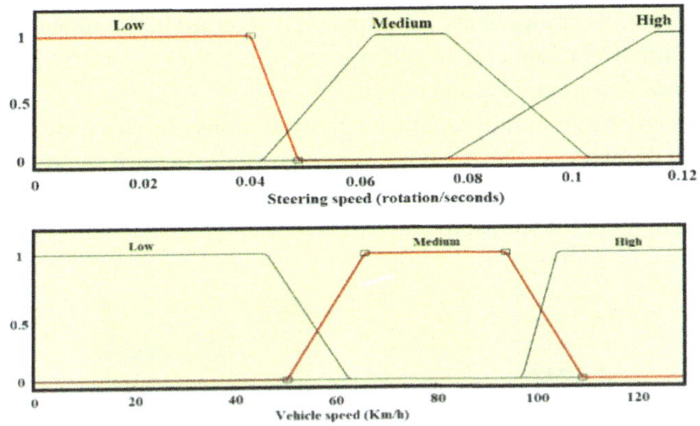

Fig. 6. Fuzzy logic membership function.

Similarly, vehicle speed has 3 variables, namely low, medium, and high with a value of 0–63, 50–110, and 90–120 km/h, respectively. Fuzzy logic in the steering control system is shown in Fig. 6.

The decision statement of the fuel system mode control that considers the steering and vehicle speed factors is shown in Table 1.

Table 1. Fuel system mode control based on fuzzy logic.

No.	Average steering speed	Vehicle speed	Fuel mode system
1	Low	Low	Economizer
2	Low	Medium	Stoichiometry
3	Low	High	Stoichiometry
4	Medium	Low	Stoichiometry
5	Medium	Medium	Stoichiometry
6	Medium	High	Stoichiometry
7	High	Low	Stoichiometry
8	High	Medium	Stoichiometry
9	High	High	Stoichiometry

2.3 Drivetrain System

The drivetrain system is a vehicle drive model, which functions to transmit and regulate rotation and torque from the engine to the wheels as required. This system has 3 components, namely the clutch, transmission, and differential. Furthermore, the clutch is a component that serves to disconnect and connect the engine speed to the transmission. The assumption in this study is that the clutch does not slip, and continues on a 1 to 1 rotation. Therefore, its components were modeled together in the transmission. In this study, the transmission modeled is 5-speed gears, i.e. 1, 2, 3, 4, and 5. The ratio of the designed transmission model is shown in Table 2.

The differential is a drivetrain component that functions to divide the rotation from the transmission to the wheels. Furthermore, it is used to distinguish the rotation between the left and right wheels as the vehicle turns and has a ratio of 0.04. The developed drivetrain system model is shown in Fig. 7.

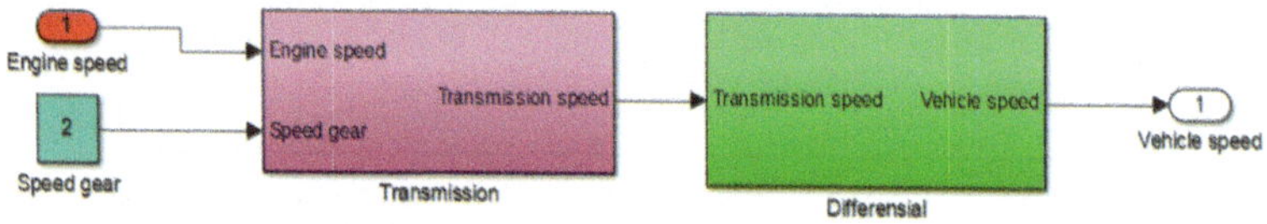

Fig. 7. Drivetrain system model.

Table 2. Transmission gear ratio

No.	Speed gear	Gear ratio
1	1	3.345
2	2	1.904
3	3	1.310
4	4	0.969
5	5	0.815

The gas engine dynamics has 3 components, namely mixing and combustion dynamics, throttle and intake manifold dynamics, and engine speed. Mixing and combustion dynamics are models of internal combustion engines. Meanwhile, throttle and intake manifold dynamics are a model of the dynamics of airflow in the intake manifold. The gas engine dynamics model is shown in Fig. 8.

Engine speed is the rotational motion of the crankshaft when the engine is in operation, and it is modeled from 2200–2900 rpm (medium crankshaft rotation). In this

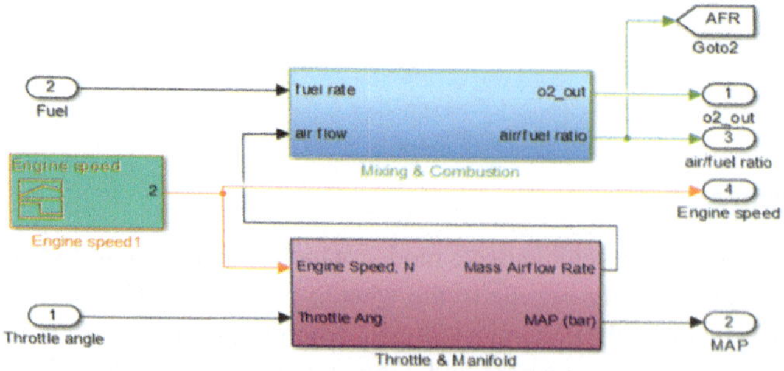

Fig. 8. Gas engine dynamics models.

rotation, the engine is in medium operation (not full power). The engine speed model is shown in Fig. 17.

Throttle and intake manifold dynamics have 2 components, namely throttle valve and intake manifold dynamics. The throttle valve controls the dynamics of air movement as the valve opens. Meanwhile, the intake manifold controls the dynamics of air movement while the engine is operating. The throttle and intake manifold dynamics model is shown in Fig. 9.

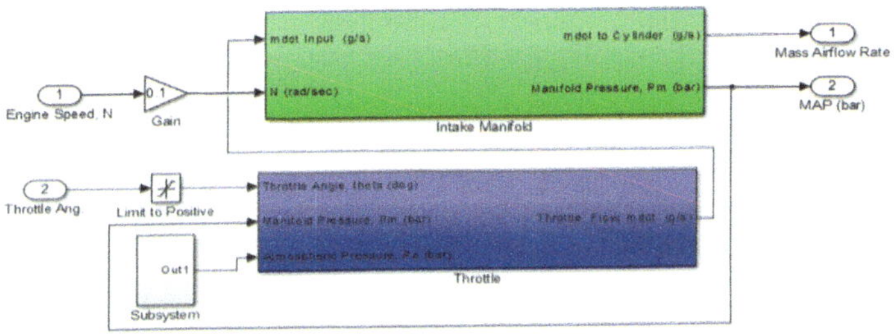

Fig. 9. Throttle and intake manifold dynamics model.

Intake manifold dynamics have 2 inputs and outputs each and are calculated based on engine speed and airflow into the intake manifold. The output is in the form of intake manifold pressure and airflow into the engine. Furthermore, the airflow is calculated based on the intake manifold pressure and engine speed which refers to Eq. 7 from the previous study [26]. The intake manifold dynamics model is shown in Fig. 10.

$$y = -0.366 + 0.08979 \times u1 \times u2 - 0.0337 \times u2 \times u1^2 + 0.0001 \times u1 \times u2^2 \quad (7)$$

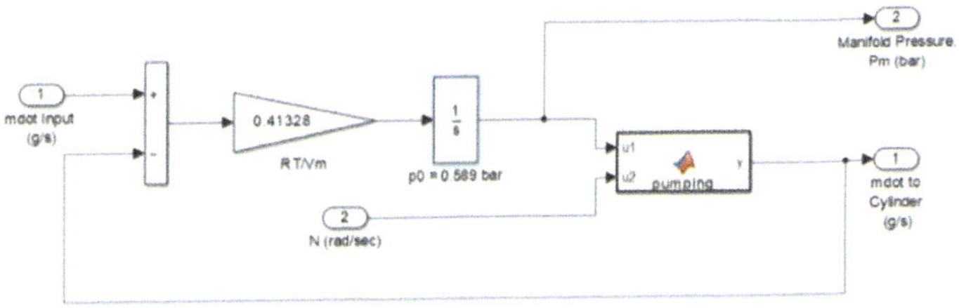

Fig. 10. Intake manifold dynamics pressure.

Throttle dynamics has 3 inputs and 1 output. These inputs include throttle valve opening angle, intake manifold pressure, and atmospheric pressure. The throttle valve opening angle was calculated based on Eq. 8.

$$y = 2.821 - 0.05231 \times u + 0.10299 \times u^2 - 0.00063 \times u^3 \quad (8)$$

The intake manifold and atmospheric pressures were calculated based on the smallest values. Throttle dynamics output is in the form of airflow into the intake manifold. The throttle dynamics model is shown in Fig. 11.

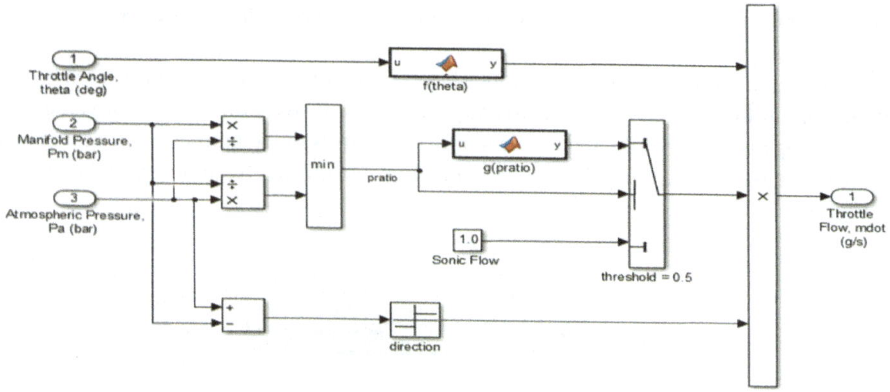

Fig. 11. Throttle dynamics models.

3 Result and Discussion

The results on the design of the steering operation control system model in the fuel control system based on fuzzy logic were in the form of simulations. These include the Air To Fuel Ratio (AFR) dynamics, fuel consumption, vehicle speed, the introduction of mode fuel, throttle angle dynamics, engine and steering speed.

3.1 Steering Speed

The data on steering speed was carried out using data acquisition. The speed sensor was installed in the steering system, and the microcontroller was connected to a computer as a signal reader and a sensor. The results of the displayed signal were stored in the computer and modeled as the steering operation control system for the fuel control system. The data collection process is shown in Fig. 12.

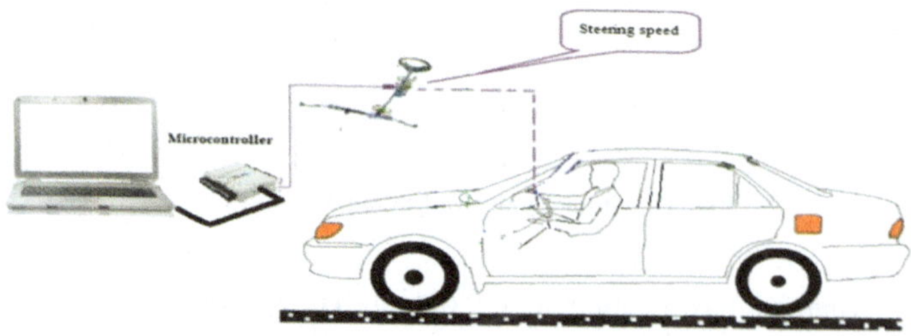

Fig. 12. The concept of data collection.

The sensor used to measure the steering speed is shown in Fig. 13.

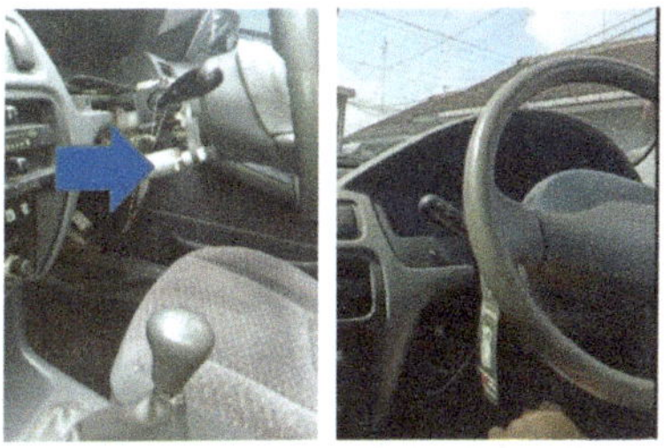

Fig. 13. Sensors used to measure steering speed.

The results of the steering speed measurement are shown in Fig. 14. Furthermore, the data collection process was carried out on a winding road. In the first, second, and third periods, the vehicle was driven with an average steering speed of 0.02, 0.04, and 0.02 rotation/seconds, respectively. The results of the average steering speed process were further designed with fuzzy logic to create a fuel mode cluster.

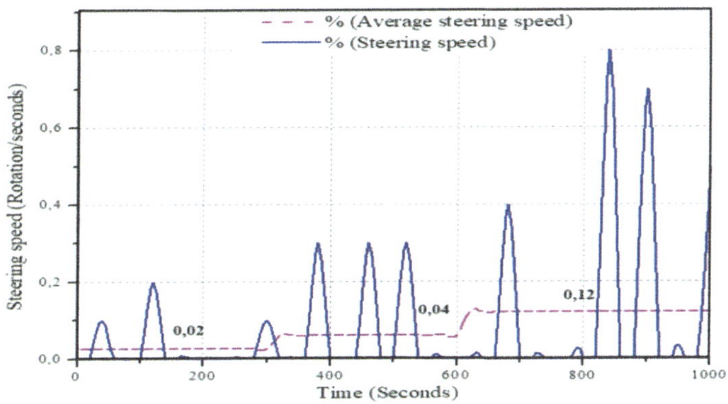

Fig. 14. Data created in model design.

3.2 Fuel Dynamics

The design of the steering operation system model for the fuel control system was made with the MATLAB Simulink software and is shown in Fig. 1. This model requires input and produces several outputs. The outputs include the Air To Fuel Ratio (AFR) dynamics. In the period 0–300 s, the steering speed control system made a decision, after which the fuel mode entered the economizer area. During this period, the increase in fuel economy was quite good. AFR in this period had a value of 15.4, but after 15 s, the fuel mode was in the stoichiometry area (14.7). Furthermore, during this period, the AFR returned to the vehicle's normal system. The AFR dynamics with the steering control system to manage the fuel control system are shown in Fig. 15.

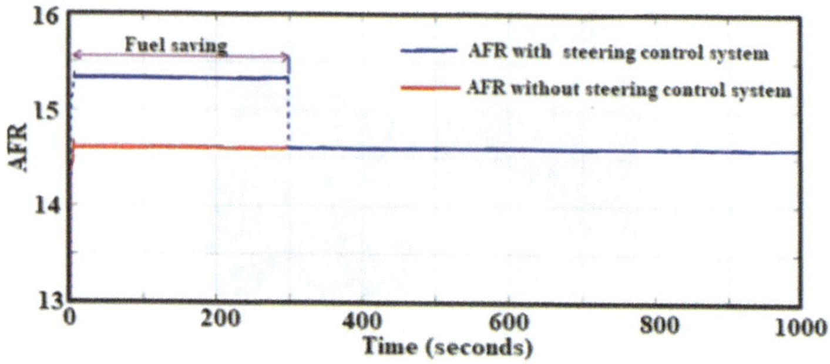

Fig. 15. AFR simulation dynamics with the steering control system.

The fuel flow in the steering operation control system model for the fuel control system is shown in Fig. 16. This simulation displays the fuel flow when the mode is in the economizer area. It was smaller compared to the fuel flow without the steering control system. Therefore, the increased economy was seen when the steering and vehicle speeds were in a low position, which is in the period 0–300 s. Meanwhile, from 300 s and above, the fuel flow returned to the normal standard vehicle system.

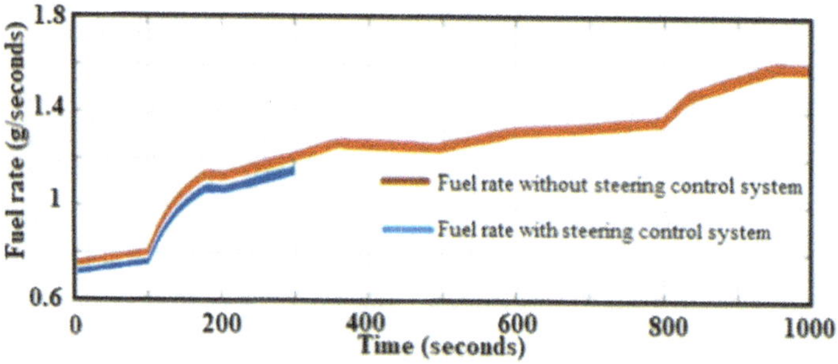

Fig. 16. Fuel rate with the steering control system.

3.3 Vehicle Dan Engine Dynamics

Engine speed is simulated at 2200–2900 rpm. Modeled speed, depicts the engine in normal operation (not at full power). The simulation results were further used for modeling operations in the next step. The engine speed model is shown in Fig. 17.

The simulation of vehicle speed was obtained by converting engine speed with transmission and differential. It was simulated from a speed of 46 km/h to 60 km/h. Generally, this speed is used when the vehicle is being operated on a busy road in an urban area, or on a road that requires the driver to operate at a medium speed. The results of the vehicle speed simulation are shown in Fig. 18.

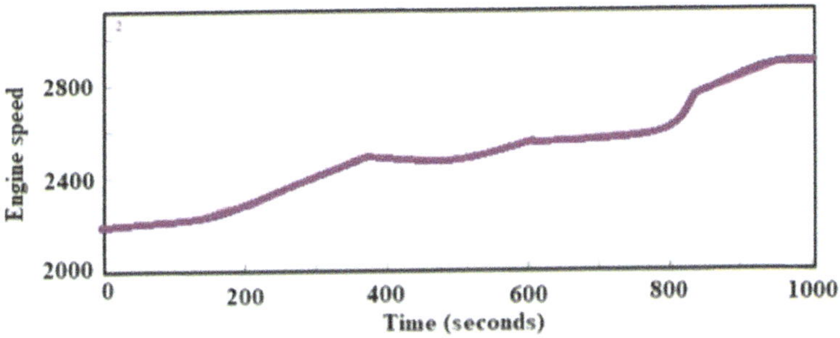

Fig. 17. Engine speed models.

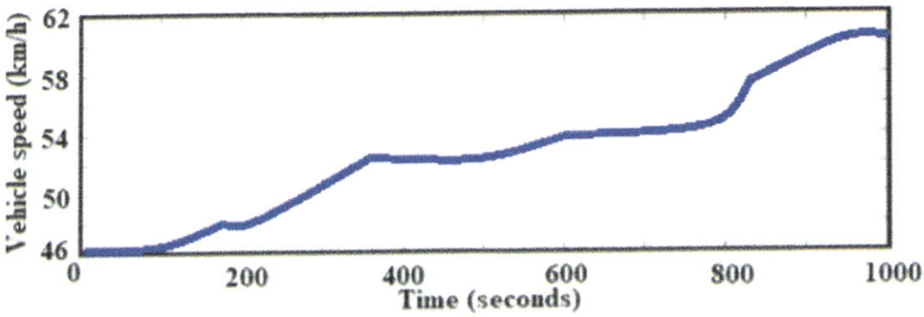

Fig. 18. Vehicle speed.

The throttle valve opening position was simulated at an angle of 10–24° shown in Fig. 19. It produces an analog signal that can be used for the control system of the engine control unit. This component is driven by the driver's feet to recognize the desired driving pattern when operating at a high load/speed.

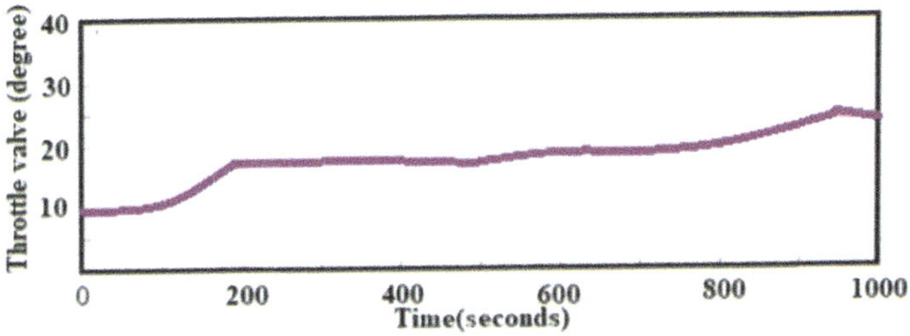

Fig. 19. Throttle valve opening position.

4 Conclusion

Based on the results of this study on the fuel control system, using a steering operation control model based on fuzzy logic, several conclusions were obtained, including that fuel economy can be obtained when the vehicle speed and steering operation are in a low position. This indicates that the driver is comfortable with slow conditions, therefore less fuel power is required. Furthermore, this condition can be used to adjust the scale of lean mixture fuel. The simulation results also illustrate that this model has the potential of being used in real vehicles.

Acknowledgment. The author expresses gratitude to the Robotics Laboratory of Mechanical Engineering, Diponegoro University, Semarang and PRVI, the Muhammadiyah University of Magelang for assisting this study.

References

1. Alkidas, A.C.: Combustion advancements in gasoline engines. Energy Convers. Manag. **48**(11), 2751–2761 (2007)
2. Kiencke, U., Nielsen, L.: Automotive Control System for Engine, Driveline and Vehicle, 2nd edn. Springer, Heidelberg (2005). https://doi.org/10.1007/b137654
3. Al Al-Fattah, S.M.: Non-OPEC conventional oil Production decline, supply outlook and key implications. J. Pet. Sci. Eng. **189**, 107059 (2020)
4. Lee, Y., Oh, S., Kim, C., Lee, J., Lee, K., Kim, J.: The dual-port fuel injection system for fuel economy improvement in an automotive spark-ignition gasoline engine. Appl. Therm. Eng. **138**, 300–306 (2018)
5. Li, Y., Khajepour, A., Devaud, C., Liu, K.: Power and fuel economy optimizations of gasoline engines using hydraulic variable valve actuation system. Appl. Energy **206**, 577–593 (2017)
6. Rajagopalan, S.S.V., Midlam-Mohler, S., Yurkovich, S., Dudek, K.P., Guezennec, Y.G., Meyer, J.: A control design and calibration reduction methodology for AFR control in gasoline engines. Control Eng. Pract. **27**(1), 42–53 (2014)
7. Triwiyatno, A., Sinuraya, E.W., Munahar, S., Setiyawan, J.D.: Smart controller design of air to fuel ratio (AFR) and brake control system on gasoline engine. In: International Conference on Information Technology, Computer, and Electrical Engineering (ICITACEE) (2015)
8. Khajorntraidet, C., Shen, T.: Adaptive lean air-fuel ratio control and analysis of commercial gasoline engines. IFAC-PapersOnLine **51**(31), 423–428 (2018)
9. Stone, R., Ball, J.K.: Automotive Engineering Fundamentals (2018)
10. Robertson, D., Prucka, R.: Evaluation of autoignition models for production control of a spark-assisted compression ignition engine. Int. J. Engine Res. **22**, 2732–2744 (2020)
11. Wang, Y., Shi, Y., Cai, M., Xu, W.: Predictive control of air-fuel ratio in aircraft engine on fuel-powered unmanned aerial vehicle using fuzzy-RBF neural network. J. Franklin Inst. **357**, 8342–8363 (2020)
12. Sardarmehni, T., Aghili Ashtiani, A., Menhaj, M.B.: Fuzzy model predictive control of normalized air-to-fuel ratio in internal combustion engines. Soft. Comput. **23**(15), 6169–6182 (2018). https://doi.org/10.1007/s00500-018-3270-2
13. Barghi, F., Safavi, A.A.: Experimental validation of recurrent Neuro-Fuzzy Networks for AFR estimation and control in SI engines. In: IEEE IEEE International Conference on Computational Intelligence for Measurement Systems and Applications Proceedings, pp. 107–112 (2011)
14. Du, K.L.: Clustering: a neural network approach. Neural Netw. **23**(1), 89–107 (2010)
15. Hebboul, A., Hachouf, F., Boulemnadjel, A.: A new incremental neural network for simultaneous clustering and classification. Neurocomputing **169**, 89–99 (2015)
16. Xu, R., Wunsch, D.: Survey of clustering algorithms. IEEE Trans. Neural Netw. **16**(3), 645–678 (2005)

17. Arsie, I., Pianese, C., Sorrentino, M.: Development and real-time implementation of recurrent neural networks for AFR prediction and control. SAE Tech. Pap. **1**(1), 403–412 (2008)
18. Mansfield, L.R., Guros, F., Truxillo, D.M., MacArthur, J.: Individual and contextual variables enhance transfer for a workplace eco-driving intervention. Transp. Res. Part F Traffic Psychol. Behav. **37**, 138–143 (2016)
19. Wang, H., Fu, L., Zhou, Y., Li, H.: Modelling of the fuel consumption for passenger cars regarding driving characteristics. Transp. Res. Part D Transp. Environ. **13**(7), 479–482 (2008)
20. Grove, K., Soccolich, S., Engström, J., Hanowski, R.: Driver visual behavior while using adaptive cruise control on commercial motor vehicles q. Transp. Res. Part F **60**, 343–352 (2019)
21. Kohl, J., Gross, A., Henning, M., Baumgarten, T.: Driver glance behavior towards displayed images on in-vehicle information systems under real driving conditions. Transp. Res. Part F Psychol. Behav. **70**, 163–174 (2020)
22. Martinelli, F., Mercaldo, F., Orlando, A., Nardone, V., Santone, A., Kumar, A.: Human behavior characterization for driving style recognition in vehicle system R. Comput. Electr. Eng. **83**, 102504 (2020)
23. Hong, Z., Chen, Y., Wu, Y.: A driver behavior assessment and recommendation system for connected vehicles to produce safer driving environments through a 'follow the leader' approach. Accid. Anal. Prev. **139**, 105460 (2020)
24. Sharma, A., Zheng, Z., Bhaskar, A., Haque, M.: Modelling car-following behaviour of connected vehicles with a focus on driver compliance. Transp. Res. Part B **126**, 256–279 (2019)
25. Vaezipour, A., Rakotonirainy, A., Haworth, N.: A simulator evaluation of in-vehicle human machine interfaces for eco-safe driving. Transp. Res. Part A **118**, 696–713 (2018)
26. The MathWorks Inc., Modeling a Fault-Tolerant Fuel Control System. MathWorks (2021). https://www.mathworks.com/help/simulink/

Hybrid ANN-AHP to Optimize the Inventory with a 5th Polynomial Model

Edy Fradinata[1(✉)], Muhamad Mat Noor[2], and Zurnila Marli Kesuma[3]

[1] Industrial Engineering, Universitas Syiah Kuala (USK), Banda Aceh 23111, Indonesia
edy.fradinata@unsyiah.ac.id

[2] Faculty of Mechanical and Manufacturing Engineering, Universiti Malaysia Pahang, 26600 Pekan, Pahang, Malaysia
muhamad@ump.edu.my

[3] Natural Science, Statistica Department, Universitas Syiah Kuala (USK), Banda Aceh 23111, Indonesia

Abstract. The study try to solve the general problems in the inventory to maximize some variables in the inventory system with the hybrid methods to get the fitting data then polynomial model was applied to create the model. The data is simulated from the generate dataset, then it uses the algorithm of hybrid *Artificial Neural Networks* and *Analytical Hierarchy Process* (ANN-AHP). Moreover, *fifth polynomial model* is applied. The results shown that the model contribute the maximum revenue from some variables in the inventory system.The analysis demonstrated that the result from optimization method start to implement in to the real company. The analysis demonstrated that the result from optimization method start to implement in to the real company This is good for warehouse manager to maximize the profit when managing the inventory system. The paper provides a critical perspective on the minimize the risk through maximization the inventory's variables when managing the inventory system. The result, the fifth polynomial mathematical model $g(x) = -f(x) = -0.00078*x^5 + 0.027*x^4 - 0.33*x^3 + 1.7*x^2 - 2.4*x + 27$. With the result profit is \$.5070.

Keywords: The average inventory level · Hybrid ANN-AHP · Optimization

1 Introduction

We see a major problem in warehouse management related to inventories. The inventory crisis affects almost all sectors and supply chains around the world. In the factory system, inventory continues to be a major expense. In most cases, correctly managing inventory levels is very difficult. To protect against the high risk of quantity, the effective inventory level must be controlled to the bare minimum. A risk arises when inventory is unable to control inventory costs at the lowest possible level. When the right location and time for the decision aren't available, this appears.

As a consequence, we expected to use forecasting techniques to correct predictions for a specific period of time in order to reduce the risk of cost while maintaining customer

Akhyar et al. (Eds.): ICECME 2021, LNME, pp. 74–81, 2023.
https://doi.org/10.1007/978-981-19-3629-6_8

demand fill rates and satisfaction. Unfortunately, the fill rate of consumer demand where it is met by immediate stock availability is a difficult decision to relate to since it has a direct effect on customer service and supply chain expenses. We tried to use the hybrid ANN-AHP with the polynomial model to solve the problem [1].

One of the techniques for determining the price of products in a competitive market is to use a management system for lowering costs and improving service levels in the warehouse. Customer demand is needed in order for the warehouse manager to forecast the exact situation in the future and improve the warehouse's control system. The warehouse manager is expected to be familiar with three forms of forecasting *behaviours*: patterns, cycles, and seasonal. The cost of protecting the storage room for the stockpile could be a concern due to a lack of prediction. The level of volatility and variability of conditions must be anticipated by the safety blanket.

Despite the fact that many businesses are aware of inventory management today, manufacturers and retailers do not. To forecast demand fluctuations, safety stocks were used. Some of the variables are the random output capability, optimum buffer inventory, and opportunistic preventive maintenance. The phenomenon of supplementary demand from suppliers could cause manufacturers to overreact, causing them to store supply inventory higher than distributor demand and increase orders to suppliers [2]. That case necessitates warehouse inventory level management, which improves performance.

An Analytical Hierarchy Process (AHP) is a decision-making system focused on requirements and alternatives for selecting a target. The approach can be applied to a variety of situations, according to the author of this procedure. The value of the criterion is graded on a scale of one to nine. The alternatives vary in terms of choice for each criterion used to make *trade-offs*. The problem's goal was established at the start, and the problem was covered in order to respond to it. The weight of the product will be determined by the measurement using AHP software [3].

Inventory must be managed in proportion to the amount of demand from customers. Because of certain factors, such as the difficulty of predicting consumer demand, the life cycle of a growing number of goods, the introduction of new competitive products into the marketplace, the complexity of cost and time of distribution, the delivery of lead time, and the economic scale of delivery by special transportation on large quantities, the inventory level must be held at a high level. Inventory management would be more successful if the warehouse manager planned how much to order, when to order it, and how to get it. With the inventory policy scheme, the technique removes the inventory's instability situation in the warehouse [4].

These issue statements often have the aim of identifying certain variables in order to achieve the best possible condition from the average inventory level to benefit. Average demand, reorder point, order quantity, factor service level, safety stock, and average inventory level are the variables in the warehouse that need to be identified. The type of inventory system will then be chosen to ensure that inventory management is as risk-free as possible. Finally, as a result of the inventory level's relationship to average demand, we'd like to observe the order quantity in order to achieve the best benefit.

2 Method

2.1 The Variation Demand

When handling inventory in a factory, there are two types of policies. Continue to use the policy method of review and the policy type of periodic review. Both would work well in the warehouse if used in the right situation. Customer demand in the function of *t*, D*t*, where D*t* = + D*t*-1 + t, was observed in the dataset, and an order, qt, was placed with the warehouse. After identifying the autocorrelation at the beginning of the pattern, the model tends to the Moving Average (MA) demand characteristic [5]. The variation Eq. 1 gives a normal distribution to the customer demand imposed on the warehouse. It is about the variance of Dt, i.e., with a variance of production, where it increases in variability, where it must assess the variance of q*t*. it can be represented by the symbol. Since qt has a negative value, the excess inventory is returned to the customer free of charge.

The target inventory level was determined using the equation $y_t = 0$, in which is the estimated mean of lead time demand, is the estimated standard deviation of forecast errors above the lead time, and *z* is chosen to meet a desired service level. To enforce this inventory strategy, the warehouse calculates the mean and standard deviation of demand based on customer lead time demand, which is calculated as = L*t*, where t is the average of previous p demand observations with Eq. 1.

Var(D) = and cov(yield) the second equation. To further evaluate Var(d), the following formula is required: follow from Var(D) $=\frac{\sigma^2}{1-\rho^2}$ and $\mathrm{cov}(D_{t-1}, D_{t-p-1}) = \frac{\sigma^2}{1-\rho^2}\sigma^2$ to further evaluate Var(q_t), it need the following formula.

If a warehouse uses a simple moving average forecast for a p request and the request reaches $D_t = \ldots + D_{t\text{-}1+t}$, then Cov(D (t-i,) _$t_L$) = 0 for all I = 1, ……, p. As a result, when the variability from warehouse usage to producer increases, we consider setting a lower cap.

$$\frac{Var(q^{MA})}{Var(D)} \geq 1 + \frac{2L}{p} + \frac{2L^2}{p^2}(1 - \rho^p) \tag{1}$$

where var *(D)* is a consumer-demand version. From these instances, various observations can be drawn. To begin with, these parameters' three functions are *p, l*, and there will be an increase in variability from producer to warehouse. *l denotes the lead time between the warehouse and the manufacturers, and p denotes the number of observations used in the moving average method.* Furthermore, there is a connection. When using the observation that it was used in the moving average, the *p* function showed an increase in the variability with a function of p values as well as an increase in the function of *l*, which is the lead variable [6].

2.2 ANN

The algorithms use a backpropagation algorithm neural network if it is highly efficient, a training network to update the bias, and a weight to achieve optimization. The input is where the transfer functions measure the output from the sheet [7]. The gradient descent with momentum weight and bias learning function with Eq. 2 is the function of *traingdm*.

$$E = \frac{1}{2}\sum_{h}^{M} E_h = \frac{1}{2}\sum_{h}^{M}\sum_{i}^{N}(t_{hi} - O_{hi})^2. \tag{2}$$

where E represents the total error pattern, Eh represents pattern h with its error, and the index h ranges over the set of the input pattern, and I represents the output neuron's i*th* iteration. When pattern h is introduced, this is the ideal output for the i*th* output neuron. When data is entered into the network, the relevance transfer function must be selected; this transfer function keeps the training process in the network more consistent and speeds up the training process. Since it generated the smallest MSE, the data will be run with Sigmoid as the dataset's transfer feature in an ANN network [8].

2.3 Hybrid of ANN-AHP

If sales do not meet expectations, there is an excess of inventory in the warehouse. When the products are already in storage in the factory, the opportunity exceeds the profit, and less profit appears. The accurate forecasting information provided by the forecasting tool would aid management in making inventory investments. To offer an overview of the approach, consider the following:

The hybrid model from ANN and AHP. will be tested in this analysis. The output of the AHP from the six variables of the determinant of demand (population, number of buyers, price, GDP, expectation of future price, promotion, and demand) should be the input variable in the input layer, and some data should be the input weight in the hidden layers ANN scheme. With the AHP, the normalized data process was used to make data on the scale of 0 to 1, and it rendered data between the ranges of less than 0.01. From the output layer, the ANN uses a backpropagation network based on *Levenberg-Marquardt*. In *trimln*, the network uses the backpropagation training feature to minimize the error value. The stimulation was designed to reduce the error value from the previous error. After the network has completed its training and testing, the model of the combination system will be reached. Furthermore, production performances are calculated by mean square error (MSE). Finally, it calculated the fitting value to obtain the model's result [9].

The neural network's learning algorithm was split into two sections [7]. Propagation and weight update are the two. The following is the.

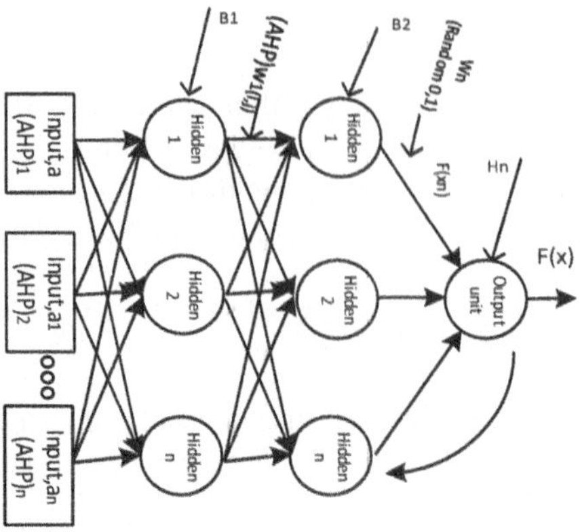

Fig. 1. Combination methods of ANN with Analytical Hierarchlly Process

The changed weight from the Analytical Hierarchy Process is shown in Fig. 1. This weight would take the place of the network's default weight in the input layer. The error is then produced using the data from the input layer in the hidden layers, which is better than the initial input.

Quantity of Order

The formula for calculating order quantity is as follows:

As before, the order quantity Q satisfies

$$Q = \sqrt{\frac{2K * AVG}{h}} \tag{3}$$

The warehouse checks inventory every period, location, and order in order to get its inventory level up to a target level. Note that the review period is one in this case. As a result, the base stock level is determined as follows:

$$L * AVG + \ z * SD * \sqrt{L} \tag{4}$$

The average and standard deviation of regular (or weekly) consumer demand are AVG and STD, respectively. The safety factor is represented by the constant z. Q/2 + z is the average Inv standard.

Polynomial

Polynomial models are generalized to any number of predictor variables x_i (i = 1,…, N) as follows:

$$y(x) = a_0 + \sum\nolimits_{i=0}^{N} a_i x_i + \sum\nolimits_{i=0}^{N} a_{ij} x_i x_j + \sum\nolimits_{i=0}^{N} a_{ij} x_i^2 + \ldots.. \tag{5}$$

The model includes, from left to right, an intercept, linear terms, quadratic interaction terms, and squared terms. Higher order terms would follow, as necessary.

The Revenue

The benefit of the result of the ANN-AHP combination is to optimize warehouse profit inventory. Follow the overfitting formula from the model moving average in this situation.

The formula determined that x_1 (demand), x_2 (selling price), x_3 (production cost), and x_4 (fixed cost) were variables. It can be shown as follows:

$$\text{Profit}(x) = x_1(x_2 - x_3) - x_4 \tag{6}$$

Maximize:

$$g(x) = -f(x) = -a * x^4 + b * x^3 - c * x^2 + d * x + e \tag{7}$$

The profit formula identifies the x_1 (demand), x_2 (selling price), x_3 (production cost), and x_4 (Fixed cost) with the maximization of g(x).

This experiment is being carried out in order to achieve the study's goals. It should be noted that the ANN-AHP run dataset. The result for 15 weeks, and the trend is smoothed by fitting it with a polynomial graph. There are three peaks on the dataset, indicating that the pattern is moving average. Commercial revenue may also be referred to as sales or as turnover. Some companies receive revenue from interest, royalties, or other fees. Revenue growth is the income in general, or it may refer to the amount, in a monetary unit, earned during a period of time, as in "Last year, Company X had revenue of $42 million." Profits or net income generally imply total revenue minus total expenses in a given period. In accounting [10].

3 Result and Discussion

The determinant of demand dataset affected the neural network training. The network reduces mean square error (MSE) feature errors in the overall iteration process on demand results. The demand, on the other hand, was determined by the weight and the input data, which were combined by iteration and evaluation of the smaller error from the initial error in the network using nonlinear data, weight, and transfer functions. When the training happened in the overfitting situation, the iterating was then halted. Since continuously differentiable distributions have properties for the network learning process and take the place of the mean and standard deviation value of output, the nonlinear transfer function, especially *trainlm*, has a large function in forming patterns.

Moreover, the model calculated the best validation state, which is depicted in the ANN-AHP network performance validation process. It is 0.85285 (between production and target), 1 for training, and 0.99911 for testing, indicating that the validation is performing well since the line and points are similar to 1 (one). The perfect outcome – outputs = goals – is represented by the line in the axis. The solid line represents the best fit for the linear regression condition between outputs and goals.

The error is small, indicating that the dataset has a small error when it was predicted to the original dataset. This means this dataset from the generalized data is stable enough to calculate some variables in the inventory system.

Using the model maximization from the fitted polynomial model, estimate the maximum benefit and average profit on the inventory position in the warehouse to reduce the bullwhip effect [11]. As an example, consider the following:

$$y(x) = a_0 + \sum_{i=0}^{N} a_i x_i + \sum_{i=0}^{N} a_{ij} x_i x_j + \sum_{i=0}^{N} a_{ij} x_i^2 \tag{8}$$

A specific examination of the function is generally needed to see whether the point is a minimum or a maximum. This study gets the mathematical fifth polynomial model formula as follows:

Maximize with the fifth mathematical polynomial model:

$$g(x) = -f(x) = -0.00078 * x^5 + 0.027 * x^4 - 0.33 * x^3 + 1.7 * x^2 - 2.4 * x + 27. \quad (9)$$

When the order quantity is 20.2 and the value for benefit optimization is \$5070. drawn the condition of residual from the data where most of them closed to the center line. Correlation between residuals and input for negative lags, is not necessarily an indication of an inaccurate model. When current residuals at time *t* affect future input values, there might be feedback in the system. In the case of feedback, concentrate on the positive lags in the cross-correlation plot during model validation. In the previouse journal with run with the fouth mathematical polynomial model.

4 Conclusion

In this analysis, the hybrid ANNAHP model was used to estimate the optimization of average inventory level and benefit with a factor service level of 97%. Demand changes might heighten the risk in the warehouse. It may be due to an inability to monitor inventory, a quality control problem, or a lack of customer service. When more inventory is spent in the warehouse, this risk will increase the cost. The demand must be filled as soon as possible to fully satisfy the customer's needs. The ANNAHP combination method was found to be a viable alternative model for predicting average inventory levels and benefits. This research came to three conclusions. First, the accuracy of prediction can be determined by the determinant of demand variables, which is affected by the dataset's typical customer demand. Second, the continuous review policy type is the type of inventory policy system that minimizes risk, and benefit optimization from the average inventory level can be calculated. They are as follows: The data has a mean of 39.2 and a standard deviation of 12.9. Furthermore, the order quantity is 20.2 units, with an average demand of 57.3 units over the duration. In this condition, z's service level is 97%. While the average inventory level is 20.2 units, the optimal order quantity is \$5070.The fifth of five mathematical models based on polynomials. $g(x) = -f(x) = -0.00078*x^5 + 0.027*x^4 - 0.33*x^3 + 1.7*x^2 - 2.4*x + 27$.

References

1. Al-Barqawi, H., Zayed, T.: Infrastructure management: integrated AHP/ANN model to evaluate municipal water mains' performance. J. Infrastruct. Syst. **14**, 305–318 (2008)
2. Giri, B.: Managing inventory with two suppliers under yield uncertainty and risk aversion. Int. J. Prod. Econ. **133**, 80–85 (2011)
3. Saaty, T.L.: The analytic hierarchy process (AHP). J. Oper. Res. Soc. **41**, 1073–1076 (1980)
4. Muller, M.: Essentials of Inventory Management. HarperCollins Leadership (2019)
5. Gu, G.-F., Zhou, W.-X.: Detrending moving average algorithm for multifractals. Phys. Rev. E **82**, 011136 (2010)

6. Lee, H.L., Padmanabhan, V., Whang, S.: The bullwhip effect in supply chains. Sloan Manage. Rev. **38**, 93–102 (1997)
7. Buscema, M.: Back propagation neural networks. Subst. Use Misuse **33**, 233–270 (1998)
8. Fradinata, E., Suthummanon, S., Sirivongpaisal, N., Suntiamorntuthq, W.: ANN, ARIMA and MA timeseries model for forecasting in cement manufacturing industry: Case study at lafarge cement Indonesia—Aceh. In: 2014 International Conference of Advanced Informatics: Concept, Theory and Application (ICAICTA), pp. 39-44 (2014)
9. Fradinata, E., Suthummanon, S., Suntiamorntut, W., Noor, M.M.: Compare the forecasting method of artificial neural network and support vector regression model to measure the bullwhip effect in supply chain. J. Mech. Eng. Sci. **13**, 4816–4834 (2019)
10. Fradinata, E., Kesuma, Z.M.: Numeric model to predict the location of market demand and economic order quantity for retailers of supply chain. In: IOP Conference Series: Materials Science and Engineering, p. 012010 (2018)
11. Fradinata, E., Suthummanon, S., Suntiamorntut, W.: Reducing the bullwhip effect from signal demand of hybrid artificial neural network models of supply chain in Indonesia. Int. J. Adv. Appl. Sci. **4**, 64–75 (2017)

Business Process Model Design as a Basis for Determining the Price of Coffee Beans in Tanah Karo

Meilita Tryana Sembiring(✉), Andri Nasution, Tania Alda, Sawaluddin, and Letno Gurusinga

Universitas Sumatera Utara, North Sumatera, Indonesia
meilita@usu.ac.id

Abstract. Karo Regency is one of the coffee producing areas in North Sumatra. In 2015–2019, the total annual coffee production in Karo district was 4,808, 5,270, 6,877, 7,379, and 7,402 tons. Coffee growth in Karo has increased by around 648.25 tons/year. However, the growth of coffee in Karo is not supported by good coffee management. After making observations at 3 collectors, the researchers found a problem, namely the poor flow of information between farmers and collectors, namely the absence of quality standardization in determining prices, so that coffee prices were not uniform and varied. For this reason, it is necessary to improve the flow of information and pricing mechanisms. Improving the flow of information is done by making a coffee quality determination form based on SNI, use case diagrams and business process model and notation (BPMN). The result of this research is an electronic-based information flow model interface design platform. The conclusion of this study is to determination of prices at the collector level has standardization based on the quality that has been determined.

Keywords: Information flow · Use case diagram · Business Process Model and Notation (BPMN)

1 Introduction

Currently, the coffee commodity is the prima donna of international trade and Indonesia is the 4th largest producer in the world with an average income of 639 thousand tons/year. The types of coffee that dominate the production are robusta and arabica coffee with a production percentage of 72.84% for the robusta type and 27.16% for the arabica type [1].

Karo Regency is one of the regencies in North Sumatra which is one of the coffee producers. The area for developing Arabica coffee commodities is 40,893.1 Ha. The total production of arabica coffee in Karo Regency is 25,546.81 tons. The prospect of the coffee market is quite large in Karo Regency, but it is not supported by coffee management in Karo Regency. Based on observations, the problem of coffee supply chain management in Karo Regency is the poor flow of information, causing problems in determining the price of coffee that varies. Farmers do not know the mechanism for determining the price

Akhyar et al. (Eds.): ICECME 2021, LNME, pp. 82–92, 2023.
https://doi.org/10.1007/978-981-19-3629-6_9

by the collectors due to the lack of information on the standardization of determining the price of coffee that has been set by the collectors [2].

Research on the use of information technology to improve the Supply Chain Management model conducted by Varma and Khan (2014) produces an electronic information-based information flow model that significantly replaces information in written reports, so that it can control daily supply chain activities more effectively [3].

Research conducted by Ferantino and Koten (World Bank Group, 2019) To find out the impact of supply chain 4.0, the use of technology in the supply chain will transform the supply chain model by integrating the flow of information into a supply chain center [4]. Research conducted by Himanshu et al. (2013) explains that to be able to compete globally, information management must be carried out. The use of information technology provides benefits in making decisions by considering the quality of information to the company and can integrate between stakeholders in the supply chain [5].

Research conducted by Sutri Handayani (2018) explains that the design of information Systems e-commerce is done by making use case diagrams to describe what the system should do. Use case diagrams provide a way of describing an external view of the system and its interactions with the outside world [6].

Research conducted by Ismanto (2020) explains that modeling business processes in an organization in detail with the flow of information in the form of messages conveyed between related parties can be done using the Business Process Model and Notation (BPMN) [7].

Based on several studies that have been carried out regarding the importance of supply chain management, this study will focus on a design of information flow model to control the condition of the coffee supply chain in Karo Regency.

2 Method

The object under study is the flow of information in the Arabica coffee supply chain between collectors and farmers.

The independent variable in this study is the quality of coffee, while the dependent variable of this study is the price of coffee. The data used in this study are primary data, namely the price of coffee, the actual flow of information by direct observation, and secondary data, namely the total coffee production in Karo Regency.

Data processing is carried out by making a model of the information flow system between collectors and farmers, taking into account user needs such as coffee quality standards, coffee prices, the total weight of coffee needed by each collector. The modeling of this system is carried out by conceptualizing the model by making use case diagrams, Business Process Model and Notation (BPMN), then verifying use case diagrams and BPMN, as well as designing visualization of the system model design by making system mock-ups.

The stages of the research can be seen in Fig. 1.

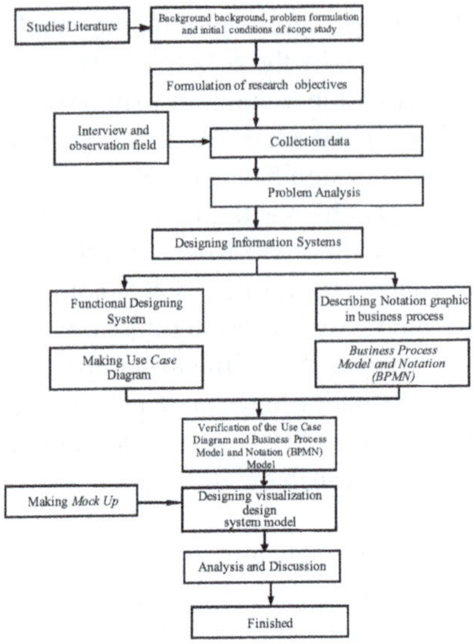

Fig. 1. Research stages

3 Results and Discussion

Varied prices are caused by differences in the quality of coffee sold by farmers and the strategy of collectors to win the market. During the observation, the researcher noticed that there were no clear standards for determining price and quality in each collector. Sampling is still done descriptively without following the quality standards that have been set by SNI. The following is a model of the coffee buying and selling system in Tanah Karo that has been happening at the farmer level as shown in Fig. 2 below.

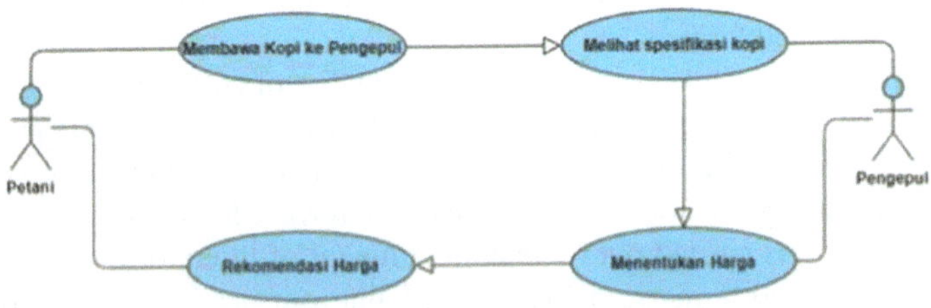

Fig. 2. Use case diagram of the coffee buying and selling model in Tanah Karo

Through the use case diagram above, it can be seen that the flow of information is not good because to find out the price of coffee, farmers have to bring coffee to each collector. Farmers in this case too, do not know the standard of price determination made by collectors. For this reason, it is necessary to improve a system by integrating the needs of collectors and the wishes of farmers into a system.

1. Making Coffee Quality Standards Based on SNI

The standard of determining the quality and price of coffee beans at the collector level is carried out by descriptive sampling, causing farmers not to know how to determine the quality of their coffee. In addition, collectors do not provide information to farmers about the steps in determining the quality of coffee and determining the price of farmers' coffee beans. So far, the advice given by collectors to farmers is to get a high price, then the farmer must separate/discard the defective coffee beans from the pile of coffee to be sold, without any explanation to farmers regarding the division of coffee quality categories based on the level of defect. For this reason, it is necessary to improve coffee quality standards based on National Standardization Agency (BSN) 2017. The standardization of the quality of coffee beans in accordance with the Indonesian National Standard (SNI) can be seen in Table 1.

Table 1. Coffee bean quality requirements

Quality	Condition
Quality 1	The maximum number of defects is 11
Quality 2	Total defect value 12 to 25
Quality 3	Total defect value 26 to 44
Quality 4a	Total defect value 45 to 60
Quality 4b	Total defect value 61 to 80
Quality 5	Total defect value 81 to 150
Quality 6	Total defect value 151 to 225

Notes: For Arabica coffee quality 4 is not divided into sub quality 4a and 4b. The determination of the value of the defect from each defective seed is listed in Table 2

To determine the value of coffee bean defects can be seen in Table 2.

Table 2. Determination of the value of coffee bean defects

No	Type of defect	Defect value
1	1 (one) black seed	1 (one)
2	1 (one) black seed partially	(half)
3	1 (one) broken black seed	(half)
4	1 (one) cup of coffee	1 (one)

(continued)

Table 2. *(continued)*

No	Type of defect	Defect value
5	1 (one) cocoa	(quarter)
6	1 (one) large size coffee	1 (one)
7	1 (one) medium size coffee	(half)
8	1 (one) small size coffee	1/5 (one fifth)
9	1 (one) seed with horn skin	(half)
10	1 (one) large size horn skin	(half)
11	1 (one) medium sized horn skin	1/5 (one fifth)
12	1 (one) small horn skin	1/10 (tenth)
13	1 (one) broken seed	1/5 (one fifth)
14	1 (one) young seed	1/5 (one fifth)
15	1 (one) seed with one hole	1/10 (tenth)
16	1 (one) seed with more than one hole	1/5 (one fifth)
17	1 (one) spotted seed	1/10 (tenth)
18	1 (one) twig, soil, or large stone	5 (five)
19	1 (one) medium-sized twig, soil, or stone	2 (two)
20	1 (one) twig, soil, or small stone	1 (one)

By applying standardization according to the Indonesian National Standard (SNI) it is hoped that it can help collectors and farmers in determining prices based on quality categories.

2. Making a Calculation Form for Coffee Quality Standards
Determination of the quality category is done by calculating the total number of defects and then categorizing the quality according to the value of defects in Table 1. Thus, collectors and farmers have the same quality determination standard, so that farmers can estimate the quality of their coffee, and know the basis for determining the price made by collectors. The formula for calculating the total number of defects is as follows.

Total defect value for each type of defect = defect value x number of defects the form for determining the value of defects in Arabica coffee beans can be seen in Table 3.

3. Making Use Case Diagram System Model
Use case diagramsis a UML model that is used to describe the expected functional requirements of a system. Use case diagrams are use cases that are used to briefly

Table 3. *Form* determination of the value of coffee bean defects

No	Type of defect	Defect value	Number of defects	Total defect value
1	1 (one) black seed	1 (one)		
2	1 (one) partial black seed	(half)		
3	1 (one) broken black seed	(half)		
4	1 (one) cup of coffee	1 (one)		
5	1 (one) cocoa	(quarter)		
6	1 (one) large size coffee	1 (one)		
7	1 (one) medium size coffee	(half)		
8	1 (one) small size coffee	1/5 (one fifth)		
9	1 (one) seed with horn skin	(half)		
10	1 (one) large size horn skin	(half)		
11	1 (one) medium sized horn skin	1/5 (one fifth)		
12	1 (one) small horn skin	1/10 (tenth)		
13	1 (one) broken seed	1/5 (one fifth)		
14	1 (one) young seed	1/5 (one fifth)		
15	1 (one) seed with one hole	1/10 (tenth)		
16	1 (one) seed with more than one hole	1/5 (one fifth)		
17	1 (one) spotted seed	1/10 (tenth)		
18	1 (one) twig, soil, or large stone	5 (five)		
19	1 (one) medium-sized twig, soil, or stone	2 (two)		
20	1 (one) twig, soil, or small stone	1 (one)		

describe who uses the system and what can be done. Making use case diagrams is generated with the help of Power designer software.

A. Use Case Diagrams Visitors

Use case diagrams visitors are used to see the interaction between actors and the available system. The visitor use case diagram can be seen in Fig. 3.

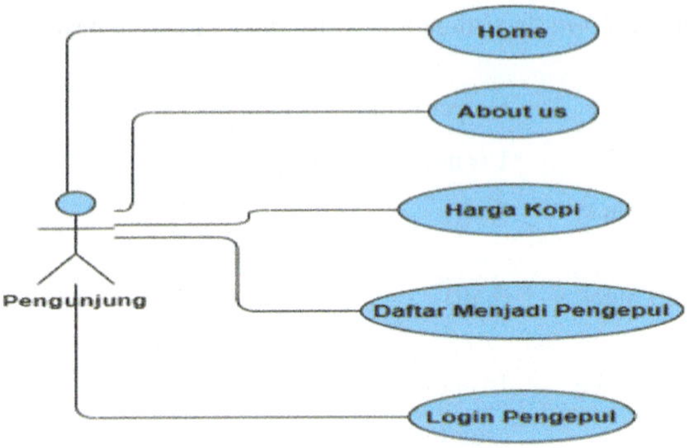

Fig. 3. Visitor use case diagram

In the use case diagram above, there is 1 actor, namely the visitor. Visitors can visit all available menus. Home is used to see an overview of coffee and its characteristics. About us is used to view information about the usability of the system along with general information about the system designer. Coffee Prices are used to see the price of coffee inputted by collectors for each quality. Visitors can also register as a collector by selecting the Register to be a Collector menu. As for access to the system, collectors can log in to collectors. The collector login menu is used by collectors to edit price data for each quality and quantity of coffee needed. To see the conceptual truth of the system being built, the model verification is carried out with the help of the power designer application. The results of the model verification are shown in Fig. 4.

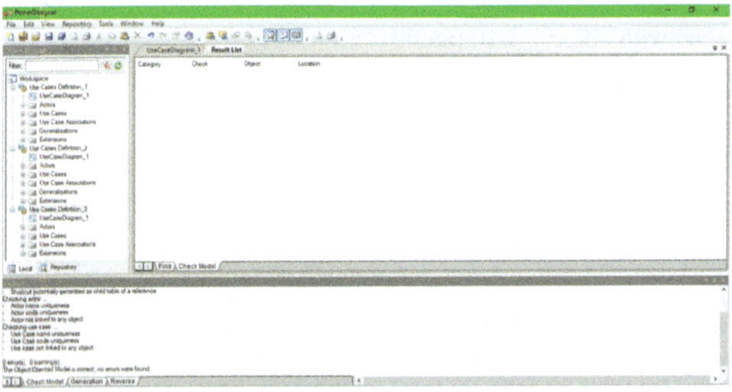

Fig. 4. Verification of the visitor diagram use case

From the picture above, it is explained that the verification results show that the model formed is correct. The parameters that indicate the model has been formed correctly are the absence of error comments in the display of the tests performed.

B. Use Case Diagrams Register as a Collector

Use case diagrams this is used to see the steps that must be taken by visitors to become collectors. To become a collector, visitors must prepare the owner's identity, company name, and business location. The display of the use case diagram for the list of collectors can be seen in Fig. 5.

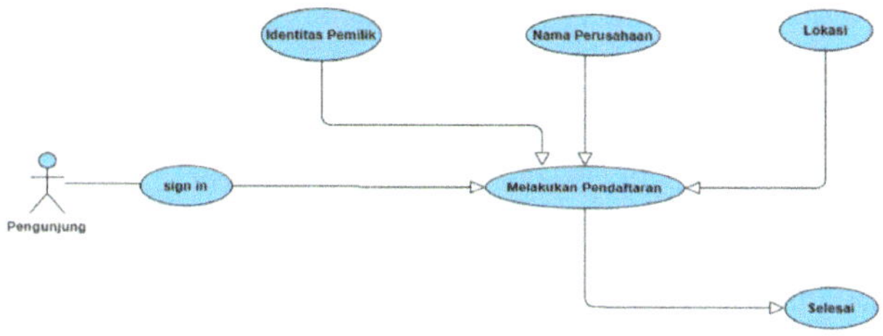

Fig. 5. Use case diagram register to be a collector

To see the conceptual truth of the system being built, a model verification is carried out with of the power designer application. The results of the model verification are shown in Fig. 6.

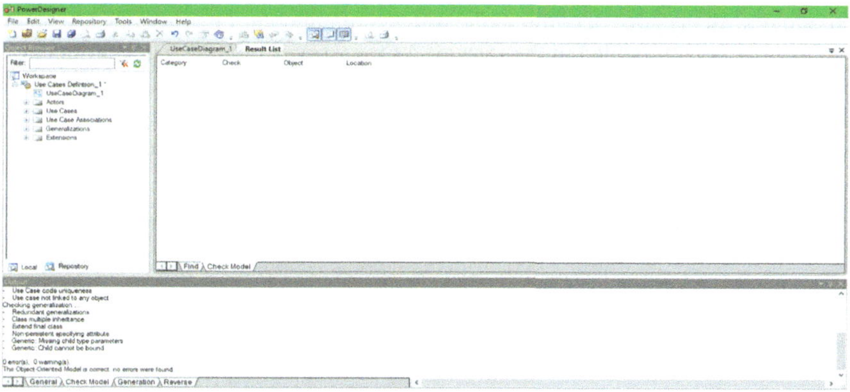

Fig. 6. Verify use case diagram register to be a collector

From the picture above, it is explained that the verification results show that the model formed is correct.

C. Use Case Diagram System Model

Use case diagramsit displays the overall interaction between the actor and the system. Actors who play a role in this system are collectors, farmers, and admins. These three actors have their respective roles in running the system model that is built. To find out the system model, it can be seen through the use case diagram of the system model in Fig. 7.

In the use case diagram above, there are 3 actors, namely farmers, collectors and admins. Collectors have a role in inputting data on the need for coffee orders and coffee prices based on coffee quality specifications. Farmers have a role in inputting the location

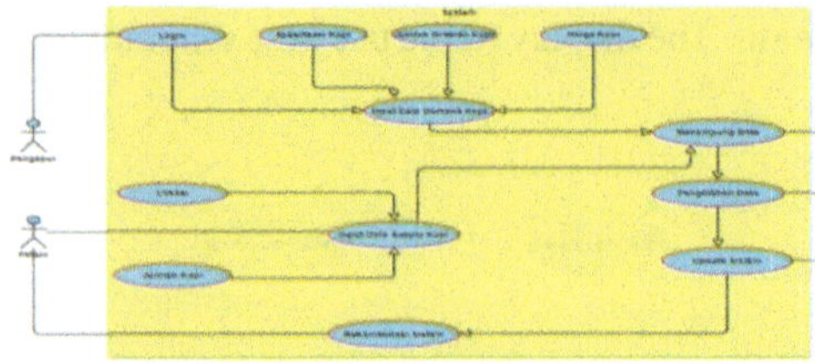

Fig. 7. Use case diagram system model

of coffee and the amount of coffee owned and in making recommendations for buying and selling coffee systems. The admin has a role to collect data, then from the data inputted by collectors, data processing is carried out. The data that has been processed is then entered into the system for system updates. The results of the system update will be displayed in the application as the latest rule for coffee buying and selling transactions. To see the conceptual truth of the system being built, the model verification is carried out with the help of the power designer application.

All the activities of collectors and farmers can be seen in a system. This certainly benefits both users. The advantage of the collector is that the collector can tell what he needs along with the price of coffee to the wider community just by entering data into the menu that has been provided in the system. As for farmers, after this it is profitable for farmers because they know how to know the price of their coffee, the quality of coffee is done based on the value of defects. In addition, farmers can also see the difference in prices at the level of collectors, so that farmers get a reference for the collectors who are the purpose of buying and selling coffee.

The results of the model verification are shown in Fig. 8.

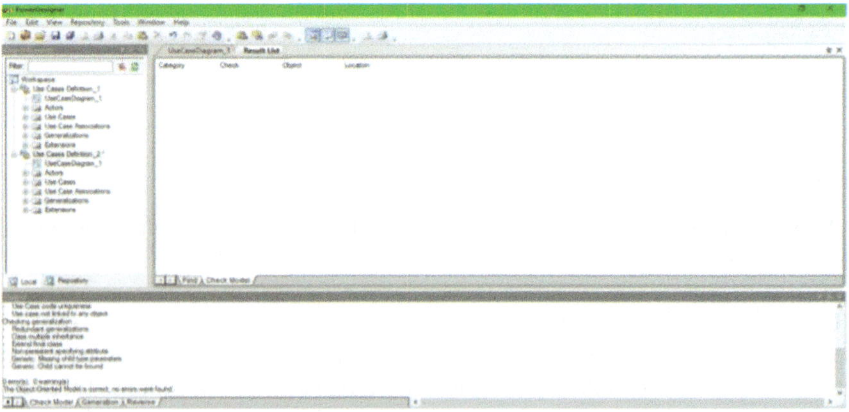

Fig. 8. Verification use case diagram system model

From the picture above, it is explained that the verification results show that the model formed is correct.

4. Making Business Process Model and Notation (BPMN)

Business Process Model and Notation (BPMN) is a graphical representation for specifying business processes in the business process model. The main objective of BPMN is to provide a standard notation that is easily understood by all business stakeholders. This includes the business analysts who create and refine processes, the technical developers responsible for implementing them and the business managers who monitor and manage them. As a result BPMN serves as a common language, bridging the frequent communication gap between business process design and implementation. BPMN generated with the help of Power designer software. The Business Process Model and Notation of this system can be seen in Fig. 9.

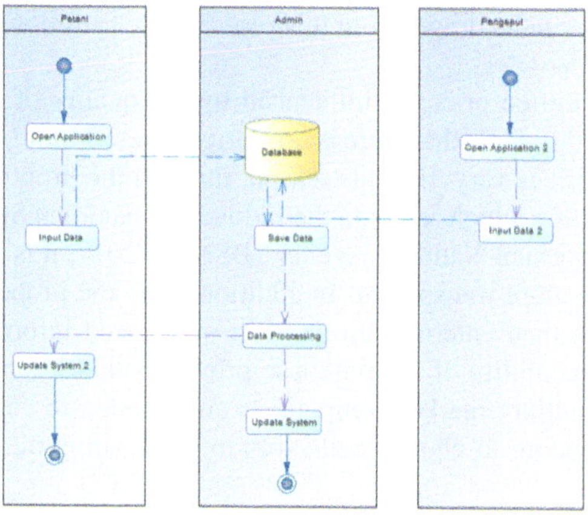

Fig. 9. BPMN system

To see the conceptual truth of the system being built, it is done Verify the model with the help of the Power Designer application. The results of the model verification are shown in Fig. 10 below.

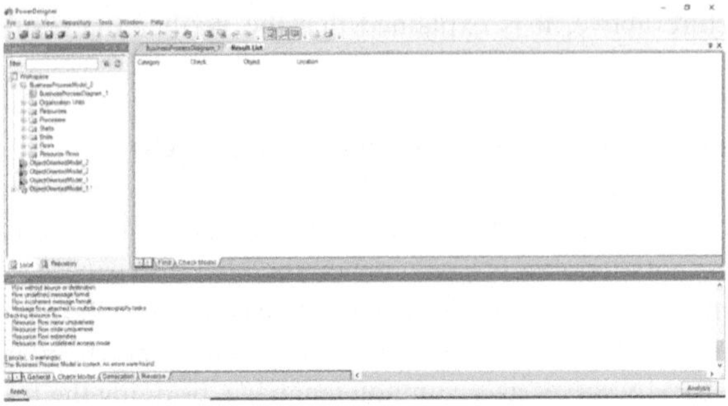

Fig. 10. System BPMN verification

4 Summary

The conclusions that can be drawn from this research are as follows.

- Determination of coffee prices is influenced by the quality of coffee farmers. The problem in the field so far is that there is no clarity in determining the quality of coffee by collectors, so prices vary. For this reason, through the proposed improvement of coffee standardization which refers to the Indonesian National Standardization (SNI) issued by the Indonesian National Agency (BSN) in 2017, it is hoped that the price issue will become more transparent. In addition, with the proposed standardization that refers to SNI, then entered into an electronic-based information flow system, farmers can see the ability of the purchase price inputted by the collectors, so that this becomes a healthier market competition and is open to collectors and farmers more have the freedom to choose collectors to carry out coffee buying and selling transactions.
- Improvements to coffee standardization referring to the Indonesian National Standardization (SNI) issued by the National Standardization Agency (BSN) in 2017 can help solve the problem of determining coffee quality and price by collectors.

References

1. http://indonesiabaik.id/
2. https://disbun.sumutprov.go.id/
3. Varma, T.N., Khan, D.A.: Information Technology in Supply Chain Management. National Institute of Technology, India (2014)
4. Ferrantino, M.J., et al.: Understanding Supply Chain 4.0 and Its Potential Impact On Global Value Chains. World Bank Group, Switzerland
5. Moharana, H.S., et al.: Importance Of Information Technology for Effective Supply Chain Management. International Journal of Modern Engineering Research (IJMER), India
6. Handayani, Sutri. 2018. Design of an E-Commerce-Based Sales Information System Case Study of Toko Kun Jakarta. STMIK Nusa Mandiri Jakarta, Jakarta
7. Ismanto, et al.: Business Process Modeling Using Business Process Modeling Notation (BPMN). Jurnal.unublitar.ac.id, Blitar (2020)

Thermal Performance of the Heat Pipe and Thermoelectric Sterilization Face Mask

Wayan Nata Septiadi[1(✉)], Komang Manik Marianti[2], Made Nara Pradipta Adi[2], Muhamad Alim[2], Cheto Rizkiantoro[2], Dandi Ramadhani[2], and Anak Agung Gde Agung Krisnanta Dwipayana[2]

[1] Department of Mechanical Engineering, Udayana University, Bali, Indonesia
wayan.nata@unud.ac.id

[2] Bachelors Program in Mechanical Engineering, Udayana University, Bali, Indonesia

Abstract. The spread of the coronavirus has been the focus the world's attention. The government has issued a new regulation on the coronavirus that obligates the entire community to always wear a face mask and implement social distancing to prevent the spread of viruses. These issues have been impacted by an increase in the use of face masks throughout the world and in Indonesia. Moreover, the World Health Organization (WHO) suggested using medical face masks. Furthermore, in the health care sector, the possible loading of pathogens in sub-micrometer sizes and properties such as splash resistance and the ability to prevent infection by reducing the concentration of inhaled particulates has created limitations on the types of face masks that can be used. Therefore it is necessary to design a face mask sterilizer device that can extend the life span of masks, thereby decreasing the masks consumption. An investigation on mask sterilizer devices based on heat pipes and thermoelectrics was conducted in this study. The objective of this study was to investigate the thermal performance of the mask sterilizer device. The method used was an experimental method using variations in the input voltage on the thermoelectric, namely 9 V, 10 V, and 11 V, and the microcontroller was also adjusted to control the temperature by 70 °C, 80 °C, and 90 °C. The results showed that the thermoelectrics can generate heat with a temperature difference between the hot side and the cold side that can reach 80.11 °C. In terms of voltage, the greater voltage that was given led to a greater resulting temperature of the sterilization device and reduced the amount of time to achieve the minimum sterilization temperature.

Keywords: Sterilization · Face mask · Heat pipe · Thermoelectric

1 Introduction

The COVID-19 pandemic is now an issue and the center of attention throughout the world because the number of cases is increasing day by day, both positive and negative cases [1]. COVID-19 was first confirmed in December 2019 in Wuhan, Hubei, China [2, 3]. The pathogen of this outbreak was later identified as Severe Acute Respiratory Syndrome Coronavirus 2 (SARS-CoV-2) [4]. Indonesia first confirmed SARS-CoV-2 on March 2

Akhyar et al. (Eds.): ICECME 2021, LNME, pp. 93–102, 2023.
https://doi.org/10.1007/978-981-19-3629-6_10

2020, in Jakarta, with overtime the pandemic has spread to all provinces in Indonesia [5]. The spread of coronavirus is urgently needed to break the chain of transmission and protect the population from risk [6]. Its spread includes droplets, aerosols, and physical contact, so on March 11, 2020, WHO declared this outbreak a global pandemic [7]. Breaking the chain of transmission of the coronavirus can be done individually by doing personal hygiene, washing hands, in groups by doing social distancing, and the WHO also issued a recommendation to use face masks for all people, both healthy and sick, in early April 2020 [8, 9].

With the recommendation from WHO, the Indonesian government requires the use of face masks for all citizens [10]. However, this has led to a shortage of medical face masks because briefly, the use of face masks has increased to reach 12,740 tons in 60 days in Indonesia [11], the shortage of medical face masks can be overcome by using alternatives such as using cloth face masks. WHO also recommends that countries affected by the COVID-19 pandemic recommend the use of non-medical face masks for their citizens, including cloth face masks [12]. However, in the health care term, the possible loading of pathogens in sub-micrometer sizes and properties such as splash resistance and the ability to prevent infection by reducing the concentration of inhaled particulates has created limitations on the types of face masks that can be used. The use of medical face masks or disposable face masks causes a new problem, namely the waste of face masks [13]. The infectious disease protocol set by the government requires doctors to dispose of N95 face masks after contact with patients exposed to COVID-19 and replace face masks within a certain time [8, 14], this is also a trigger for the increase in face mask waste in Indonesia, thus it is important to develop a face mask sterilization protocol without compromising its efficacy [11, 14, 15].

Sterilization is the process of destroying microorganisms including vegetative microorganisms or spores such as germs, viruses, rickettsiae, and fungi. Sterilization can be done physically and chemically with sterilants in the form of liquids, gases, or electromagnetic radiation and heat [16]. Sterilization by heating can be done in 2 ways, namely moisture sterilization and dry heat sterilization. dry heat sterilization is one of the methods of sterilizing face masks without reducing the filtration efficiency of face masks. Anand Kumar, et al. conducted a study on decontamination of N95 face masks using sterilization technology with a peak temperature of 121 °C heating for 15 min and a total cycle time of 40 min [17]. High temperatures can certainly destroy pathogens but rapidly reduce the filtration efficiency of face masks [15]. Recent studies have shown that dry heating at 70 °C for one hour can destroy six types of bacteria, one type of fungus, and maintain a face mask filtration efficiency above 95% after three cycles [18].

Based on this, the researchers tried to conduct a study on the thermal management of heat pipe and thermoelectric-based face mask sterilization. Utilization of heat pipes because they have high thermal conductivity by transferring heat and maintaining the temperature difference between the heated and cooled parts so that it helps the healing of the face mask more optimally [19, 20] and thermoelectric is a direct conversion process where if there is an electric voltage, it will create a temperature difference or vice versa. The input current will be converted into 2, namely heat and cold [21]. In this study, the hot side of the thermoelectric will be used as a heater which then distribute the heat to the heat pipe and the heat pipe will distribute the heat evenly on the copper plate that clamps the face mask to be sterilized. The purpose of this face mask sterilization research is to understand the concept of good heating to sterilize facial masks and to develop thermal management of heat pipe and thermoelectric-based facial mask sterilizers.

2 Experimental Set-Up

This research was conducted using experimental methods, which were conducted in several stages of research. The first stage was the design stage of the sterilization model based on heat pipes and thermoelectrics. The model was designed using Autodesk Inventor 2018 software. The design of the device was made to have portable dimensions. Fritzing and Arduino IDE software are used to design a temperature control system based on the Arduino microcontroller, which is designed with a semi-automatic concept for controlling the temperature of the thermoelectric to hold it at a certain temperature so the temperature will stay in the set temperature range. The case of the sterilizer was made from filament material, which was processed using a 3D printer by the Ultimaker Cura software. Device manufacturing using CNC machine is also carried out for the less complicated case parts using acrylic. In the Fig. 1, the sterilizer system assembly consists of a heatsink, two thermoelectric TEC-12706, several heat pipes, copper plates, polyurethane, an Arduino microcontroller and battery.

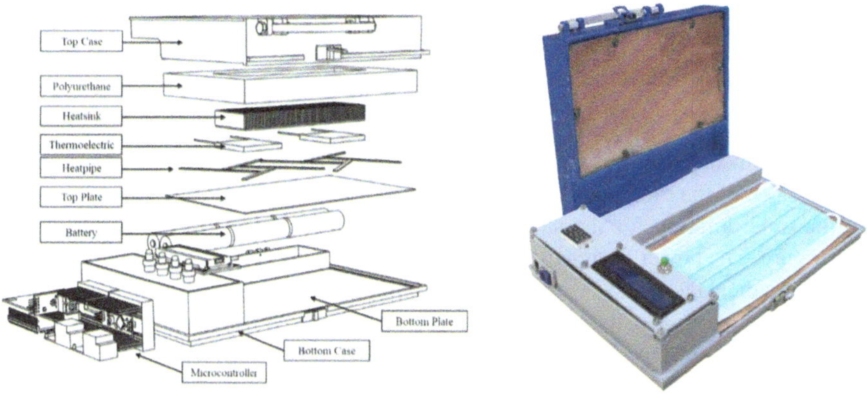

Fig. 1. The component and schematic of the portable mask sterilization device

The hot side of the thermoelectric is connected to eight heat pipes with a size of 7 × 70 mm. The heat pipes were used to distribute heat to the sterilization media in the form of a copper plate, while the cold side of the thermoelectric will be given a heatsink to dissipate heat into the environment. This sterilization device also uses insulation with low thermal conductivity made from polyurethane to increase heating efficiency and prevent the heat from dissipating into the environment [22]. The power source of this research comes from a 3S2P battery with a 12V voltage, which is controlled using an Arduino microcontroller, so the temperature does not exceed the set temperature. The Arduino-based temperature control and timer system in the Fig. 2 consists of several components such as: pushbutton, LCD display, Arduino Nano, temperature sensor, and relay. The control system is designed in the form of a PCB circuit so it can occupy the available space. The system will run semi-automatically when pressing the power on button, where the temperature and duration of mask sterilization have been previously set. After the timer is finished, it can be re-sterilized by pressing the reset button.

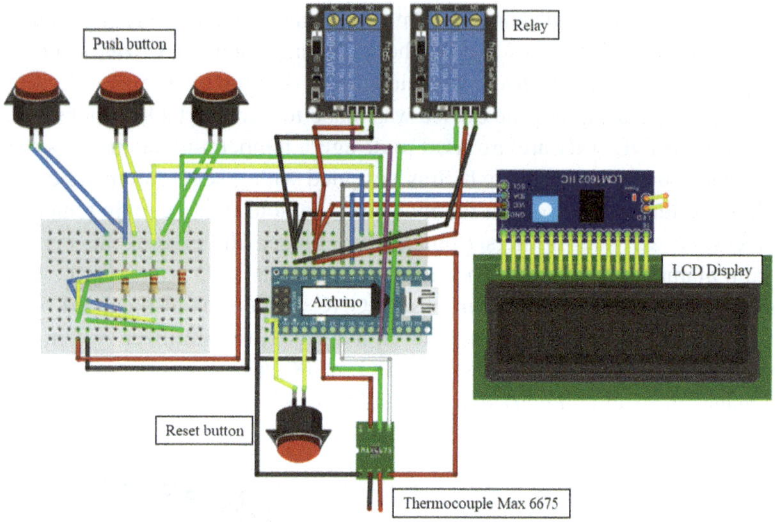

Fig. 2. Schematic Arduino microcontroller system

The sterilization device was observed to know the temperature difference between the two sides of the thermoelectric surfaces, the resulting temperature for sterilization, and the time duration for the sterilization device to reach the minimum sterilization temperature. The power source used in this test is a DC Power Supply SP-305 30 V/5 A. Then the medical face masks were used as specimens or sterilization media. The research was conducted with an experimental method with several thermocouples affixed to the upper copper plate, lower copper plate, hot side of the thermoelectric, and cold side of the thermoelectric. The temperature data read by the sensor would be detected by the computer system using the NI 9123 and C-DAQ 9174 modules. The results of the temperature data can be viewed using the NI LabView 2017 software [21]. The data temperatures and time that were used in the experiment to observe the temperature on the hot and cold sides of the thermoelectric and the duration of the sterilization device to reach the minimum sterilization temperature were the result of the variation of input voltage of 9 V, 10 V, and 11 V. For variation temperature data on the copper plate, it was held at a temperature of 70 °C, 80 °C and 90 °C. The temperature range was chosen based on previous research indicating that bacteria or viruses would be destroyed at temperatures above 70 °C [18].

The experimental scheme used can be seen in Fig. 3. The thermal performance of the sterilizer is observe using heat pipes and thermoelectrics that are insulated with polyurethane. Where the battery's electrical energy is supplied using a DC Power Supply

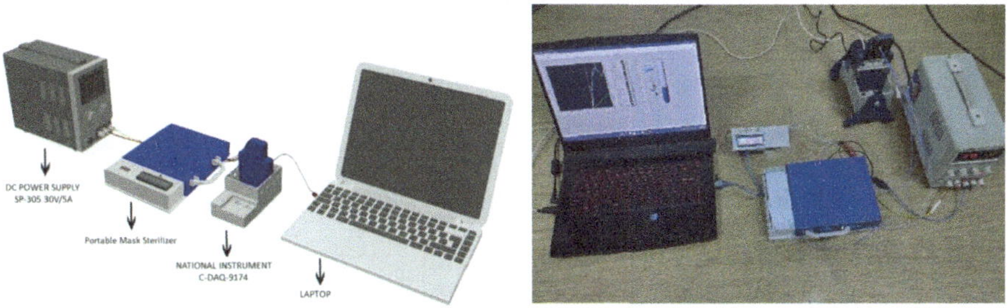

Fig. 3. Experimental set-up of the heat pipe and thermoelectric thermal performance

SP-305 30 V/5 A. Thermal performance on heat pipe and thermoelectric-based sterilizers was observed by attaching a K-type thermocouple to the bottom plate, top plate, hot side and cold side of the thermoelectric.

3 Result and Discussion

Based on the tests that have been carried out, data is obtained in the form of temperature distributions for each temperature variation carried out. The test results obtained a comparison of temperatures on the hot side and the cold side of the thermoelectric. Figure 4 exhibits a comparison of the temperature difference between the cold side and the hot side of the thermoelectric. The test was conducted with voltage variations of 9 V, 10 V, and 11 V voltage. This investigation is necessary to study, thus the performance and the characteristic of the thermoelectric is well known [23]. First, when the thermoelectric was given 9 V voltage, the temperatures of the hot and cold sides were 108.60 °C and 49.31 °C respectively, thus the temperature difference was 59.29 °C. When the thermoelectric was given 10 V voltage, the hot and cold sides of the thermoelectric were 131.20 °C and 57.83 °C respectively. Furthermore, the result of the temperature difference was 73.37 °C. Finally, when the thermoelectric was given with an 11 V voltage, the temperature of the hot and cold sides was 135.14 °C and 55.03 °C respectively. Therefore the 80.11 °C was achieved as the value of the temperature difference. The temperature difference between the 2 sides of the thermoelectric was caused by the Seebeck effect, which occurred because there is a temperature difference between 2 different conductors. Thus, the higger the voltage given, the bigger the temperature difference between the two sides of the thermoelectric [24]. Therefore, the results showed that the highest temperature difference was achieved when the thermoelectric was given with 11 V voltage. In contrast, the lower temperature difference was achieved when a 9 V voltage was given to the thermoelectric.

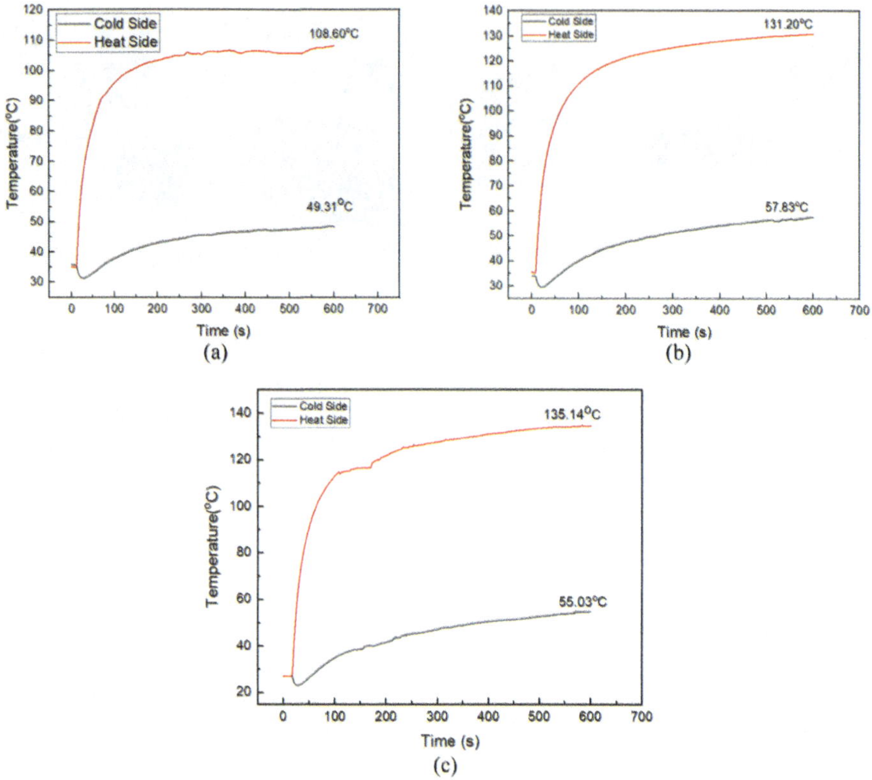

Fig. 4. Graph of the difference in temperature sterilization device (a) at 9 V voltage (b) at 10 V voltage (c) at 11 V voltage

An investigation of temperature distribution on sterilization components has also been conducted. The main focus of the investigation was to measure the temperature between the top plate side and the bottom plate side (Fig. 5) and also the time difference of the sterlization device to reach 70 °C at voltage variations. The objective of this investigation was to assess the amount of heat that can be transferred from the sterilization device to the mask surface and how long the sterilization device can reach the minimum sterilization temperature 70 °C at 9 V, 10 V, and 11 V voltage.

Figure 6 shows that when the device was set at 9 V voltage, it took 30 min to achieved 70 °C as the minimum sterilization temperature. It took 16 min to achieve 70 °C when the device was set to 10 V voltage, and 9 min when the device was set to 11 V voltage. Thus, the sterilization device showed the best performance when an 11 V voltage was given, because the time duration needed to achieve the sterilization temperature was faster than at 9 V or 10 V voltage. These results are also similar to the previous research where the higher the voltage given to the device would affect the increase in temperature duration in the device [25]. Therefore, the 11 V voltage was chosen for the input voltage in the portable mask sterilizer.

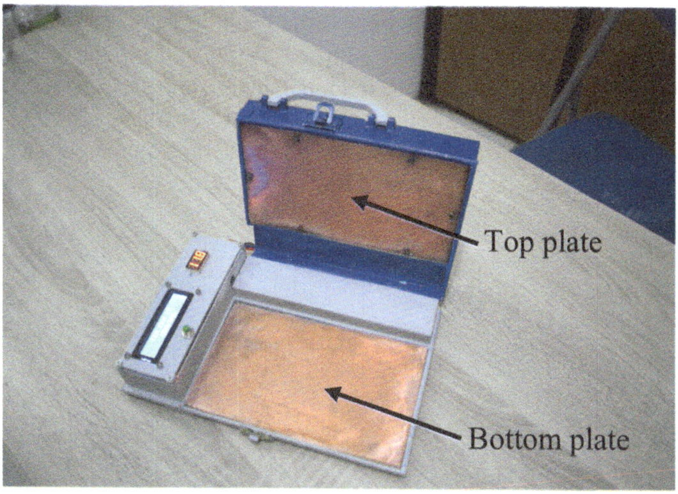

Fig. 5. Top plate and bottom plate sterilization device

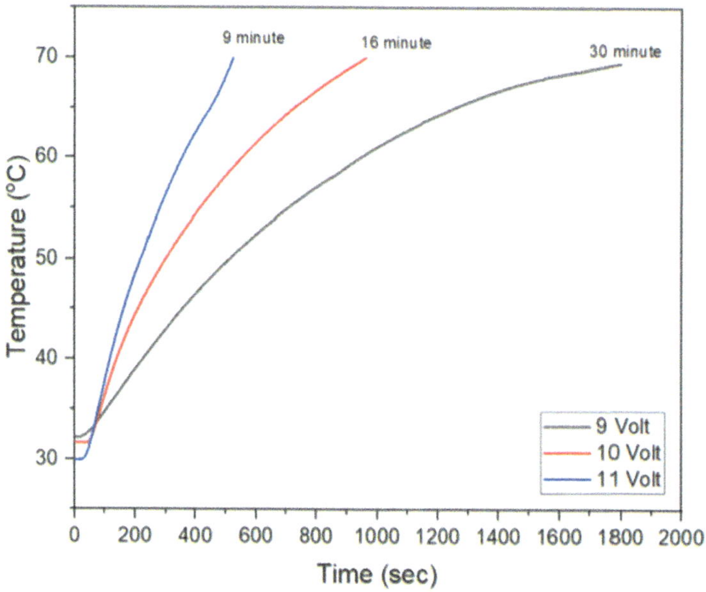

Fig. 6. Graph of the time difference sterilization device reaching 70 °C at 9 V, 10 V, and 11 V voltages

Figure 7 depicts the results of the temperature difference with an application of a holding temperature mechanism. The holding temperature is the condition when the microcontroller is adjusted to maintain the temperature of the thermoelectric thus, temperature of the thermoelectric will remain stable. This investigation was used to hold temperature variations of 70 °C, 80 °C, and 90 °C. When the holding temperature was adjusted to 70 °C, the top and bottom plate could reach 72.67 °C and 58.18 °C. Subsequently, when the holding temperature is adjusted at 80 °C, the top and bottom plates can reach 82.74 °C and 64.14 °C. When the holding temperature was adjusted at 90 °C, the temperature between the top and bottom plates was 91.67 °C and 70.80 °C. Moreover, the temperature difference with these various holding temperature at 70 °C, 80 °C, and 90 °C was 14.49 °C, 18.6 °C, and 21.08 °C respectively.

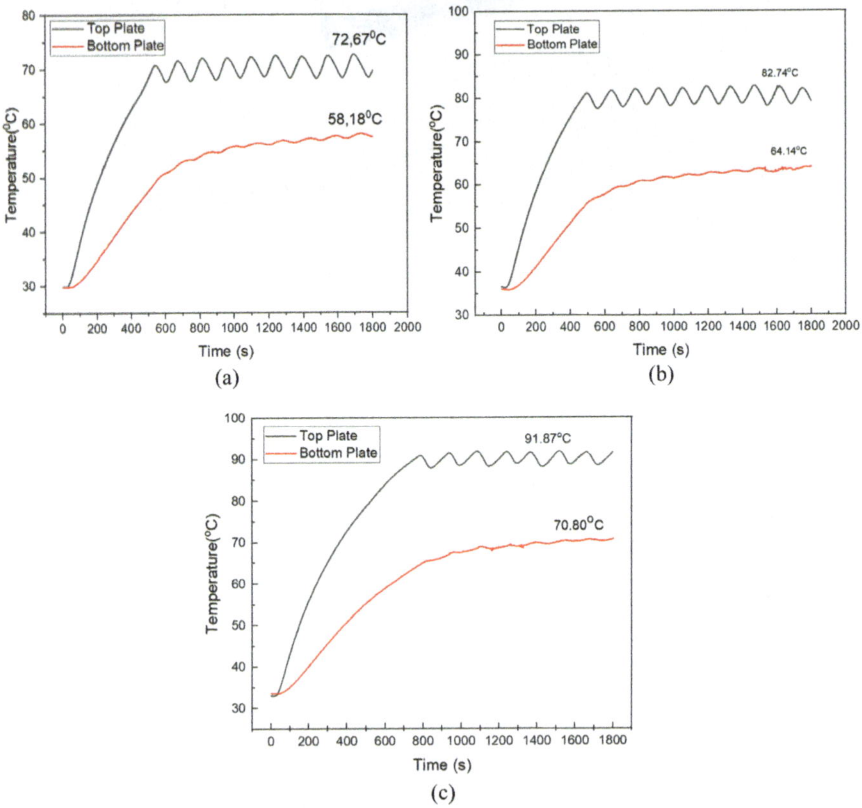

Fig. 7. Graph of the difference temperature sterilization device (a) at a holding temperature of 70 °C (b) at a holding temperature of 80 °C (c) at a holding temperature of 90 °C

These results showed that the sterilization components could transfer the heat to the mask until 91.87 °C when the holding temperature was adjusted at 90 °C. This temperature was effective enough to sterilize the mask from microorganisms. However, when the holding temperature was adjusted at 90 °C, it could cause more power consumption because there is a correlation between the increase of thermoelectric temperature with the power consumed by the thermoelectric [26]. Subsequently, for the application of this sterilization device, the holding temperature of 80 °C is chosen as the optimal temperature that can generate 82.74 °C on the top plate and 64 °C on the bottom plate. Whereby, based on previous research, these ranges of temperatures were effective for destroying microorganisms with the dry heat sterilization method [27]. Therefore this sterilization device can sterilize the microorganism effectively.

4 Conclusion

Based on the research and analysis conducted, a thermoelectric can distribute heat from the hot side with a temperature ratio between the hot side and the cold side of the thermoelectric reaching 80.111 °C. In terms of voltage, the greater voltage that is given leads to a greater resulting temperature of the sterilization device and reduces the amount of time it takes to achieve the minimum sterilization temperature. Based on this research, an 11 V voltage and a holding temperature of 80 °C were choosen to achieve the best performance of the device while maintaining the optimal temperature for destroying microorganisms.

References

1. Hira, H., Amelia, T.: HEALTHCARE WORKERS SECURITY: Jaminan, Regulasi, dan Sanksi. Khatulistiwa Law Rev. **1**(4), 109–129 (2020)
2. Wu, Y.C., Chen, C.S., Chan, Y.J.: The outbreak of COVID-19: an overview. J. Chinese Med. Assoc. **83**(3), 217–220 (2020). https://doi.org/10.1097/JCMA.0000000000000270
3. Fl,L.: Stability of SARS-CoV-2 in different environmental conditions(Supplementary Appendix), *lancet diabetes Endocrinol.*, **3**(18), 925–927 (2020)
4. Yang, X., et al.: Clinical course and outcomes of critically ill patients with SARS-CoV-2 pneumonia in Wuhan, China: a single-centered, retrospective, observational study. Lancet Respir. Med. **8**(5), 475–481 (2020). https://doi.org/10.1016/S2213-2600(20)30079-5
5. Abdullah, D., Susilo, S., Ahmar, A.S., Rusli, R., Hidayat, R.: The application of K-means clustering for province clustering in Indonesia of the risk of the COVID-19 pandemic based on COVID-19 data. Qual. Quant. 0123456789 (2021). https://doi.org/10.1007/s11135-021-01176-w
6. Zhang, Y., Zhao, Q., Hu, B.: Community-based prevention and control of COVID-19: experience from China. Am. J. Infect. Control **48**(6), 716–717 (2020). https://doi.org/10.1016/j.ajic.2020.03.012
7. Olivia, S., Gibson, J., Nasrudin, R.: Indonesia in the time of Covid-19. Bull. Indones. Econ. Stud. **56**(2), 143–174 (2020). https://doi.org/10.1080/00074918.2020.1798581
8. Amalia, V., Hadisantoso, E.P., Wahyuni, I.R., Supriatna, A.M.: Penanganan Limbah Infeksius Rumah Tangga Pada Masa Wabah COVID-19. Lp2M **2**, 7 (2020)
9. Maharaj, S., Kleczkowski, A.: Controlling epidemic spread by social distancing: do it well or not at all. BMC Public Health **12**, 679 (2012). https://doi.org/10.1186/1471-2458-12-679
10. Prasetiawan, T.: Permasalahan Limbah Medis Covid-19 Di Indonesia. Info Singk. **12**(9), 13–18 (2020)
11. Sangkham, S.: Face mask and medical waste disposal during the novel COVID-19 pandemic in Asia. Case Stud. Chem. Environ. Eng. 2, 100052 (2020). https://doi.org/10.1016/j.cscee.2020.100052
12. Donasi, G., Kain, M., Edukasi, D.A.N., Masker, P., Upaya, D., Penyebaran, M.: Universitas Serang Raya urusan Tata Busana, Alumni SMKN 3 Malang **3**, 53–57
13. Kumar, A., et al.: N95 Mask Decontamination using Standard Hospital Sterilization Technologies (2020). https://doi.org/10.1101/2020.04.05.20049346

14. Hirotsu, Y., Maejima, M., Nakajima, M., Mochizuki, H., Omata, M.: Environmental cleaning is effective for the eradication of severe acute respiratory syndrome coronavirus 2 (SARS-CoV-2) virus in contaminated hospital rooms: a patient from the Diamond Princess cruise ship. Infect. Control Hosp. Epidemiol. **41**(9), 1105–1106 (2020). https://doi.org/10.1017/ice.2020.144
15. de Man, P., van Straten, B., van den Dobbelsteen, J., van der Eijk, A., Horeman, T., Koeleman, H.: Sterilization of disposable face masks by means of standardized dry and steam sterilization processes; an alternative in the fight against mask shortages due to COVID-19. J. Hosp. Infect. **105**(2), 356–357 (2020). https://doi.org/10.1016/j.jhin.2020.04.001
16. Ma'at, S.: Sterilisasi dan Disinfeksi. Airlangga University Press, Surabaya (2009)
17. Kumar, A., et al.: N95 mask decontamination using standard hospital sterilization technologies. medRxiv, p. 2020.04.05.20049346 (2020). https://doi.org/10.1101/2020.04.05.20049346
18. Xiang, Y., Song, Q., Gu, W.: Since January 2020 Elsevier has created a COVID-19 resource centre with free information in English and Mandarin on the novel coronavirus COVID-19. The COVID-19 resource centre is hosted on Elsevier Connect, the company's public news and information, no. January 2020
19. Tio, A.O., Komposisi, H.O., Yonathan, F., Tamba, M., Septiadi, W.N., Astawa, K.: Koefisien Perpindahan Panas Sumbu Kapiler Pipa Kalor berbasis Sintered Powder Tembaga pada Fluida Kerja Hybrid Nanofluida **5**(1), 20–25 (2019)
20. Saleh, R., Putra, N., Prakoso, S.P., Septiadi, W.N.: Experimental investigation of thermal conductivity and heat pipe thermal performance of ZnO nanofluids. Int. J. Therm. Sci. (2013). https://doi.org/10.1016/j.ijthermalsci.2012.07.011
21. Septiadi, W.N., Murti, M.R., Arliyandi, Pristha Arvikadewi, I.G.A., Putra, I.P.Y. P.: Output voltage characteristic in system lighting road based on heat pipe and thermoelectric. E3S Web Conf. **67**, 10–13 (2018). https://doi.org/10.1051/e3sconf/20186702058
22. Carriço, C.S., Fraga, T., Carvalho, V.E., Pasa, V.M.D.: Polyurethane foams for thermal insulation uses produced from castor oil and crude glycerol biopolyols. Molecules (2017). https://doi.org/10.3390/molecules22071091
23. Gandi, F., Yusfi, M.: Berbasis mikrokontroler ATmega8535. J. Fis. Unand **5**(1), 35–41 (2016)
24. Alashkar, A., Alami, A.H.: Overview of Thermoelectric Materials, no. March. Elsevier Ltd. (2022)
25. Pranita, N.H., Azura, K., Ismardi, A. T. A. Ajiwiguna, Handayani, I.P.: Implementing thermoelectric generator on CPU processor. In: ICCEREC 2015 - International Conference on Control Electronics Renewable Energy Communities, vol. 2, no. 2, pp. 108–111 (2015)., doi: https://doi.org/10.1109/ICCEREC.2015.7337026
26. Thiangchanta, S., Do, T.A., Tachajapong, W., Mona, Y.: Experimental investigation of the thermoelectric cooling with vacuum wall system. Energy Rep. **6**, 1244–1248 (2020). https://doi.org/10.1016/j.egyr.2020.11.048
27. Xiang, Y., Song, Q., Gu, W.: Decontamination of surgical face masks and N95 respirators by dry heat pasteurization for one hour at 70°C. Am. J. Infect. Control **48**(8), 880–882 (2020). https://doi.org/10.1016/j.ajic.2020.05.026

Risk Analysis of Domestic Egg-Laying Chicken Farms Using Z-Score and VAR (Value at Risk) Methods

"A Case Study: Chicken Farms in Nagan Raya Regency"

Jumaidi Umran[1], Iskandar Hasanuddin[2(✉)], and Hasan Yudie Sastra[2]

[1] Master of Industrial Engineering, Department of Machinery and Industry, Faculty of Engineering, Universitas Syiah Kuala, Banda Aceh, Indonesia

[2] Department of Machinery and Engineering, Engineering Faculty, Universitas Syiah Kuala, Banda Aceh, Indonesia

iskandarhasanuddin@unsyiah.ac.id

Abstract. The high risk of mortality is one of the main problems faced by domestic egg-laying chicken farmers. This study examines the risks facing domestic egg-laying chicken farmers and the economic loss impacts contributed by those risks. The analysis methods used in this study were z-score and VaR (Value at Risk). The results show four types of production risks facing a group of farmers in Nagan Raya Regency: stocking density, changes in weather, predatory pests, and diseases. The most significant sources of risk come from diseases and changes in weather. The risk of disease has a probability of 17.8%, while the change in the weather is 12.7%. Stocking density and predatory pests, on the other hand, are the least significant sources of risk with the probability of only 4% and 2.2%, respectively. The risk of disease brings the most significant economic loss impact amounting to Rp.362,617.00, - and farmers lost Rp.117,926.00,-due to the weather changes.

1 Introduction

The owners of domestic egg-laying chicken farm businesses in Nagan Raya Regency often face challenges in achieving the success of their business. The challenges could be in the form of different levels and types of risks. The most common type of problem is production risk. Production risk is the probability of the occurrence of events that may cost economic loss impact. Such a problem may be caused by the conditions of sheds, weather, predatory pests, or disease.

Akhyar et al. (Eds.): ICECME 2021, LNME, pp. 103–115, 2023.
https://doi.org/10.1007/978-981-19-3629-6_11

2 Domestic Chickens

Domestic chickens, the local or native chickens from Indonesia, are originally from the Red Junglefowl that has been through the taming process. Domestic chickens successfully adapt to the local environment through evolution and domestication to be more resistant to the disease and weather than broiler chickens [1]. Domestic chickens can be found in almost all places in Indonesia.

In Indonesia, they are various breeds of domestic chickens, some of which are still not identified. The definition of domestic chicken may vary from one place to another. In general, however, domestic chicken has various feather colors (black, white, brown, yellow, or combinations). It also has a relatively small body size with long black, white or yellow legs. Pelung, Kedu, Merawang, and Sentul are probably some most widely recognized Indonesia's domestic breeds. As a result of the cultivation process and interbreeding naturally or wildly, as well as different environmental influences, various types of chickens with various physical appearances and varieties are formed [2].

Consuming domestic chicken meat and eggs is healthier because the cholesterol content is lower than broiler chicken, and the meat is also tastier and drier. Domestic chicken eggs are also much preferred because they are believed to increase stamina or vitality. The advantages of raising domestic chickens include ample and sustainable market opportunities; high and relatively stable selling price; the longer the maintenance, the more expensive the selling price; relatively resistance to disease and stress; and pride for raising local poultry [3].

3 Phases of Domestic Egg-Laying Chickens

The cultivation of egg-laying chickens is generally divided into three phases based on age, namely: [2].

1. The starter phase or the initial phase starts from the age of 0–8 weeks, where the shape, size, and uniformity are the goals.
2. The grower phase starts from the age of 8–20 weeks, and the chickens need to be kept under very carefully controlled feed management to prevent the chickens from having inappropriate body weight.
3. The layer phase begins after the chickens are 20 weeks old. In this phase, the growth of the chickens must be accelerated for sexual development and to achieve optimal body weight uniformity.

4 Production Factors

1. Day Old Chick (DOC)

a. Selection of Day Old Chick

Selection of the type of chicken breed is a significant factor in a laying hens farming industry. Day Old Chick (DOC) is a leading poultry commodity gained from crosses of high-productive chicken species with high economic value. One of the characteristics of this commodity is that it has very fast growth. Some of the characteristics of a good quality DOC include being free from disease, weighing not less than 37 grams, DOC looks active, has bright hair, big and wet legs, looks fresh, has no physical defects, and has no pasty vent [4].

b. Delivery process of DOC
Although the quality of Day-Old Chickens produced (DOC) is excellent, they are still at risk if they are stressed and do not want to eat and drink, leading to death. In this case, the handling technique during the shipping process becomes a very critical point. The following are deaths caused by transportation factors.

a). Delivery Time of Doc
Delivery time is also a concern in the shipping process. The preparation generally begins in the afternoon until delivery at night. This is done to avoid unpredictable hot weather conditions during the day throughout the trip. The most common case related to DOC transportation is where the DOC condition is dehydrated (lack of fluids) when they arrive at the farm. This condition can occur because the DOC is too long in transit [5].

2. Feed
Feed or ration is a collection of food ingredients suitable for eating chickens and arranged according to specific rules. The feed has the value of nutritional needs for chickens and the value of the nutritional content of the food ingredients used. Feed for egg-laying chickens in Indonesia is mostly divided into two types according to the period of maintenance: feed for the early period of laying and feed for the late period of laying. These two types of feeds have different nutritional content. Therefore, it is necessary to pay attention to the age of the chicken. Chicks less than four-week old are given early feed, whereas they are given late feed when the chickens are four weeks old [4].

Feeds have three physical forms: complete flour, granules (pellets), and (crumble). Complete form feed can be used for all ages, from day-old laying chicks to laying hens. Pellets or granular feed are only used for laying hens, namely pellets with a diameter of 3.2 mm. Meanwhile, crumble is widely used for laying hens during the starter period [3].

Domestic chickens aged 1–2 months need rations ranging from 25–45 g/day/head, age 2–3,5 months the ration needs 45–60 g/day/head, age 3.5–5.5 months 60–80 g/head/day [6]. The amount of feed consumption depends on the size of the animal's body, genetic characteristics, temperature, environment, size of the sheds, feeding place per head, condition of drinking water, quality and quantity of feed, and disease [7].

3. Drugs and Vaccines
Drugs and vaccines in this study refer to the drugs used to treat disease in livestock. Vaccines are used for the prevention of diseases originating from viruses, as well as antibiotics and vitamins, and can support the growth of chickens so that they can grow optimally.

Vaccines, vitamins, and antibiotics should be given regularly. This is very important to prevent disease outbreaks in chickens, especially against ND disease, which is very dangerous for livestock, and Infectious bursal disease (IBD), known as Gumboro. Administering the vaccine can be done through eye drops, nasal drops, injection, or spray method (fine spraying) [8].

4. Shed

Just like humans, chickens also need a proper place to live that can guarantee their health. Cleanliness of the shed and the ideal shed environment is an absolute requirement so that chickens can avoid various diseases. Some models of chicken sheds have different variations depending on the purpose of raising chickens, for example, egg-laying chickens, broilers, or chickens for decoration. [9]. The following are the requirements that must be considered when building a chicken shed.

1. Shed Location

   The following are essential items that must be considered when building a chicken shed,

   1. The shed must be built in an ideal location that will not disturb the environment, especially if it is built for larger chickens.
   2. The position of the cage should be higher than its surroundings so that water does not stagnate when it rains.
   3. The minimum distance is at least 10 m from the residential area.
   4. The shed must be designed for easy cleaning.
   5. The shed design must have good ventilation so that air exchange in the shed can work properly.
   6. The shed design must let sunlight into the chicken shed, especially in the morning.

2. Shed Density

   The ideal density standard for grower egg-laying chicken is 15 kg/m^2, equivalent to 6–8 broilers and 12–14 grower layer hens [6]. The size of the shed depends on several factors, such as the type of the shed, size of the chicken, temperature, environment, and ventilation conditions. The floor area recommended by the European Commission for medium-sized domestic chickens is 45 cm^2/head. The shed density of 8 birds /4050 cm^2 equivalent to 506 cm^2/head in young Wareng-Tangerang chickens provides a comfortable enough space for optimal production life [10].
3. Shed Construction

   The shed can be made of economically strong material, but it must have easy access for maintenance such as cleaning and disinfection. In addition, the material and construction of the cage must provide chickens protection from accidents and physical damage. The sheds should also have sewerage.

The shed's position should be built facing east so that it can get enough direct sunlight in the morning. The chickens are protected from the harmful heat of the sun during the day. The location must be separated from the residential area and is at least 500 m from the outer fence. The distance between chicken breeding farms and other farms (cows/buffaloes, goats/sheep, and horses) is at least 500 m and at least 1000 m away from animal waste shelters. The shed must not be located in the exact location as the hatchery or at least 500 m away.

4. Shed equipment
   Egg-laying chicken farms should have some shed equipment depending on the capacity/number of chickens. The equipment must be easy to use and easy to clean, and not easy to rust. The equipment needed are brooder, feeder for various ages, waterer for various types of age, lighting equipment, cleaning tools, egg trays and nests that function for laying eggs are usually made on the side of the shed in the litter/postal cage, the exhaust fan is a tool to remove CO2 gas and ammonia gas from inside the cage out of the shed.
5. Shed Lighting
   Lighting is crucial to help the chicken egg production process. The functions of lighting are:

   1. Stimulates appetite
   2. Stimulates the FSH hormone for the formation of egg yolks
   3. Stimulates the hormone LH for the release of egg yolk cells.
   4. The duration of lighting should not be reduced.

For egg production, the maximum duration for lighting is 16 h (12 h of sunlight + 4 h of light at night), or the addition of light is 4 h/day.

5 Research Method

The method used to examine the probability of risk occurrence is the standard value method or z-score. This method can be used if there is historical data and is in continuous (decimal) form. In this study, what will be calculated is the probability of risk in production activities (Fig. 1).

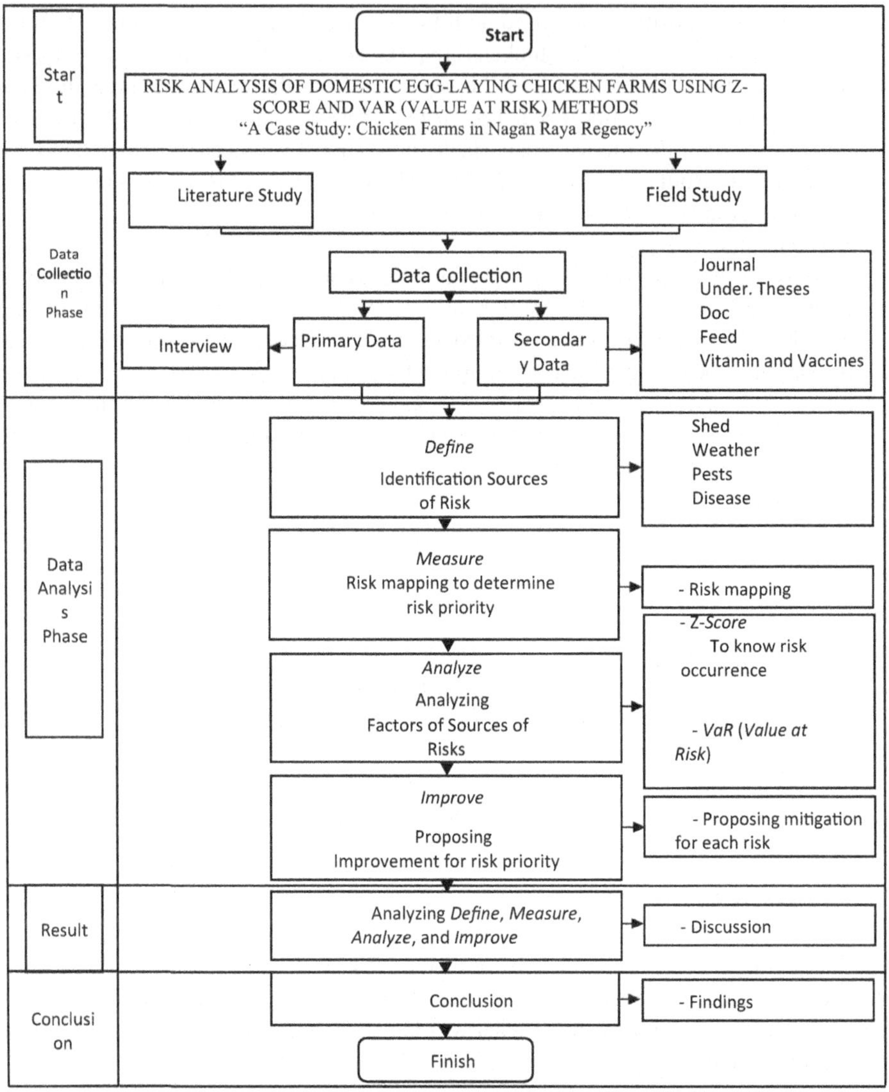

Fig. 1. The flowchart of the research

6 Discussion

6.1 Domestic Egg-Laying Chicken Mortality Rate

The data of mortality rate of domestic egg-laying chickens obtained from the group farmers in Nagan Raya Regency can be seen in Table 1.

Table 1. Data of chicken mortality in Domestic Egg-Laying Chicken Farms in Nagan Raya

No	Month	Initial number	Disease	Weather change	Predatory pest	Shed	Total mortality	Mortality percentage
1	January	4,000	10	5	0	0	15	0.38
2	February	3,985	0	0	0	0	0	0
3	March	3,985	0	5	0	0	5	0.13
4	April	3,980	20	10	5	0	35	0.88
5	May	3,945	0	0	10	5	15	0.38
6	June	3,930	0	10	3	0	13	0.33
7	July	3,917	0	0	0	0	0	0
8	August	3,917	28	0	0	5	33	0.83
9	September	3,884	0	25	5	0	30	0.75
10	October	3,854	0	15	0	0	15	0.38
11	November	3,839	37	20	0	8	65	1.63
12	December	3,774	15	20	0	0	35	0.88
13	January	3,739	40	7	5	0	52	1.30
14	February	3,687	15	13	0	16	44	1.10
15	March	3,643	50	0	0	0	50	1.21
16	April	3,593	40	20	5	5	70	1.75
17	May	3,523	0	5	0	10	15	0.38
18	June	3,508	20	0	0	0	20	0.50
Total		**3,488**	**275**	**155**	**33**	**49**	**512**	**12.80**

Source: Group of Domestic of Egg-Laying Chicken Farmers in Nagan Raya

6.2 Probability Analysis of Production Risk

The results of identifying the sources of production risks in the domestic egg-laying chicken farming industry in Nagan Raya Regency provide information that there are four types of production risk sources. The following is a probability analysis of each source of risk to examine their probability of occurrence.

This probability analysis was carried out to determine which sources of risk are most and least likely to occur so that the priority of handling can then be determined. The data used to carry out this probability were obtained from interviews with the farm owners and production reports from January 2020 to June 2021. Meanwhile, the determination of conditions, limits, and quantities used to calculate probability analysis based on farmer calculations referring to the reality that occurred in previous periods. The calculation of this analysis was performed using the z-score method.

6.3 Probability Risk Analysis of Changes in Weather

The probability of the risk from weather changes in the domestic egg-laying chicken farms is shown as follows.

The z-score method calculation showed that the z value for the source of production risk in the form of weather changes is −1.14. The negative (−) z value indicates that the value is on the left of the average value of the normal distribution curve. When mapped on the z distribution table, the z value for the source of the risk of weather changes will show a value of −1.14 with a probability of weather change of 12.7%.

6.4 Probability Risk Analysis of Predatory Pests

The following is the probability of the source of predatory pest risk in the domestic egg-laying chicken farms.

The z value for this source of risk obtained from calculations using the z-score method was −2. The z value, which was also negative (−), indicates that the value is on the left side of the average value of the normal distribution curve. The z value for the source of predatory pest risk, when mapped on the z distribution table, will show a value of −2 with a probability of predatory pest risk of 2.2%.

6.5 Probability Risk Analysis of Shed Density

The following is the probability of Shed Density Risk in the domestic egg-laying chicken farms.

The z value for shed density risk obtained from calculations using the z-score method is −1.75. A negative (−) z value indicates that the value is on the left side of the average value of the normal distribution curve. The z value for the source of shed density risk, when mapped on the z distribution table, will show a value of −1.75 with a spatial density probability of 4%.

6.6 Probability Risk Analysis of Disease

The z value for risk of disease obtained from the calculations using the z-score method is −0.92. A negative (−) z value also indicates that the value is on the left side of the average value of the normal distribution curve. When mapped on the z distribution table, the z value for the disease risk shows a value of −0.92 with a disease risk probability of 17.8%.

From the probability analysis results of the sources of production risk in domestic egg-laying chicken farms in Nagan Raya, the probability of each source of production risk can be seen in Fig. 2.

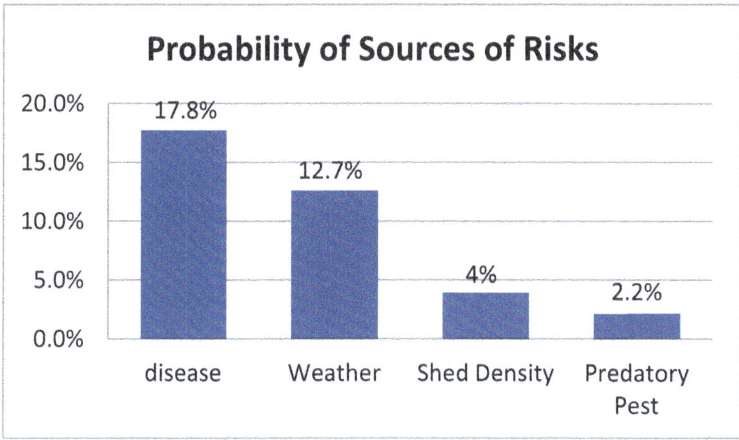

Fig. 2. Pareto diagram of probability Source of Risk

6.7 Production Risk Mapping

The probability and impacts of each source of production risk on domestic egg-laying chicken farm groups in Nagan Raya have been analyzed and calculated. Before formulating a risk management strategy, it is vital to carry out risk measurements to produce a risk status and risk map. Risk status is a measure that shows the level of risks from several sources of production risk that have been previously identified. The value of the risk status is obtained from the result of the multiplication of the risk probability and the risk impact of each risk source. The risk status of each source of production risk can be seen in Table 2.

Table 2. Status of the Risk Source in Egg-Laying Chicken Production

Source of production risk	Probability (%)	Impact (Rp)	Risk status
Disease	17.8	2,037,179	362,617
Weather	12.7	928,555	117,926
Shed Density	4	999,553	39,982
Predatory Pest	2.2	303,029	6,666

6.8 Analyzing the Sources of Risks in Domestic Egg-Laying Chicken Farms

Based on the data simulation analysis results of risk sources in domestic egg-laying chicken farms, the next step is to analyze the sources of domestic egg-laying chicken farms in Nagan Raya. The following is an analysis and mitigation of risk sources in domestic egg-laying chicken farms:

1. Changes in Weather

Extreme weather conditions significantly affect the mortality of domestic egg-laying chickens. In general, the change in weather from rain to heat is the primary source of risk impacting the chickens. Such a condition often affects the condition of the chicken's body. The immune system in the chicken body will significantly decrease in case of extreme weather changes. The following is a precaution that can be taken when extreme weather occurs (Table 3):

Table 3. Mitigating sources of weather change risk

No	Activity/Action	Objectives
1	Vitamin Supplementation (Especially Vitamin C)	To increase the body's immune, so that the chicken is more resistant to extreme weather changes
2	Paying attention to the opening and closing on the curtain	To maintain the temperature in the shed
3	Improving biosecurity	To prevent disease transmission. For example, controlling entry, spraying sheds with disinfectants (sterile infectious agents), maintaining the shed's cleanliness, feeder, and the surrounding environment

2. Disease

Gumboro or Infectious bursal disease (IBD), omphalitis (inflammation of the ovaries), Newcastle disease, and chronic respiratory disease are the common diseases attacking domestic egg-laying chickens. Viruses and bacteria cause this disease. Table 4 is the prevention that can be done against the disease:

Table 4. Management of sources of disease risk

No	Activity/Action	Objectives
1	ND Hitcner B1/ND Lasota vaccination on 4 day old chickens	To increase chicken's antibody. The antibody of chickens aged 4 decreased, making them prone to disease, especially ND
2	Vaccination of Gumbroro A on chickens aged 4–8 weeks	To increase immunity to Gumboro disease, because Gumboro disease generally attacks in the first and second weeks
3	Vaccinate ND Lasota/ND Clone 45 on chickens aged 4 weeks and so on every 3 months	To maintain antibody titers preventing other diseases

If chickens have been vaccinated three times but still die, they must be diagnosed by a veterinarian. If the chickens are tested positive for avian influenza (AI), then the AI vaccine should be given in the next vaccination.

3. Predatory Pests

Pests are predatory organisms. The most common predatory pests for chicken farms are civets and snakes. Many predators are active at night, making them difficult to detect. Table 5 shows the prevention that can be done against predatory pests:

Table 5. Mitigation of Predator Pest Risk Sources

No	Activity/Action	Objectives
1	Maintaining shed cleanliness	To keep pests out of the shed
2	Performing checks during the night	To prevent the attack of the pests during the night
3	Installing security such as wire on the floor and the wall	To prevent the pest from entering the shed

4. Analyzing the sources of the risk of shed density

Under normal conditions, the ideal number of chickens in the cage is 8–10 birds/m^2. The impact of the density of a shed is an increase in the temperature that may lead chickens to experience a critical condition, such as unstable (snoring), which occurs due to heat evaporation. The chickens need to stabilize the heat in their body. If the heat stress continues and the density of the shed is not controlled, it may make chickens weak or die because their heart fails to function. Usually, the chicken will die when the body temperature reaches 38–42°. However, the shed density in the chicken farms in Nagan Raya did not significantly affect the chickens because the size of the shed and the number of chickens in each shed were ideal.

The measures that farmers can put in place when observing the signs of heat stress in their chickens or when knowing that the temperature in the shed is hot can be seen in Table 6.

Table 6. Mitigation of shed density risk

No	Activity/Action	Objectives
1	Opening the curtain during the day	To ensure optimal air circulation in the shed

(*continued*)

Table 6. (*continued*)

No	Activity/Action	Objectives
2	Installing cooling aids, such as fans, blowers, etc	To facilitate air circulation in the shed
3	Performing the grading of the chickens based on their size	To prevent competition between big and small
4	Replacing drinking water with cold water	To help stabilize the body heat
5	Do not feed during the day	To prevent the increase of body heat
6	Multivitamin supplementation	To increase the immune system of chicken To increase the immune system of chickens

7 Conclusion

Based on findings of this research regarding the analysis of production risk analysis on domestic egg-laying chicken farms, the conclusion that can be made is as follows:

1. The results of the analysis using the z-score method showed that there are 4 sources of production risk, namely disease risk, which possess the highest probability of 17.8%, changes in weather with a probability of 12.7%, followed by shed density with the probability of 4% and predatory pest with the probability of only 2.2%. Risks of disease and weather changes brought the most significant impact, while the shed density and predatory pests had minor impacts.
2. The VaR (value at risk) method showed that disease and weather changes were the most significant source of risks. The calculation showed that the impact losses of disease risk were Rp. 362,617,00, and weather changes were Rp. 117.926.00, -
3. Mitigation strategies are carried out based on the sources of risk that have the most significant impact on the group of domestic egg-laying chicken farmers in Nagan Raya Regency.

   Strategies for mitigating sources of disease risk:

   a. Doing ND Hitcner B1/ND Lasota vaccination on four day old chickens
   b. Perform Gumboro A vaccination on chickens aged 4–8 weeks.
   c. Vaccination against ND Lasota/ND Clone 45 on chickens aged four weeks and so on for tree months.

   Strategies for mitigating sources of risk of changes in weather:

   a. Opening the shed curtain in hot weather
   b. Closing the shed curtain in cold or rainy weather

c. Installing cooling aids
d. Giving colder drinking water, especially in high-temperature conditions
e. Giving multivitamins
f. Improving biosecurity.

References

1. Sarwono, B.: Beternak Ayam Buras [Breeding Free-range Chicken] Cetakan ke 3. Penebar Swadaya, Jakarta (1991)
2. Sutirto, A.B.: Pengaruh Substitusi Konsentrat Dengan Tepung Silase Keong Mas (Pomacea Canaliculata) Terhadap Kualitas Fisik Daging Ayam Kampung [The Effects of Concentrate Substitution with Silage Flour Mas Conch (Pomacea Canaliculata) on the Physical Quality of Domestic Chicken Meat]. Thesis, Universitas Mercu Buana Yogyakarta (2019)
3. Krista, B., Harianto, B.: Buku Pintar Beternak dan Bisnis Ayam Kampung" [Smart Book on Domestic Chicken Breeding and Business]. Agro Media Pustaka, Jakarta, (2010)
4. Rasyraf, M.: Beternak ayam pedaging [Broiler Chicken Farming]. Jakarta (2004)
5. Fadillah: Modeling ynamicinteractionin supply chain sing agent base simulation. journal dition, iberoamericanjournal of industrial enggineering, lorian polis, SC, Brasil (2006)
6. Ariyanti, F., Aji, M.B., Budiono, N.: Pengaruh Pemberian Air Gula Merah Terhadap Performa Ayam Kampung Pedaging" [The Effect of Giving Brown Sugar Water on the Performance of Broiler Chickens]. Jurnal Sains Veteriner (2013)
7. Suprijatno, E., Atmomarsono, U., Kartosudjono, R.: Ilmu Dasar Ternak Unggas. [Basic Foundation of Poultry]. Penebar Swadaya. Jakarta (2005)
8. Solihin, M.: Risiko produksi dan harga serta pengaruhnya terhadap pendapatan peternakan ayam broiler CV. AB Farm Kecamatan Bojonggenteng Sukabumi, [Production risk, price, and their impacts on broiler farm income at CV. AB Farm, Bojonggenteng Sub-district, Sukabumi] Skripsi, Institut Pertanian Bogor, Bogor, Indonesia (2009)
9. Ibniyah, S.: kajian terhadap efisiensi saluran tataniaga ayam broiler pada PT. Nusanto Agheng [a study of the efficiency of the broiler trading system at PT. Nusanto Ageng]. Departemen sosial ekonomi peternakan, fakultas pertanian, institunpertanianbogor (2000)
10. Iskandar, S., Setyaningrum, S.D., Amanda, Y., Iman, R.H.S.: Pengaruh Kepadatan Kandang Terhadap Pertumbuhan dan Perilaku Ayam Wareng-Tangerang Dara [The Effect of Cage Density on the Growth and Behavior of Wareng-Tangerang Dara Chickens]. JITV **14**(1), 19–24 (2009)

Hazmat Cooling Vest Temperature Using Heat Pipe and Thermoelectric

Wayan Nata Septiadi[1(✉)], Anak Agung Gde Agung Krisnanta Dwipayana[2], Komang Manik Marianti[2], I Ketut Arya Aditya Nugraha[2], I Made Pujanu Ary Shantika[2], I Gede Riski Ananta[2], I Komang Wahyu Suarsana[2], Ni Nengah Dwi Fatmawati[3], Ni Made Dian Sulistiowati[4], and Nandy Putra[5]

[1] Department of Mechanical Engineering, Udayana University, Badung-Bali, Indonesia
wayan.nata@unud.ac.id

[2] Bachelors Program in Mechanical Engineering, Udayana University, Bali, Indonesia

[3] Department of Microbiology, Faculty of Medicine, Udayana University, Bali, Indonesia

[4] Mental Health Nursing Department, Faculty of Medicine, Udayana University, Bali, Indonesia

[5] Department of Mechanical Engineering, Universitas Indonesia, Kampus Baru UI-Depok, Indonesia

Abstract. Corona Virus (COVID-19) is a virus that is endemic almost all over the world, including Indonesia. COVID-19 was first confirmed by the World Health Organization (WHO) on December 31, 2019, in Wuhan City, Hubei Province, China, and then rapidly expanded outside of China. To suppress the Covid-19 case, medical volunteers are needed as the main actors in efforts to handle Covid-19 patients. This makes health care facilities also need to focus on the principles of health worker safety, not only focus on the principles of patient safety. This also makes health care facilities also need to focus on the principles of health worker safety, not only focus on the principles of patient safety. The use of hazmat clothes is one of the efforts to protect health workers when in contact with Covid-19 patients. Hazmat clothes are technically referred to as "encapsulated waterproof protective clothing" which is PPE that must be used for officers from the risk of contracting the Covid-19 virus through airborne droplets and contact with patients and patient body fluids. Although hazmat clothing is an important PPE for health workers to stay protected, the use of hazmat clothing for a long time often makes medical personnel feel uncomfortable when providing services. Based on the problems above, the researchers conducted a study on the heat pipe - thermoelectric hazmat suit cooling vest. This technology can absorb more heat than other methods by simply applying the principle of capillarity to the wicks on the pipe walls. schematic of testing a cooling vest on a hazmat suit. The loading on the thermoelectric is given through the DC - Power supply. The temperature data read by the sensor will be detected by the computer system using the NI 9123 and C-DAQ 9174 modules. The test results can be viewed using the NI LabView 2017 software. The temperature used in this experiment is the result of tests carried out for 30 min. Based on the tests that have been carried out, the heat pipe-based thermoelectric hazmat suit cooling vest has been able to reach the lowest thermoelectric temperature of 24,42 °C, which is distributed through heat pipes to body parts.

Akhyar et al. (Eds.): ICECME 2021, LNME, pp. 116–123, 2023.
https://doi.org/10.1007/978-981-19-3629-6_12

Keywords: Heat pipe · Thermoelectric · Cooling vest · Hazmat

1 Introduction

Corona Virus (COVID-19) is a virus that is endemic almost all over the world, including Indonesia. COVID-19 was first confirmed by the World Health Organization (WHO) on December 31, 2019, in Wuhan City, Hubei Province, China, and then rapidly expanded outside China [1, 2]. This virus was declared a pandemic in Indonesia on March 11, 2020, after 2 positive cases of COVID-19 appeared on March 2, 2020, and continue to increase (Ministry of Health of the Republic of Indonesia, 2020a). Of the 50,563 people examined, Indonesia reported 8,211 positive cases, 689 cases died, 1,002 cases recovered and 42,352 were declared negative on April 25, 2020 (Ministry of Health of the Republic of Indonesia, 2020) [3]. To suppress the COVID-19 case, medical volunteers are needed as the main actors in efforts to handle Covid-19 patients. This makes health care facilities also need to be focused on the principles of health worker safety, not only focusing on the principles of patient safety [4].

The use of hazmat clothes is one of the efforts to protect health workers when in contact with Covid-19 patients. Hazmat clothes are technically referred to as *encapsulated waterproof protective clothing* which is PPE that must be used for officers from the risk of contracting the Covid-19 Virus through air droplets and contact with patients and patient body fluids [4, 5]. The use of hazmat clothing is effective protection in handling cases of COVID-19 patients because it can protect health workers from contamination of blood, body fluids, respiratory secretions, and aerosols. Although hazmat clothes are important PPE for health workers to stay protected, the use of hazmat clothes for a long time often makes medical personnel feel uncomfortable when providing services.

JaHyun Kang and his team have researched the effectiveness of the use of personal protective equipment among health workers using a descriptive observation method which is divided into 4 stages where clinical observations, simulations, surveys, and follow-up evaluation simulations are carried out. The results of the study and observations were obtained for 1 h of simulation, the level of PPE contamination was 79.2% and the obstacles experienced were in the form of impracticality and the effectiveness of PPE in protecting the body [6]. Long-term use causes discomfort and heat, difficult verbal communication and loss of the normal sense of smell, reduced touch sensitivity, difficulty moving and difficulty seeing due to moisture deposition on PPE glasses [7]. Heat on protective equipment can result in death if no action is taken [8]. The human body temperature must be maintained at 37 °C, if the body temperature drops 1 °C then the immune system decreases by 30% [9, 10]. This is certainly a problem for health workers when in contact with Covid-19 patients who still have to protect themselves with PPE but also have to be able to endure uncomfortable conditions. Pamela JM Tay, Zhi X. Koh, and MKF Liong have researched the half body cooling system in hazmat suits using a prospective clinical evaluation method to determine the effectiveness of the Carboncool cooling system, obtained from 23 healthy participants who provided qualitative feedback on safety, mobility and cooling are mostly positive. This explains the cooling system developed on the hazmat suit is very effective and tolerable [11].

Therefore, it is necessary to develop a cooling system as a solution so that in the use of hazmat clothes, health workers still feel comfortable even though they have to use it for a long time while on duty.

Based on the problems above, the researchers conducted a study on the heat pipe - thermoelectric hazmat suit cooling vest. The heat pipe is a metal container containing a capillary axis as an alternative heat delivery system to regulate thermal conditions with the concept of passive cooling, allowing the heat pipe to transfer heat without the influence of external energy in the working fluid circulation [12]. This technology can absorb more heat than other methods by simply applying the principle of capillarity to the wicks on the pipe walls [13].

Thermoelectric is a technology that can convert heat energy into electrical energy or vice versa from electrical energy into heat energy. With its compatibility properties, it can optimize the cooling of the thermoelectric [14, 15]. In this case, thermoelectric as a cooling system is an alternative technology with practical size and lightweight, compact, quiet (not vibrating) and cooling rate is controlled using the supplied voltage (DC) [16]. Thermoelectrics utilize the principle of the Peltier-Seebeck effect where the applied voltage produces a temperature difference on the thermoelectric side on the cold side and the hot side so it needs to be combined with a heat sink to reduce heat from the thermoelectric hot side so that it can be a good cooling system [17].

In this study, researchers used heat pipes and thermoelectrics as the cooling system of the vest for temperature distribution of the Hazmat suit cooling vest technology. Judging from the importance of the cooling system for hazmat clothes, it is necessary to evenly distribute the cooler. The distribution of the cooling system temperature can be done by designing a thermoelectric system and considering the specifications to ensure the cooling system can optimally cool the hazmat suit [18].

2 Method

This research was conducted using an experimental method which was carried out in several stages. The first stage is the design stage of the heat pipe and thermoelectric-based hazmat suit cooling vest model. The cooling system used consists of a heat pipe sink fan and two thermoelectric TEC1–12706. The thermoelectric is connected to a heat pipe that has a diameter of 5 mm as a cooler designed on the backside with a length of 350 mm, back-chest with a length of 600 mm, and waist - abdomen 600 mm. The heat pipe is used to distribute the temperature to the plate and then to the vest. The power source used for the heating system is connected to a 12-V battery circuit which is controlled by an Arduino microcontroller so that the temperature will remain within the temperature setting range. Then the cooling vest model was observed to determine the effectiveness of heat pipe-thermoelectric cooling on the hazmat suit cooling vest. The human body simulator is treated with the use of Hazmat clothes by isolating the simulator with polypropylene material (Hazmat clothing material). Type-K thermocouple is used to measure temperature at several points of the body such as the chest, back, waist, and abdomen as well as the arms. Several points of the thermocouple are also placed on the absorber plate as well as the cold side and hot side of the thermoelectric part.

Figure 1 shows a schematic of a cooling vest test on a hazmat suit. The loading on the thermoelectric is given through the DC-Power supply. The temperature data read by

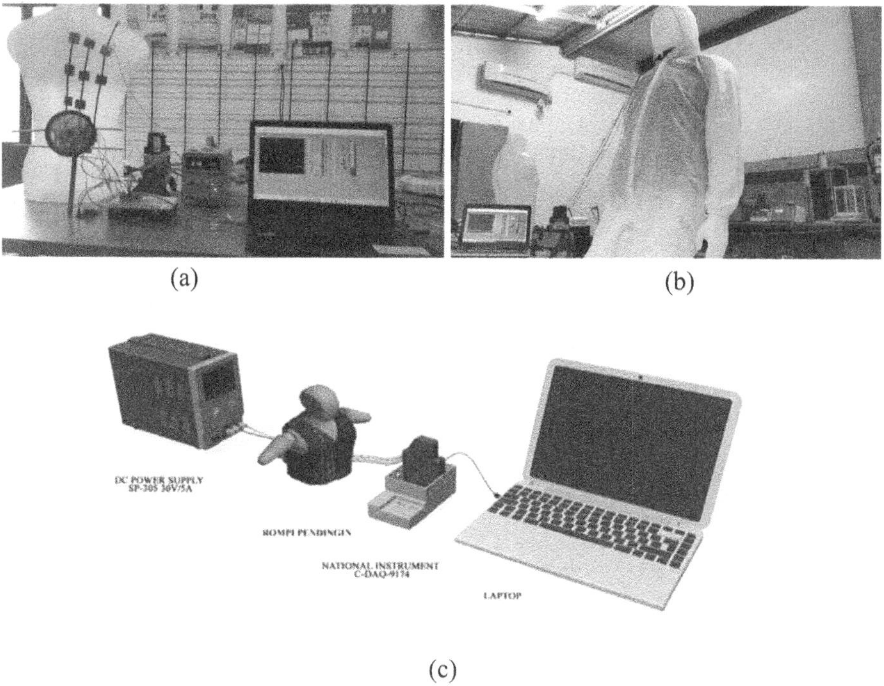

Fig. 1. Experimental setup

the sensor will be detected by the computer system using the NI 9123 and C-DAQ 9174 modules. The test results can be viewed using the NI LabView 2017 software [19]. The temperature data taken in this study are the temperature for the use of a hazmat without a cooling vest, the temperature for the use of a hazmat with a wish vest, and temperature data for the cold side and the hot side.

3 Result and Discussion

Based on the tests that have been done, data obtained in the form of temperature distribution is tested. The test results obtained the distribution of temperature from the thermoelectric -Heatpipe to the chest, back, and abdomen. The output voltage generated by the thermoelectric pends on the temperature difference on the hot and cold sides of the thermoelectric. Figure 2 indicates the temperature of hazmat use without a cooling vest. The temperature conditions of the vest on the hazmat suit. From the results of testing the use of hazmat suits without a vest, the back temperature reached 39,594 °C, the right 39.153 °C, the left chest reached 38.893 °C, and the abdomen reached 39,129 °C, Based on the test results that the temperature will continue to rise directly proportional to the time of use of hazmat without refrigeration.

Figure 3 tested the use of hazmat suits with cooling vests. The temperature on the back reaches 25.915 °C, the temperature in the right chest reaches 25.969 °C, the temperature on the left reaches 25.969 °C, the temperature in the Abdomen reaches 25.969 °C, and

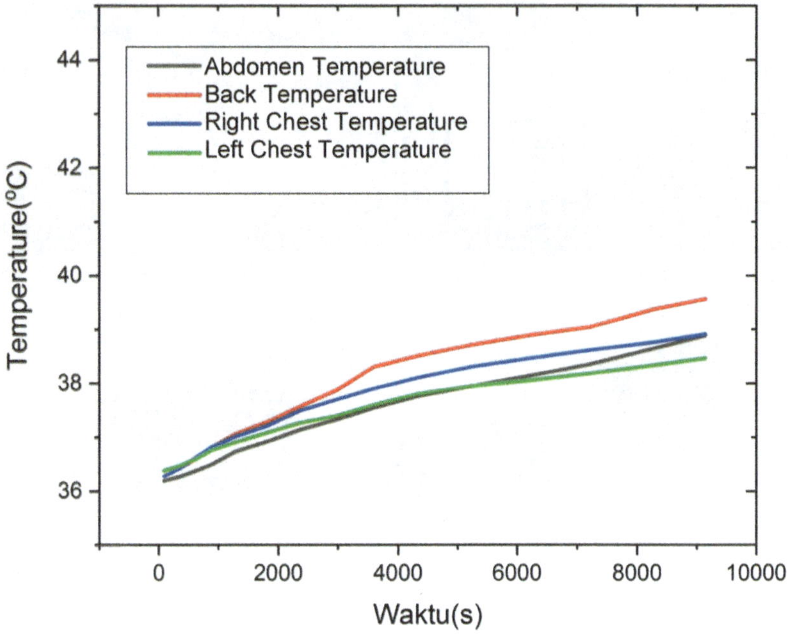

Fig. 2. Body temperature chart when wearing hazmat without cooling vest

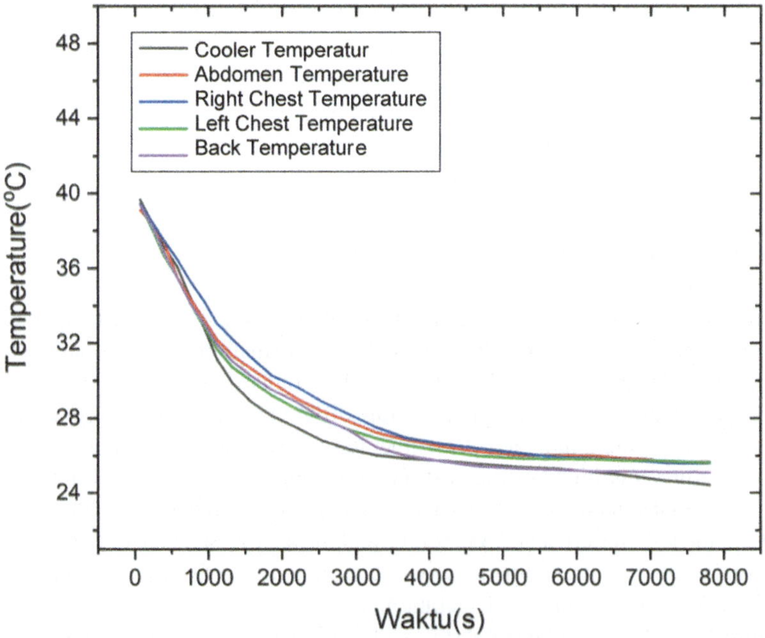

Fig. 3. Body temperature chart when wearing hazmat with cooling vest

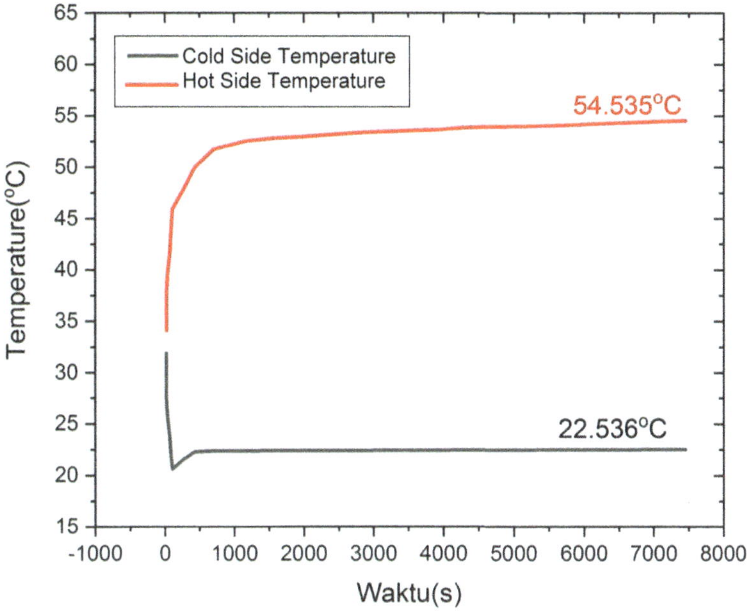

Fig. 4. Thermoelectric temperature graph

the temperature of cooler reaches 24.057 °C. From the results of tests that have been done that the temperature of using hazmat suits will decrease when using a cooling vest.

Figure 4 shows the results of the thermoelectric test where there is a temperature difference on the cold side of the thermoelectric and the hot side of the thermoelectric. The temperature on the cold side of the thermoelectric reaches 22.536 °C while the temperature on the hot side of the thermoelectric reaches 54.535 °C. Based on these data, the temperature on the cold side of the thermoelectric is lower than that on the hot side of the thermoelectric. between the hot side of the thermoelectric and the hot side of the thermoelectric has a difference of 31.999 °C. based on research results the cooling system succeeded in dissipating the heat generated when using a hazmat without a cooling vest.

4 Calculation

Based on the test results that the temperature will continue to rise directly proportional to the time of use of hazmat without refrigeration and the temperature of using hazmat suits will decrease when using a cooling vest. The temperature on the cold side of the thermoelectric reaches 22.536 °C while the temperature on the hot side of the thermoelectric reaches 54.535 °C. Based on these data, the temperature on the cold side of the thermoelectric is lower than that on the hot side of the thermoelectric. between the hot side of the thermoelectric and the hot side of the thermoelectric has a difference of 31.999 °C. based on research results the cooling system succeeded in dissipating the heat generated when using a hazmat without a cooling vest.

Acknowledgment. Thank you to the Ministry of Technology and Higher Education and the Udayana Institute for Research and Community Service for financial support through the 2021 Penelitian Unggulan Udayana (PUU) scheme, Contract Number: B/96–141/UN14.4.A/PT.01.05/2021.

References

1. Wu, Y.C., Chen, C.S., Chan, Y.J.: The outbreak of COVID-19: an overview. J. Chinese Med. Assoc. **83**(3), 217–220 (2020). https://doi.org/10.1097/JCMA.0000000000000270
2. Chin, A.W.H., et al.: Stability of SARS-CoV-2 in different environmental conditions. Lancet Microbe **1**(1), e10 (2020). https://doi.org/10.1016/s2666-5247(20)30003-3
3. Susilawati, S., Falefi, R., Purwoko, A.: Impact of COVID-19's pandemic on the economy of Indonesia. Budapest Int. Res. Critics Inst. Humanit. Soc. Sci. **3**(2), 1147–1156 (2020). https://doi.org/10.33258/birci.v3i2.954
4. Ningsih, K.P., Hardjo, K., Purwanti, E.: The use of personal protective equipment for medical recorders and health information during covid-19 pandemic in Indonesia **514**(Icoship 2020), 64–68 (2021). https://doi.org/10.2991/assehr.k.210101.015
5. Rubenstein, C.D., DenHartog, E.A., Deaton, A.S., Bogerd, C.P., DeKant, S.: Fluid replacement advice during work in fully encapsulated impermeable chemical protective suits **14**(6), 448–455 (2017). https://doi.org/10.1080/15459624.2017.1296230
6. Kang, J.H., O'Donnell, J.M., Colaianne, B., Bircher, N., Ren, D., Smith, K.J.: Use of personal protective equipment among health care personnel: results of clinical observations and simulations. Am. J. Infect. Control **45**(1), 17–23 (2017). https://doi.org/10.1016/J.AJIC.2016.08.011
7. Vidua, R.K., Chouksey, V.K., Bhargava, D.C., Kumar, J.: Problems arising from PPE when worn for long periods **88**(1), 47–49 (2020). https://doi.org/10.1177/0025817220935880
8. Brown, D.B., Smith, M.J., Kirkland, J.B.: Monitoring heat injuries in a hazmat environment. Fed. Pract. **33**(2), 24 (2016)
9. Bin Park, E., Yazdi, S.J.M., Lee, J.H.: Development of wearable temperature sensor based on peltier thermoelectric device to change human body temperature. Sens. Mater. **32**(9), 2959–2970 (2020). https://doi.org/10.18494/SAM.2020.2741
10. Lim, C.L., Byrne, C., Lee, J.K.W.: Human thermoregulation and measurement of body temperature in exercise and clinical settings. Ann. Acad. Med. Singapore **37**(4), 347–353 (2008)
11. Tay, P.J.M., Koh, Z.X., Hao, Y., Leong, M.K.F.: Safety and efficacy of carboncool half-body vest for HAZMAT decontamination crews wearing personal protective equipment: a pilot study. Prehosp. Disaster Med. **35**(6), 645–650 (2020). https://doi.org/10.1017/S1049023X20001144
12. Vasiliev, L., et al.: Loop heat pipe for cooling of high-power electronic components. Int. J. Heat Mass Transf. **52**(1–2), 301–308 (2009). https://doi.org/10.1016/j.ijheatmasstransfer.2008.06.016
13. Putra, N., Septiadi, W.N., Irwansyah, R.: Effect of concentration and loading fluid of nanofluids on the thermal resistance of sintered powder wick heat pipe. Adv. Mater. Res. **651**, 728–735 (2013). https://doi.org/10.4028/WWW.SCIENTIFIC.NET/AMR.651.728
14. Sun, X., et al.: Experimental research of a thermoelectric cooling system integrated with gravity assistant heat pipe for cooling electronic devices. Energy Procedia **105**, 4909–4914 (2017). https://doi.org/10.1016/J.EGYPRO.2017.03.975

15. Snyder, G.J., Toberer, E.S., Khanna, R., Seifert, W.: Improved thermoelectric cooling based on the Thomson effect. Phys. Rev. B - Condens. Matter Mater. Phys. **86**(4), 1–9 (2012). https://doi.org/10.1103/PhysRevB.86.045202
16. Simons, R.E., Chu, R.C.: Application of thermoelectric cooling to electronic equipment: a review and analysis. In: Annual IEEE Semiconductor Thermal Measurement and Management Symposium, pp. 1–9 (2000). https://doi.org/10.1109/STHERM.2000.837055
17. Devireddy, S., Mekala, C.S.R., Veeredhi, V.R.: Improving the cooling performance of automobile radiator with ethylene glycol water based TiO2 nanofluids. Int. Commun. Heat Mass Transf. (2016). https://doi.org/10.1016/j.icheatmasstransfer.2016.09.002
18. Budiyanto, M.A., Nadhilah, N., Fikri, A.H., Ayuningtyas, H.: Study on the application of thermoelectric coolers inside unmanned surface vehicles. J. Adv. Res. Fluid Mech. Therm. Sci. **82**(1), 12–20 (2021). https://doi.org/10.37934/arfmts.82.1.1220
19. Septiadi, W.N., Murti, M.R., Arliyandi, Pristha Arvikadewi, I.G.A., Putra, I.P.Y.P.: Output voltage characteristic in system lighting road based on heat pipe and thermoelectric (2018). https://doi.org/10.1051/e3sconf/20186702058

Development Structure Light Projection Technique for Reconstruction 3D Shape Object

Irwansyah(✉), Muhammad Dirhamsyah, Asbar, and Agung Pratama Nasution

Department of Mechanical and Industrial Engineering,
Universitas Syiah Kuala, Banda Aceh 23111, Indonesia
{irwansyah,mdirhamsyah,asbar}@unsyiah.ac.id,
agungpratamanst@gmail.com

Abstract. Non-contact scanning technology structured light projection has become an important choice in the field of engineering. It provides potential feasibility in terms of speed data acquisition, and precision of measurements. The proposed system of structured light scanning works by projecting different grayscale lines patterns and capturing the light once it falls on the scene using a camera and projector devices. That information from captured images that contained distortion lines patterns was used to reconstruct 3D geometry. Measurement of plane displacement objects is an objective of this application. In this study, the plane object was axially punched load varying with intervals 0.5 mm to 30 mm in displacement. The results of the displacement from the punch load were recorded and saved in the format of the image. Experimental and simulation results on the plane displacements of the object test shown onto 3D topographic histogram. The proposed scanning system still has an error of 0.005 mm. However, it has the potential feasibility to reconstruct 3D shape objects and offer the plane measurement displacement.

Keywords: Structure light projection · 3D shape object · Optic measurement

1 Introduction

With modern advancements in computational methods, optics, and graphics computing, 3D scanning is rapidly becoming more prevalently adopted in society. High-density 3D scanning can be performed at rates of real time or faster, thus broadening the scope of applications to which these technologies can be applied. A structured light scanning the system projects different light patterns, or structures, and captures the light as it falls onto the scene. It then uses the information about how the patterns appear after being distorted by the scene to eventually recover the 3D geometry. The potential speed of data acquisition, noncontact nature, the availability of necessary hardware, and the high precision of measurement offered by modern 3D structured light scanning technologies are what make them highly adoptable into industries such as medicine, biology, manufacturing, security, communications, remote environment reconstruction, and consumer electronics. As the number of applications in which structured light techniques are employed

Akhyar et al. (Eds.): ICECME 2021, LNME, pp. 124–132, 2023.
https://doi.org/10.1007/978-981-19-3629-6_13

increases, more interesting and challenging problems arise. It should be noted that there is not one 3D sensing technology that solves each issue and works as a general solution. The handbook by Zhang (1), which discusses many of the major 3D acquisition technologies, can be used to identify advantages and disadvantages of each approach. This article, however, focuses on discussing structured light scanning techniques. Structured light techniques have benefited greatly from recent advancements in digital technology. To achieve real time 3D data acquisition and reconstruction, much computational power is required, yet it can now be matched by today's modern computers; even some modern tablet computers can be used for these purposes. Given this, it is clear that the barrier to entry for some applications to use these technologies is quite low. However, in the past; if a manufacturing operation, for example, wanted to use these techniques, it may have been quite difficult to obtain the minimally required hardware, let alone to deploy the actual system itself. Modern advancements have ensured that the required hardware is relatively easily available. Software techniques required to reconstruct 3D data have improved greatly as well and these will also be discussed. To be useful for a wide range of applications, a system that can capture and reconstruct 3D information in real time (online), instead of retroactively (offline), is necessary. Real time in this context typically refers to acquiring, processing, and visualizing the captured 3D data at speeds of at least 24 Hz (2). As mentioned earlier, advancements in hardware have made this a possibility, yet it is not a mere trivial task; many problems had to be solved before the technology reached this point.

As mentioned earlier, structured light scanning is only one method of 3D data acquisition. Within the area of structured light scanning, there are many different techniques, as well. One method stands out that has shown many advantages and is derived from laser interferometers: digital fringe projection (DFP). The DFP technique involves varying sinusoidal patterns; using these methods, speeds of up to 120 Hz have been realized (3). However, the DFP approach has limits when it comes to its actual implementation, due to the sinusoidal pattern itself.

Eight-bit grayscale images are required to display the sinusoidal patterns, yet modern projectors can only display 8-bit patterns at certain rates. Due to these speed limitations, among other reasons (such as nonlinearity calibration and correction), other methods have been developed. These other methods allow for increased speed while limiting the increase of system's complexity. Such improvements to DFP include the squared binary defocusing technique (4). Whereas speed barriers were approached on today's digital image projection units with 8-bit images, 1-bit structured images can be projected much faster. By displaying 1-bit patterns through a defocused projector, the bits are blended together to reproduce a natural sinusoidal pattern. Another benefit, other than speed, of the binary defocusing technique is its lack of additional complexity due to calibration. As only two intensity values are being projected, nonlinearity calibration can be ignored. The smaller data transfer rate, increased rate of projection, and lack of complexity make this technique very advantageous and 3D shape measurement speeds greater than 120 Hz can be achieved. Using binary defocusing and digital light processing (DLP) platforms, Zhang et al. have been able to successfully develop a system that performed 3D shape measurement at tens of kHz (5). This article aims to provide a review of the different principles used in structured light scanning technologies, 3D data acquisition, and, by extension, a summary of its many different fields of application. By no means does this article position itself as a comprehensive body of work covering all structured light technologies and their finer details; several other surveys have been written regarding structured light techniques that may be useful, as well (6, 7). It should be noted that the content relies on published work, either in journals or in conference proceedings, done by us based on our own experiments or by others in the field. Section 2 will cover different structured pattern encodings. Section 3 will detail the steps required to properly calibrate a structured light scanning system. Section 4 will discuss structured light scanning at real-time speeds including DLP technology, 3D data acquisition, processing, and visualization. Finally, this article concluded the discussion and future directions for 3D scanning, specifically, structured light scanning.

Recently, in rapid growth, vision computer technology has significantly assisted to reconstruct three-dimensional (3D) comminuted objects. Non-contact 3D scanning as one of the cutting technologies in computation methods has become more popular and widely adopted in industrial manufacturing, medicine, and security [1–4]. Different technologies for 3D model scanning, typically classified into active and passive techniques. Passive scanning methods refer to recording a series of photographs from various angles and finding the corresponding photos using photogrammetric. While active scanning methods reconstruct the surface object by projecting a laser beam or a light pattern to the object's surface. There is not one 3D scanning technology that could be used to solve each issue and works as a general solution. Among high levels of 3D scanning, the structured light scanning system attracted more attention due to the potential speed of data acquisition, non-contact, availability of hardware in the market, and precision of measurements. The main system of structured light scanning works by projecting different light patterns and capturing the light once it falls on the scene. Those capturing pieces of information that contain distortion light patterns are used to reconstruct 3D geometry [5]. The three-dimensional model generated by structured light techniques has certain difficulties regarding the complexity of the objects and applications. The structured-light approach for 3D reconstruction is proposed which is based on 3D triangulation of optical rays generated by a video projector and recorded by a high-resolution digital camera. The system is calibrated in one step by projecting a colored chessboard pattern on top of a targeted planar surface. Commonly, a 3D model is represented by detailed digital surface models, either in point clouds or in triangular meshes format. The rendering process employs to generate texture from real imagery [6]. The structure light projection techniques can be used as powerful methods to reconstruct three-dimensional shape objects. Several commercial structured light systems are available and the others are under research development regarding the improvement of scanning accuracy, fast speed scanning, and effectiveness and optimization of calibration issues. A recent study, reported by Nguyen H., et al. [7] proposed a hybrid structured light technique with the deep convolutional neural networks for a single-shot 3D shape reconstruction. Even though the measurement accuracy of this technique is inferior to the existing 3D imaging methods, it shows potential performance in simple and robust. Wissmann et al. [8] presented a high-speed and low-cost approach for structured light pattern sequence projection as well as assessed the error sources. Ye and Song [9] show a practical approach for the optimization of calibration parameters in a structured light system. The results show that the accuracy of the 3D reconstruction is improved by the proposed approach in comparison with the conventional calibration methods. Liu et al. [10] proposed a polynomial method for measurement fringes projection system with the results of parts geometry accurately constructed after the system calibration. The system is an optically based measurement system that can provide feedback for optimization in the manufacturing process. The objective of this study is to reconstruct the 3D shape object and to determine its deformation.

2 Materials and Methods

To reconstruct 3D shape objects, a structure light projection system is built by using a computer system with CPU i5-9300H/9th Gen RAM 8GB DDR4, a projector with a resolution of IN116xa, WXGA 1280 × 800, aspect ratio 16:9, and a camera with a resolution of SX520 HS 16.0 Megapixel 2592 × 1944 pixels. The distance between a testing object and the imaging system was about 1.3 m. The principles illustration of the structure light projection setup is shown in Fig. 1.

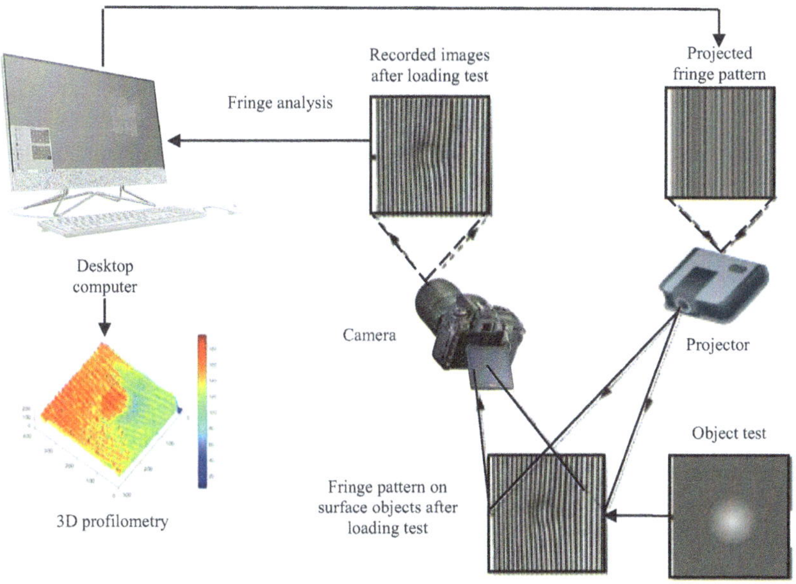

Fig. 1. Illustration of a structured light projection system

Experimental tests of 3D scanning began with the installation and setup of the light projection system concerning the triangulation principle of structured light. The scanning parameters under control include the distance from a camera to an object (L, 130 mm), the distance between the projector and a camera (D, 193 mm), and angle (α, 22°), as shown in Fig. 2. To determine scanning accuracy that equipped into the instruments and captured images deviation as well, the calibration test was carried out by comparing the results of the imaging pixels to a standard chessboard pattern.

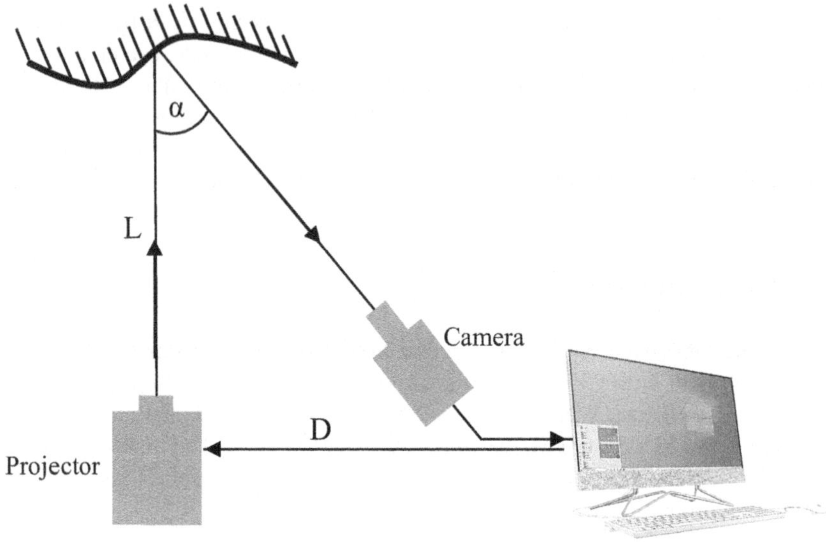

Fig. 2. Setup position of projector and camera structure light projection system

For tests, a square plate (steel, aluminum, and rubber) is mounted in a rectangular frame and penetrated horizontally by a cylindrical punch. The plane deformation is observed in the structure light scanning system. A fringe pattern is projected on the plate surface and recorded by a high-resolution camera. The changing fringe positions on the plate surface during perforation are then processed by a computer to give topography information of plane deformation. For reconstruction of the 3D model shape object, a thin material test was periodically punch-loaded 1 mm in distance for each captured image. The pressure direction of loading was pointed perpendicular to the projection planes. The loading displacement was used as representative for controlling the amount of applied pressure on the surface of the object test. A projector is used to project a grayscale pattern to the surface object. The digital camera is used to capture object deformation in time function. Overall, the scanning system was recorded for any displacements on the object surface during the pressure loading and stored in digital images format using image processing software. Images filtering, enhancement, and thresholding process were addressed to improve the quality of captured images before reconstructing the 3D shape model. Scale Invariant Feature Transform (SIFT) algorithm was employed to match images of a 3D model shape object after deformation. 3D surface models of object testing presented in the form of a Red-Blue-Green (RGB) color histogram.

3 Results and Discussion

A structure light projection system consisting of a projector and a camera was built for the reconstruction of the 3D shape objects. Three kinds of materials sheets were addressed by punch loading for identifying the feasibility of the method, as shown in Fig. 3. The pressure load is absorbed by the thin-plate surface to form a contour displacement. A plate deformation was captured and measured to refer to a profilometric histogram. The average out-of-plane surface displacement was range from 0.5–30 mm for each 30–60 repetitions. The comparison results of the recording and reconstructing 3D plane surface punch load are presented in the graph. For initial conditions before punch loading, the grayscale patterns were projected onto the test object surface and recorded using a camera. The grayscale lines pattern in the display monitor looked like a straight line on the object surface, then creating concave profiles after punch load. The comparison results between displacements of the punch loading movement onto the surface objects (3.4158 mm) to deflection measurement on the surface plane with grayscale patterned (3.4206 mm). Different results between two measurements in the range of 0.5 mm–30 that resulted from the deflection ~2.5 mm, shown in Fig. 4. Typical images of the projected fringe pattern from the camera after the punching process are presented in 3D topography, shown in Figs. 5, 6, and 7. The punch shaping after load displacement was be seen in the softened materials.

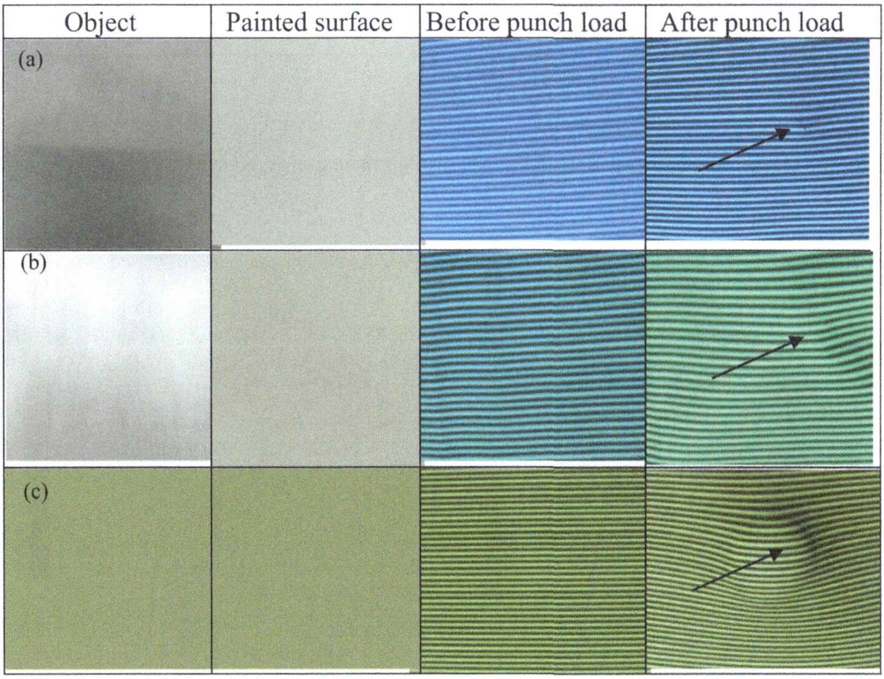

Fig. 3. Reconstruction 3D shape of the punch load on the projected plate surface, (a) rubber sheet, (b) steel plate, and (c) aluminum plate. The arrow sign shows the punch loading region on the object testing surface.

Based on the loaded reconstruction images from the cameras and the projector projection, the displacement has been measured with small errors. The out-of-plane displacement of the plate as the representative of axial distance during punch loading from the center has been plotted, It can be seen that the proposed system has potential feasibility not only to generate 3D shape objects but also offering the out-of-plane measurement displacement.

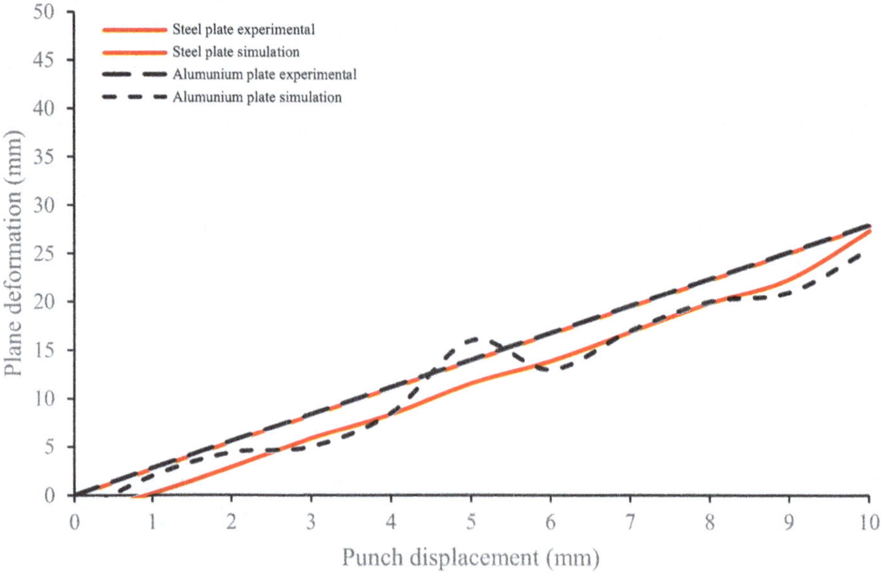

Fig. 4. Measurement of experimental punch displacement compared to simulations (polyfit)

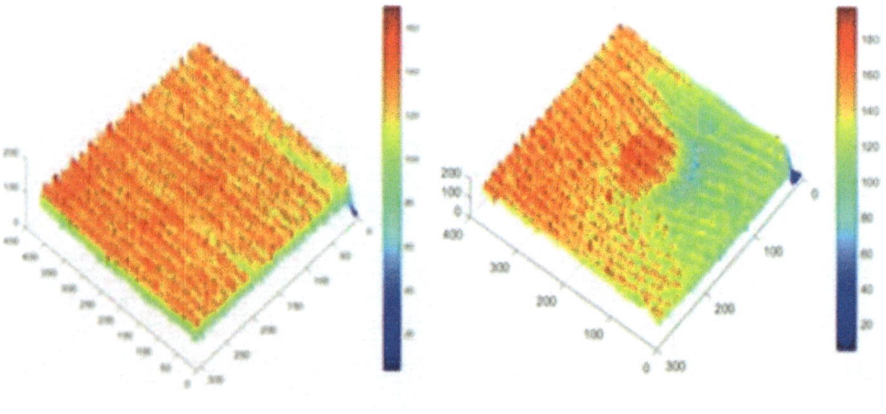

(a) (b)

Fig. 5. 3D topographic of shape reconstruction rubber sheet, (a) before (b) after punch loading

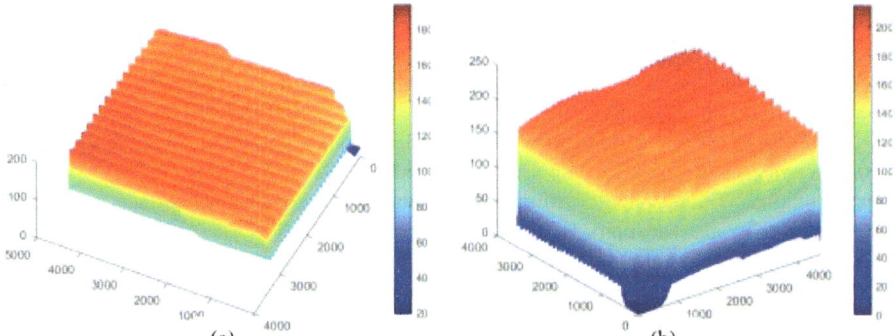

(a) (b)

Fig. 6. 3D topographic of shape reconstruction steel plate, (a) before (b) after punch loading

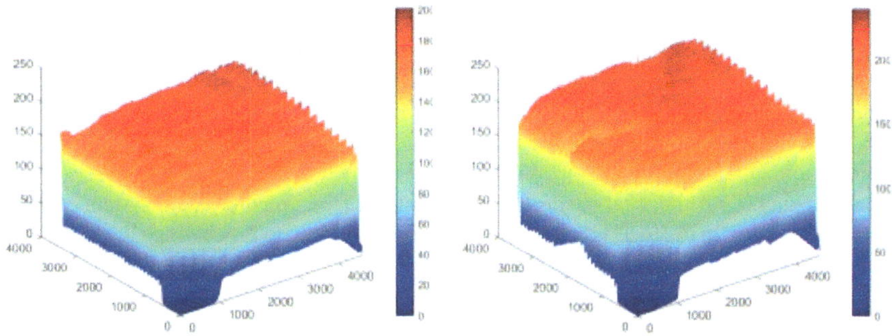

(a) (b)

Fig. 7. 3D topographic of shape model aluminum plate, (a) before (b) after punch loading

4 Summary

In this paper, a structure light projection was built based on a camera and projector for the reconstruction of a 3D shape object. The proposed system shows the feasibility as an alternative non-contact measurement to replace conventional methods. The out-of-plane displacement results include some error of ~2.5 mm for displacement range 0.5 mm–30 mm. Future research addressed to improve the scanning system performance refer to calibration accuracy, increasing the processing speed, and reducing the number of projection images.

Acknowledgments. The authors would like to express their gratitude to Direktorat Sumber Daya, Direktorat Jenderal Pendidikan Tinggi Kementerian Pendidikan, Kebudayaan, Riset dan Teknologi, Republik Indonesia (Directorate of Resources, Directorate General of Higher Education, Research and Technology, Ministry of Education, Culture, Research and Technology, Republic of Indonesia) for financial support of this research through the grant No. 154/E4.1/AK.04.PT/2021.

References

1. Javaid, M., Haleem, A., Pratap Singh, R., Suman, R.: Industrial perspectives of 3d scanning: features, roles and it's analytical applications. Sens. Int. **2**, 100114 (2021)
2. Bell, T., Li, B., Zhang, S.: Structured Light Techniques and Applications. Wiley Encyclopedia of Electrical and Electronics Engineering, pp. 1–24 (2016)
3. Shamata, A., Thompson, T.: Using structured light three-dimensional surface scanning on living individuals: Key considerations and best practice for forensic medicine. J. Forensic Leg. Med. **55**, 58–64 (2018)
4. Saha, S., Foryś, P., Martusewicz, J., Sitnik, R.: Approach to analysis the surface geometry change in cultural heritage objects. In: ElMoataz, A., Mammass, D., Mansouri, A., Nouboud, F. (eds.) ICISP 2020. LNCS, vol. 12119, pp. 3–13. Springer, Cham (2020). https://doi.org/10.1007/978-3-030-51935-3_1
5. Iizuka, K.: Structured-light pattern projection methods. In: Engineering Optics, pp. 593–626. Springer, Cham (2019). https://doi.org/10.1007/978-3-319-69251-7_18
6. Kalisperakis, I., Grammatikopoulos, L., Petsa, E., Karras, G.: A structured-light approach for the reconstruction of complex objects. Geoinform. FCE CTU **6**, 259–266 (2011)
7. Nguyen, H., Wang, Y., Wang, Z.: Single-shot 3D shape reconstruction using structured light and deep convolutional neural networks. Sensors **20**(13), 3718 (2020)
8. Wissmann, P., Forster, F., Schmitt, R.: Fast and low-cost structured light pattern sequence projection. Opt. Express **19**(24), 24657 (2011)
9. Ye, Y., Song, Z.: A practical means for the optimization of structured light system calibration parameters. In: 2016 IEEE International Conference on Image Processing (ICIP), pp. 1190–1194 (2016)
10. Liu, Y., Blunt, L., Gao, F., Jiang, X.: A simple calibration method for a fringe projection system embedded within an additive manufacturing machine. Machines **9**, 200 (2021)

The Utilization of Solar Heat with Variations of Roofing Materials in Drying Equipment

Ratna Sary, Ahmad Syuhada(✉), Dedi Afandi, and Arie H. Siregar

Department of Mechanical and Industrial Engineering, Faculty of Engineering, Universitas Syiah Kuala, Banda Aceh, Indonesia
{ratnasary_ft,ahmadsyuhada}@unsyiah.ac.id

Abstract. Solar heat is often used as a heater in the drying process, both traditional/conventional and mechanical drying with drying equipment. The objective of using a mechanical dryer is to increase the air temperature in the drying chamber by using heating from a solar collector or fuel. Heating from the solar collector commonly generates temperatures between 50 °C to 70 °C in the drying chamber. Many studies on collectors use a bottom absorber system with a glass cover. However, there are not many studies on the collector with an upper absorber system as a heater or upper heater. Solar collectors use materials with high thermal conductivity (materials with the ability to conduct heat), such as aluminum, copper and iron. In addition to high thermal conductivity, the material is also expected to function as a high absorber. This study examined the reduction in moisture content of agricultural products by utilizing heat from the roof as a solar heat absorber equipped with a heat collection channel to transfer heat into the drying room. Variations in roofing materials tested were aluminum, iron, iron coated with iron sand and zinc. To increase the temperature of the drying air in the heating channel, a sharp turn channel system was formed. Based on results obtained in this line of research, it indicated that the heater with the iron roof type was the best heater to distribute temperature, with discharge pipe of 63 °C and room temperature of 41.1 °C. The four roof variations met the requirements for standards quality of agricultural product with an average reduction in moisture content of 9.7%. The iron roof type generated the fastest reduction rate in mass of agricultural product, while the aluminum roof type generated the longest reduction rate in mass of agricultural product.

Keywords: Drying process · Solar collectors · Absorber · Roof material · Agriculture product

1 Introduction

Agricultural preservation, any of a number of methods by which food is kept from spoilage after harvest or slaughter [1]. Such practices date to prehistoric times conducted by farmers. Among the oldest methods of preservation is through the reduction of moisture content of foods. In this respect, moisture content can be reduced by drying foods to prevent damage or rotting. The process of drying agricultural products is

Akhyar et al. (Eds.): ICECME 2021, LNME, pp. 133–142, 2023.
https://doi.org/10.1007/978-981-19-3629-6_14

often carried out traditionally by the farmers through an open solar drying process. The formulated method carried out by the community has several weaknesses, including non-optimized heat. According to the BMKG in 2020, the maximum temperature in Aceh will only reach 36–37 °C [2]. Due to low temperature, the drying process is slow or takes longer, ranging from 7 to 9 days [3].

One way to increase the drying temperature is through using mechanical dryers. In this regard, to minimize the required cost, it can utilize a heat source from solar heat by collecting heat through a solar collector, then distributing it to the drying room. Solar collector is a technology to generate heat energy by utilizing solar energy as a heat source [4]. The technology is not only used by households such as solar home systems, but has also been widely applied to mechanical drying equipment in drying food [5, 6]. The technology is proven to be able to increase drying temperatures, provide an even distribution of heat, provide lower humidity and cleanliness, and provide hygiene of the dried product [7].

Several studies have been conducted using solar collectors, such as Ahmad Syuhada et al. (2018) examined cocoa drying systems using hybrid energy (solar energy and fuel). Similarly take this broad view, cocoa dryers using solar energy on 16 kg obtained a drying time of 25 h or about 3 days with energy of 23,491 kJ [8].

In general, a dryer using a solar collector as a heating source takes place by natural convection with a temperature gradient to occur in the collector. Solar radiation is collected in a solar collector using a transparent cover such as glass, fiberglass etc. One way to increase efficiency of drying equipment is to modify the existing solar collector, such as modifying the type of material or absorber pass on the solar collector [9, 10].

This research-based paper was carried out to determine the effect of heating (temperature distribution) in the drying room with variations in the types of flat plate roofing material (iron, aluminum, zinc and iron coated with iron sand materials) on the shallot drying process. With such formulated variation, it is expected to figure out the type of material with the best heat transfer rate in the drying chamber and to increase the production and quality of shallots according to SNI 01–3159-1992 by reducing the moisture content of the material with a faster drying time but not damaging the material/quality of the material [11]. An important factor in reducing the moisture content of the material is a drying rate. Quantity of moisture content in a material is a factor affecting the drying rate. The higher the quantity of moisture content in a material, the more energy is needed for the drying process [12].

The moisture content of a material can be expressed in two systematic ways, namely wet and dry basis moisture content, which is the ratio of wet basis to dry basis. Equation to determine moisture content is described as follows [13].

$$KA(bb) = \frac{Wb - Wk}{Wb} \times 100\% \tag{1}$$

where:

KA(bb) = wet basis moisture content (%)
W_b = weight of wet basis or before drying (kg)
W_k = weight of dry basis or after drying (kg)

2 Roofing Absorber Material

Collector absorber material is the most important part of a drying system using a solar collector. The function of an absorber is to absorb and store heat from solar radiation to be converted into thermal energy. When solar radiation reaches the collector, the roofing material will absorb the fraction of radiation, the fraction of radiation is reflected to the environment and most of it is transferred (conduction) to the drying chamber. In the absorber, heat absorption occurs, resulting in the increasing absorber temperature. Thus, heat transfer occurs from the absorber to the air in the room causing the air temperature to become hot [14].

3 Research Method

The present study used a flat plate roof as a solar collector by varying several types of plate material with the aim of utilizing easily available plates to be used as a heat-producing roof, and therefore the heat generated by the roof is used for the drying process to reduce the moisture content of materials or agricultural products such as shallots in order to increase the quality and reduce damage to the shallots.

In principle, the mechanism of the designed drying equipment is to utilize solar radiation as a heat source (1) on the roof with a radiant heat transfer process (Process 2) to heat the environmental air entering the gap under the roof (3) which is then heated and transferred into the drying chamber through the air pipe entering the drying cabinet (4), with the help of exhaust fan (process 5). The following is a schematic diagram of the drying equipment with 15 kg of dried shallots (Fig. 1).

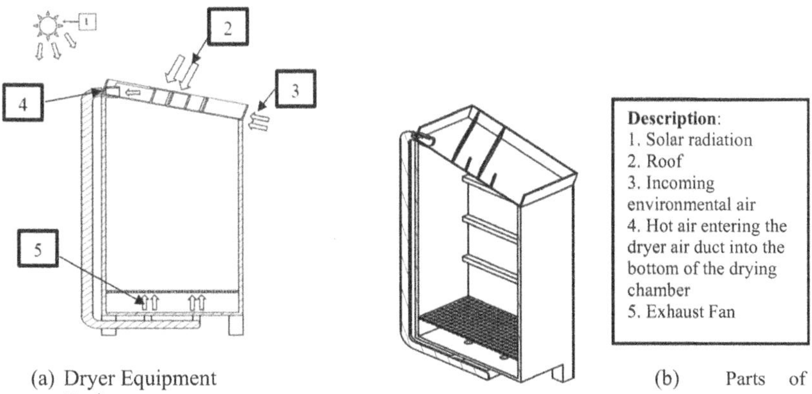

(a) Dryer Equipment Equipment

(b) Parts of Dryer

Fig. 1. Dryer equipment

The roof is designed to be separate (Fig. 2) from the drying chamber or drying cabinet so that the sharp turn channel [15–17] such as collector can be installed to increase temperature of the drying equipment. The roof has 4 variations of flat plate types as a cover or acting as an absorber in the form of a beam with a length of 1 m, a width of 1 m and a height of 0.16 m. The roof cover uses variation of the flat plate type roof (aluminum, iron, iron coated with iron sand and zinc) with a length of 1 m, a width of 1.5 m and a thickness of 2–3 mm.

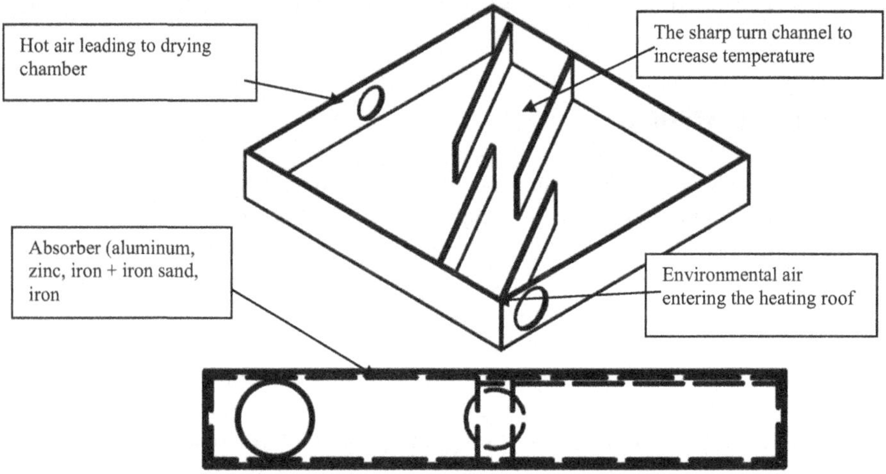

Fig. 2. Drying equipment roof design

Frame of drying cabinet (1.b) is as a shape design and load-bearing, so that the drying cabinet can be used for a long time. The frame of the drying cabinet is made of wooden blocks and the walls of the drying room are made of plywood by a machine with a dimension of 1 × 1 m. The drying rack serves as a place for placing objects to be dried. On the drying rack, the material is made of wood with a dimension of 1 × 1 m and is equipped with a base made of wire mesh for laying dried shallots so that the drying temperature can be evenly distributed. The data retrieval included the intensity of sunlight, weight of wet basis, weight of dry basis, inlet temperature of the heater, outlet temperature of the heater, inlet temperature of the room, and temperature of the drying chamber. The data were collected once in 60 min either with drying load or without drying load (Fig. 3 Measurement Point).

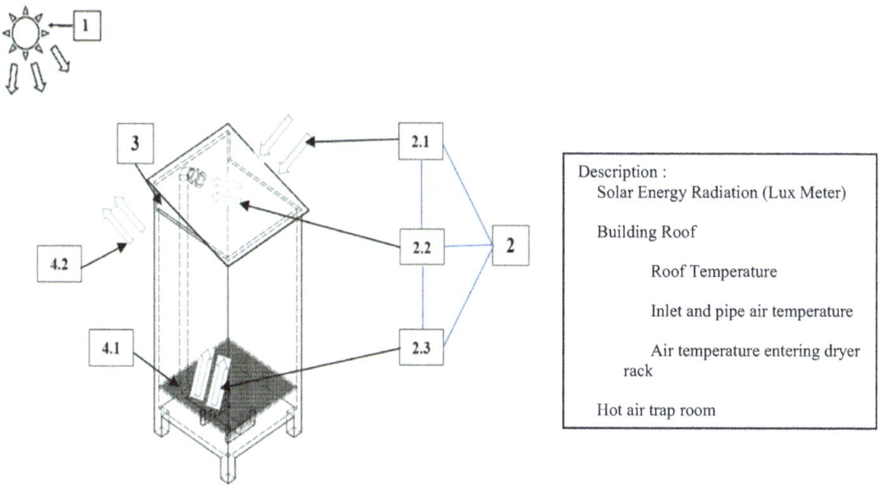

Fig. 3. Measurement point on dryer equipment

4 Result and Discussion

Figure 4 Demonstrates that the test of equipment in this report examines the heater of shallot drying system with iron coated with iron sand roof type without a load or sample, the temperature entering the drying room with sunlight intensity ranges from 856.36 W/m^2. Meanwhile, the absorber is iron coated with iron sand flat plate roof type, the highest temperature occurs at 13.00 GMT+7 with a temperature of 60 °C.

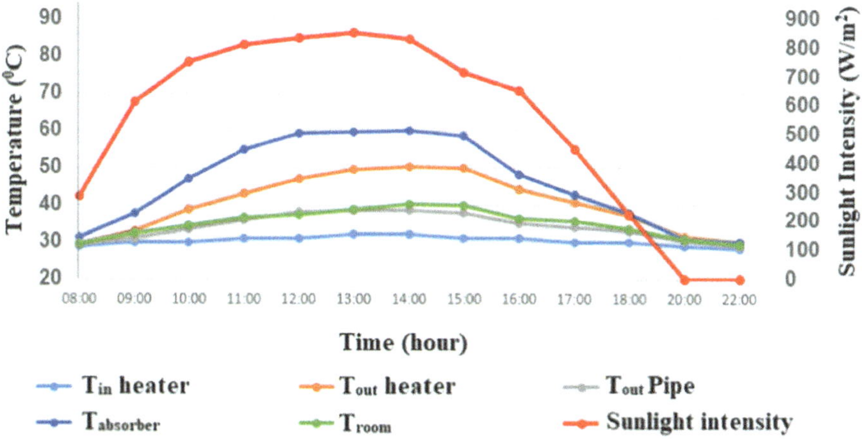

Fig. 4. Temperature distribution on drying equipment using iron+iron sand roof without load/sample

Figure 5 discloses that the test of equipment in this study examines the heater on shallot drying system with an aluminum roof type without a load (not using shallots) and the help of exhaust fan, the temperature entering the drying chamber occurs with the intensity of sunlight ranging from 856.36 W/m^2. Meanwhile, the absorber is aluminum flat plate roof type, the highest temperature occurs at 13.00 GMT+7 with a temperature of 51 °C.

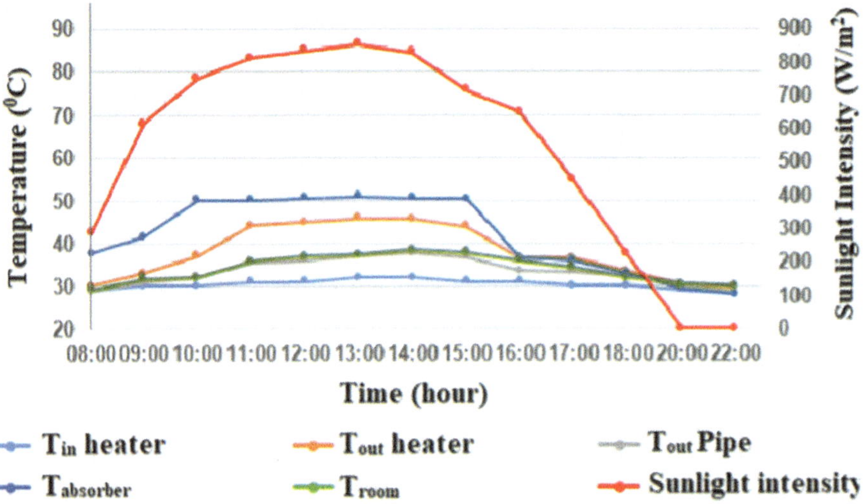

Fig. 5. Temperature distribution on drying equipment using an aluminum roof without load/sample

Figure 6 presents the findings of research that the test of equipment with a zinc roof type without a load (not using shallots) and the help of exhaust fan, the temperature entering the drying chamber occurs at 13.00 GMT+7 with the same intensity of sunlight. Meanwhile, the absorber is zinc flat plate roof type, the highest temperature occurs at 13.00 GMT+7 with a temperature of 59.1 °C.

Figure 7 appears to indicate that the test of equipment in this research-based paper examines the heater on shallot drying system either with an iron roof or without a load (not using shallots) and the help of exhaust fan, the temperature entering the drying room occurs at 13.00 GMT+7 with the intensity of sunlight ranging from 856.36 W/m^2. Meanwhile, the absorber is iron flat plate roof type, the highest temperature occurs at 13.00 GMT+7 with a temperature of 72.8 °C.

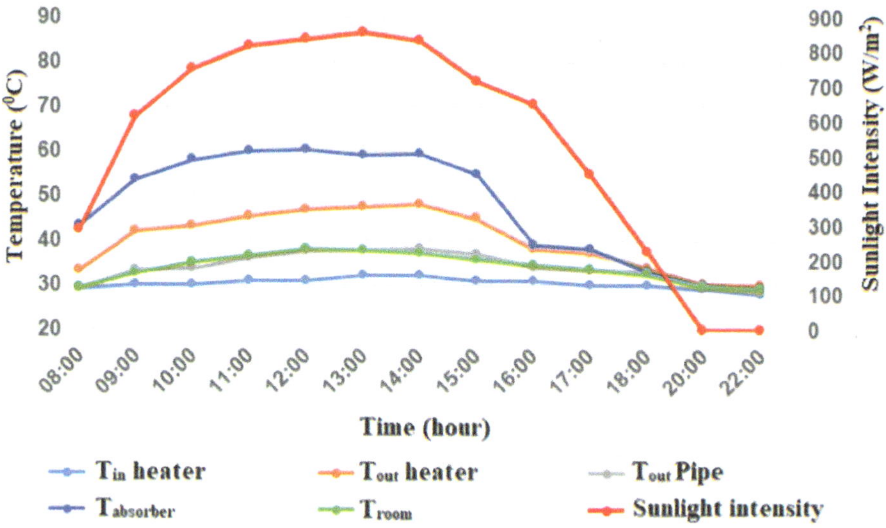

Fig. 6. Temperature distribution on drying equipment using a zinc roof without load/sample

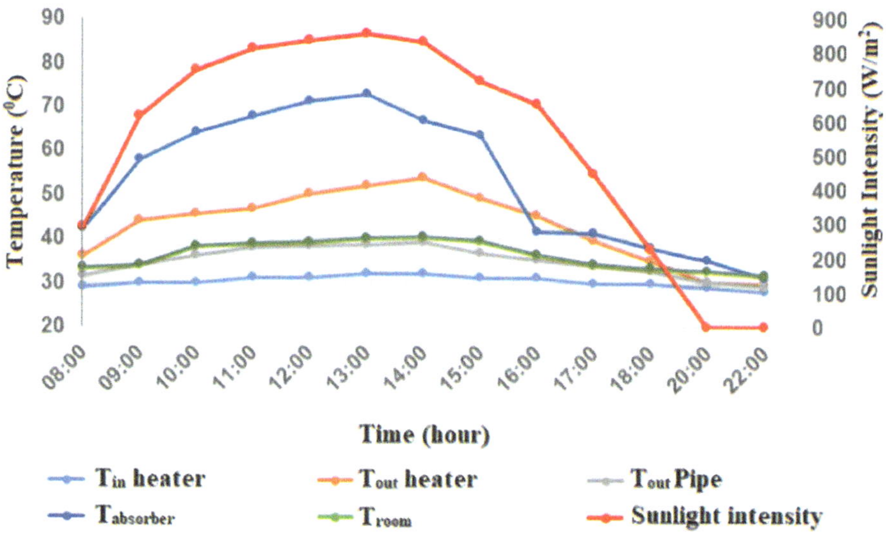

Fig. 7. Temperature distribution on drying equipment using iron roof without load/sample

5 Comparison of Reduction in Material Mass and Time with Roof Variations

Based on the results of the reduction in material/sample mass with the roof types (iron coated with iron sand, aluminum, zinc, iron), the significant reduction occurred at 09:00 to 13:00. This has drawn broad attention and was due to high sunlight intensity, hence it increased the temperature of the drying room. Eventually, the process of evaporation occurred and reduced the material mass. The highest reduction in the material mass occurred in the variation of iron flat plate roof type (Fig. 8).

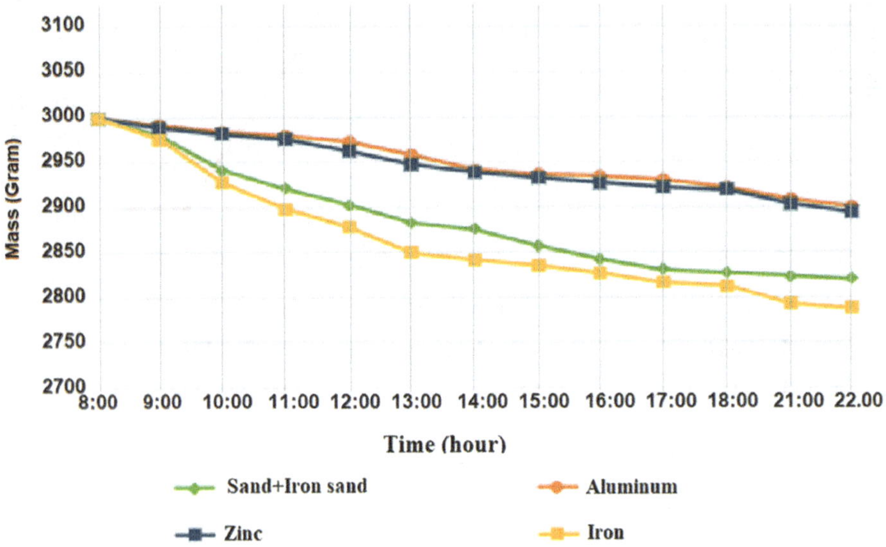

Fig. 8. Reduction in material mass with variations in the roof types

6 Calculation of Material Moisture Content

Based on the results of shallot drying/storage test using 4 variations of the roof, namely iron coated with iron sand, aluminum, zinc, and iron, it was obtained the moisture content of 80–85% and referred to the SNI standard. Based on the calculation of moisture content of each roof variation using Eq. (1), the results of iron, iron+iron sand, aluminum and zinc roof variations can be seen in the following Table 1.

Table 1. The calculation results of moisture content

Material type	Initial moisture content (Gram)	Final moisture content (Gram)	Material moisture content (%)
Aluminum	3000	2766	7.8
Zinc	3000	2700	10
Iron+iron sand	3000	2685	10.5
Iron	3000	2676	10.8

7 Conclusion

The maximum temperature obtained from the iron roofing material was 72.8 °C at 13:00 GMT+7, thus accelerating the evaporation rate of the material by 10.8%. This was because the iron material has a higher temperature, resulting in a high temperature difference (ΔT) of 21.9 °C compared to other three types of materials. In four roof variations, moisture content was obtained in accordance with the SNI standard of 80–85% with an average reduction in moisture content of 9.7%.

Acknowledgement. The researchers are grateful for the contributions of all the students at the Graduate School of Mechanical Engineering, Universitas Syiah Kuala in the recent study. Appreciation and special grateful to Universitas Syiah Kuala for funding this research.

References

1. Amit, S.K., Uddin, Md.M., Rahman, R., Rezwanul Islam, S.M., Khan, M.S.: A review on mechanisms and commercial aspects of food preservation and processing. Agricult. Food Secur. (2017). https://doi.org/10.1186/s40066-017-0130-8
2. https://www.tribunnews.com/regional/2020/07/06/suhu-udara-di-aceh-capai-35-derajat-celcius-kondisi-panas-diprediksi-berlangsung-hingga-september. Access 20 Aug 2021
3. Zamharir, S., et al.: Analisis Pemanfaatan Energi Panas Pada Pengeringan Bawang Merah (Allium ascalonicum L.) Dengan Menggunakan Alat Pengering Efek Rumah Kaca (Erk). Jurnal Ilmiah Rekayasa Pertanian dan Biosistem **4**(2), 264–274 (2016)
4. Ahmadi, M.H., et al.: Solar power technology for electricity generation: a critical review. Energy Sci. Eng. 6(5) (2018). https://doi.org/10.1002/ese3.239
5. Jayanthi, N., Suresh Kumar, R., Karunakaran, G., Venkatesh, M.: Experimental Investigation on the Thermal Performance of Heat Pipe Solar Collector (HPSC). In: Materials Today: Proceedings, diakses melalui (2012). https://doi.org/10.1016/j.matpr.2019.07.607
6. Udomkun, P., et al.: Review of solar dryers for agricultural products in Asia and Africa: an innovation landscape approach. J. Environ. Manage. **268** (2020). https://doi.org/10.1016/j.jenvman.2020.110730
7. Syuhada, A., Maulana, M.I.: Study of heat transfer characteristics on sharp turn channels for solar collectors. J. Adv. Res. Fluid Mech. Thermal Sci. **45**(I), 82–91 (2018)
8. Syuhada, A., Sary, R., Isnan, F.: Kaji sistem pengering kakao dengan mengunakan energi hybrid (energi matahari dan bahan bakar). Jurnal Teknik Mesin Unsyiah **6**(1), 17–24 (2018)
9. Phadke, C.P., Walke, P.V., Kriplani, M.V.: "Direct Type Natural Convection Solar Dryer": a review. Int. J. Adv. Res. Sci. Eng. IJARSE, **4**(2) (2015). ISSN-2319-8354. http://www.ijarse.com
10. Syuhada, A., Maulana, M.I., et al.: Pengaruh material absorber terhadap temperatru keluaran kolektor surya. Jurnal Teknik Mesin Unsyiah **6**(2), 53–56 (2018)
11. Komar, N., Rakhmadiono, S., et al.: Teknik Penyimpanan Bawang Merah Pasca Panen Di Jawa Timur. Jurnal Teknologi Pertanian **2**(2), 79–95 (2001)
12. Suherman, P.A., Aulia, M.P.: Pengaruh suhu udara dan berat sampel pada pengeringan tapioka menggunakan pengering unggun terfluidakan. Prosiding Seminar Nasional Sains dan Teknologi SNST, Fak Teknik Unviersitas Wahid Hasyim **1**(1), 45–50 (2012)
13. Winarno, F.G.: Pengantar Teknologi Bahan. PT. Gramedia, Jakarta (1993)
14. Agustina, D., Syuhada, A.: Pengaruh komposisi material absorber terhadap kinerja kolektor surya. In: Proceeding Seminar Nasional Teknologi Rekayasa (SNTR) II, pp. 92–95 (2015)

15. Razali, A.S., et al.: Analisa Pengaruh Belokan Tajam pada Pipa Besi Terhadap Perpindahan Panas pada Solar Water Heater. Jurnal Teknik Mesin Unsyiah **7**(2), 48–52 (2019)
16. Hirota, M., Fujita, H., Nakayama, H., Syuhada, A.: Heat (mass) transfer characteristics in serpentine flow passages with a sharp turn. Memoirs of the School of Engineering, Nagoya University (2001).https://doi.org/10.18999/memsenu.52.1
17. Syuhada, A., Sary, R., Zakirullah: Pengaruh Diameter Hidrolik Terhadap Perpindahan Panas Kolektor Surya Belokan Tajam. Jurnal Teknik Mesin Unsyiah **7**(2), 38–42 (2019)

Physic-Chemical Parameter of Water Around Kutaraja Fishing Port: A Preliminary Study for Shipyard Industry Development

Thaib Rizwan[1,2], Zulkarnain Jalil[1,3], Akhyar[1,4], and Husaini[1,4](✉)

[1] Doctoral Program, School of Engineering, Universitas Syiah Kuala (USK), Jl. Syech Abdurrauf No. 7, Darussalam, Banda Aceh 23111, Indonesia
{rizwanthaib,zjalil,akhyar,husainiftm}@unsyiah.ac.id
[2] Department of Capture Fisheries, Marine and Fisheries Faculty, Universitas Syiah Kuala (USK), Jl. Syech Abdurrauf No. 7, Darussalam, Banda Aceh 23111, Indonesia
[3] Department of Physics, Mathematics and Natural Science Faculty, Universitas Syiah Kuala (USK), Jl. Syech Abdurrauf No. 7, Darussalam, Banda Aceh 23111, Indonesia
[4] Department Mechanical Engineering Department, Faculty of Engineering, Universitas Syiah Kuala (USK), Jl. Syeh Abdurrauf No. 7, Darussalam, Banda Aceh 23111, Indonesia

Abstract. The shipyard is used for shipbuilding and reparation of the old ship with various damage intensities activities. Therefore, the fishing port must have a shipyard facility. This research aims to assess the water quality surrounding the fishing port and its relation with the shipyard development. The research was conducted from December 2020–March 2021 in Kutaraja Fishing Port and Marine and Fisheries Faculty Laboratory, Syiah Kuala University. Each sampling point has a distance of 120 m vertically and 310 m horizontally. Water quality data recorded were pH, temperature, dissolved oxygen, and salinity. The pH value in Lampolu waters was categorized as good, ranging between 7.13 and 8.3, with the highest pH value measured at the waters near the land. The sea surface salinity in the research area was 32–34 PPT, which gradually increased towards the coastal area. The sea surface temperature was 27.5–30 °C. Dissolved oxygen (DO) on the water surface ranges between 7.02 to 9 mg/L. Water quality parameter is one of the vital factors to determine shipyard location.

Keywords: Shipyard · Physic-chemical parameter · Kutaraja Fishing Port · Salinity · pH · Dissolved oxygen

1 Introduction

A shipyard is a place for shipbuilding and reparation of the old ship with various damage intensities. The existence of a shipyard is one of the necessary facilities in a fishing port. Current ships were designed and built with a modular approach. Each module was

Akhyar et al. (Eds.): ICECME 2021, LNME, pp. 143–147, 2023.
https://doi.org/10.1007/978-981-19-3629-6_15

developed in a specific area and transported to the development area for the final line up and attaching. In this approach, the shipyard focus only on the core area from the whole design to the launch process [1]. Modern shipyard industries have improved management systems compared to conventional shipyards. The documentation is well structured to provide specific documents with relevant information for each assembly phase [2].

Kutaraja Fishing Port, located in Banda Aceh City, is the biggest fishing port in Aceh Province [3]. It is a large-scale fishing port that is expected to provide optimal services related to operational activities, especially capture fisheries. Kutaraja Fishing Port is a fisheries business development-based fishing port used by small and large vessels to unload and load and utilize its facilities and services. Ship repairment (shipyard) installation is one of the most critical facilities that need to be made available in Kutaraja Fishing Port.

Prior to developing a shipyard in a port, it is crucial to assess the location suitability based on characteristics of the oceanographic factors, including current, seawater depth, wind, wave, tide, and sedimentation. The oceanographic aspects, such as current speed, wind, and wave height, are crucial for shipyard development. They can impede the process of exiting and entering the ship into the docking pool. These factors can also cause sedimentation, which can further impact the docking pond silting. Previous research showed that several oceanography parameters, such as current, bathymetry, wave, wind, and tide, highly influence the shipyard location [4]. This research aims to assess the water quality around the fishing port and its' relevance in shipyard development.

2 Research Method

2.1 Water Sampling

This research was conducted from December 2020 to March 2021 in the Kutaraja Fishing Port and Marine and Fisheries Faculty Laboratory, Syiah Kuala University. Each sampling point has a distance of 120 m vertically and 310 m horizontally. Water quality measured includes pH, temperature, dissolved oxygen, and salinity. Temperature and dissolved oxygen were measured using the DO meter (Lutron YK-2005WA; Taiwan), water pH was measured using a digital pH meter (ATC pH-2011; Romania), and salinity was measured using Salinometer (HANNA; USA). The data were analyzed and visualized using QGIS software [5] (Fig. 1).

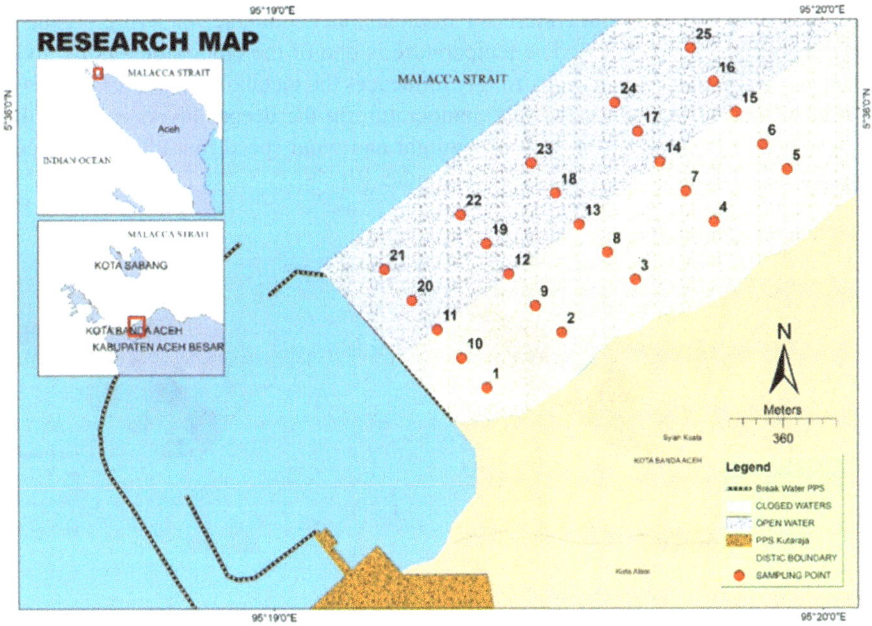

Fig. 1. Sampling sites

3 Result and Discussion

The degree of acidity (pH) value in Lampolu waters is considered a good condition ranging between 7.13 and 8.13, with the highest value near the land (Fig. 2). Generally, the ocean has a pH value ranging from 6.5 to 9.0. The degree of acidity is essential to determine the water use-value for organisms' life and other purposes. The pH value is influenced by several factors, such as photosynthesis activities, temperature, and anion cation amount in the water [6]. pH is the degree of acidity used to determine the level of acidity and basicity in a solution.

Water surface salinity in the study area ranged between 32–34 PPT, with the distribution pattern increasing towards the coastal area. Water surface temperature in the study area is 27.5–30 °C. The water surface temperature distribution varied in the study area (Fig. 2). Figure 2 shows the highest temperature is in site 5, 10, 11, 12, 13, and 25 (red color), that might be influenced by the entry of materials from various intensive economic activities in the Kutaraja Fishing Port. In conclusion, the higher the seawater salinity, the greater the corrosion rate, due to salinity in the world's waters, which are generally traversed by ships, have salt concentration around 3%–4% [7].

The temperature difference can be caused by several factors, such as sun intensity at data collection, tidal current dynamics, wind, or river flow. The study area has a dynamic characteristic with regular material inputs from land that originated from two big rivers in the eastern and western parts, thus influenced the daily temperature. Temperature is an important indicator to determine the further effect on other water parameters, such as accelerating chemical reactions, reducing gas solubility in the water, or strengthening smell and taste. Natural temperature for tropical waters suitable for living organisms ranges between 23–32 °C [6]. The temperature is one of the crucial factors for living organisms in the ocean because it strongly influences the metabolic activities and development of marine organisms [8]. The temperature in the deeper layers is lower than the sea surface because of the lack of sunlight and wind speed that affects the ocean temperature movement from one place to another [9].

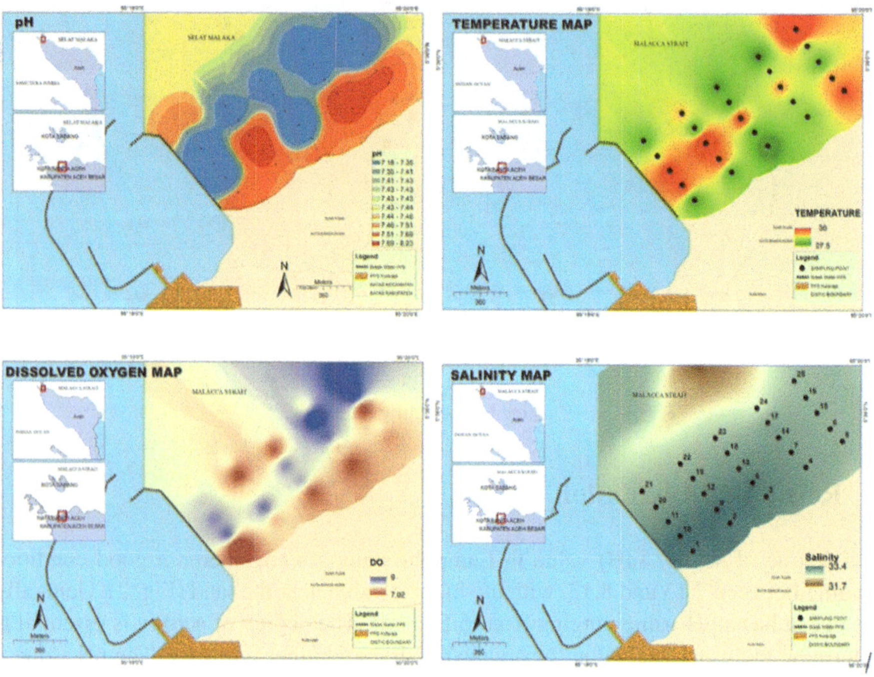

Fig. 2. Water quality parameters distribution in Lampulo Waters

The Dissolved Oxygen (DO) concentration on the surface ranged between 7.02–9 mg/L, mainly derived from direct diffusion from the air, aeration, or as a product of photosynthesis (Salmin, 2005). DO in the water can also be influenced by temperature and salinity, where the higher the temperature and salinity, the lower the DO concentration in the water. The high DO concentration during this research may be caused by the high photosynthesis activity. DO concentration that is too high or too low can threaten the marine biota, however, DO concentration in the waters still meet the quality standards for marine biota (>5 mg/L) and can be categorized as good to support the aquaculture business. DO spatial distribution varied, where high DO concentration was found in the center and eastern part of the study area. Almost all aquatic organism, including macroalgae, requires DO. The high intensity of the photosynthesis process on macroalgae decreases the DO in the area [8].

4 Conclusions

The acidity degree (pH) value in Lampulo waters is categorized as good, ranging between 7.13–8.3. The highest pH value was found in the area near the land. Sea surface salinity in the study area ranged 32–34 PPT, increasing towards the coastal area. Sea surface temperature in the study area ranged from 27.5 to 30 °C. Dissolved Oxygen (DO) in the surface area was 7.02–9 mg/L. The water quality parameter is one of the crucial factors to be assessed to determine the shipyard location.

References

1. Abdullah, A.: Managing A Shipyard From Local Perspective. Mimet's Tech. Bull. **7**, 3–8 (2016)
2. Matulja, T., Hadjina, M., Rubesa, R., Zamarin, A.: Hierarchical ranking as basis for ship outfitting process improvement. Brodogradnja **69**, 70–79 (2018)
3. Rizwan, T., et al.: The analysis of clean water need for fishing activities in Kutaraja Fishing Port, Aceh Indonesia. Austr. J. Marit. Ocean Aff. **13**, 1–11 (2020)
4. Rizwan, T., Jalil, Z., Akhyar, H.: Oceanographic factors as the indicators for shipyard industry development in Kutaraja Fishing Port: a preliminary study. J. Ecol. Eng. **22**, 237–245 (2021)
5. Farhan, A., Akhyar, H.: Analysis of tsunami disaster map by Geographic Information System (GIS): Aceh Singkil-Indonesia. IOP Conf. Ser. Earth Environ. Sci. **56**(1), 012002 (2017)
6. Siburian, R., Simatupang, L., Bukit, M.: Sea water quality analysis of activities in the port of East Sumba Waingapu- Alor. Jurnal Pengabdian Kepada Masyarakat **23**, 225–232 (2017)
7. Nova, S.M.K., Misbah, N.: Analisis Pengaruh Salinitas dan Suhu Air Laut Terhadap Laju Korosi Baja A36 pada Pengelasan SMAW. Jurnal teknik ITS **1**, 75–77 (2012)
8. Rukminasari, N., Nadiarti, K.: Awaluddin, The Effect of Acidic Level of Media on Calcium Concentration and Growth *halimeda* sp. Jurnal ilmu kelautan dan perikanan **24**, 28–34 (2014)
9. Juniarti, L., Jumarang, M.I., Apriansyah: Analisis kondisi suhu dan salinitas perairan barat Sumatera menggunakan data Argo Float. J. Phys. Commun. **1**, 74–84 (2017)

Stress Analysis on an Automotive Coil Spring Due to Speed Effect

Azhar[1,2], Husaini[1], and Teuku Edisah Putra[1](✉)

[1] Department of Mechanical Engineering, Universitas Syiah Kuala, Banda Aceh, Indonesia
{husainiftm,edi}@unsyiah.ac.id

[2] Department of Mechanical Engineering, Universitas Teuku Umar, Meulaboh, Indonesia
azharteknik@utu.ac.id

Abstract. The aim of this study is to determine the effect of car speed on the stress on a coil spring. 60-s strain signals measured on the coil spring of a car that was driving at speeds of below 20 km/h, 40–50 km/h, and above 70 km/h were utilized as the input. From the dynamic analyses, it was found that the increase in speed was directly proportional to the increase in stress. Each increase of the speed of 10 km/h was followed by an increase of stress by almost 10 kPa. The highest stress was obtained when the car was driving at a speed above 70 km/h, which was 645 kPa. This value was 43.7% of the yield strength of the material. Although the stress that occurred was relatively small, it could cause damage to the coil spring because the component is subjected to loading continuously.

Keywords: Coil spring · Speed · Strain · Stress

1 Introduction

Besides the safety and quality factors of the engine, comfort is also a factor that cannot be ignored in a car. To provide such comfort, a car is equipped with a suspension system consisting of a connecting arm for the wheels and the car body, shock absorbers, and coil spring [1]. Coil spring is used to reduce vibrations and shocks that are generated when a car is driving on an uneven or bumpy road surface [2]. These vibrations and shocks will increase when the car is travelling at a high speed.

Many works have been conducted to determine the stress on a coil spring. However, the stress analyses were carried out using a constant load which gave high stress compared to the yield strength of the material. High stress on an elastic coil spring causes the results obtained to be doubtful. For this reason, this study aims to analyze the stress that occurs on a coil spring due to dynamic loads obtained at three levels of speed, i.e. below 20 km/h, 40–50 km/h, and above 70 km/h.

Akhyar et al. (Eds.): ICECME 2021, LNME, pp. 148–153, 2023.
https://doi.org/10.1007/978-981-19-3629-6_16

2 Materials and Method

One of the newest works studying the stress on an automotive coil spring was carried out by Putra et al. [3]. In the work, the front coil spring of a minibus made of the SAE 5160 carbon steel was subjected to a static load of 3600 N. Furthermore, a strain gauge was attached at the critical point of the component and driven the car at speeds of below 20 km/h, 40–50 km/h, and above 70 km/h. The strain signals were taken over 60 s consisting of 30,000 data for each speed, as shown in Fig. 1.

In the dynamic analysis that was carried out in the current work, the strain signals in Fig. 1 were used as a load. The strain ε had converted into stress σ using the Ramberg-Osgood equation [4], which is:

$$\varepsilon = \frac{\sigma}{E} + \left(\frac{\sigma}{K'}\right)^{1/n} \tag{1}$$

where E is the modulus Young, K' is the cyclic strength coefficient, and n is the cyclic strain-hardening exponent, provided in Table 1.

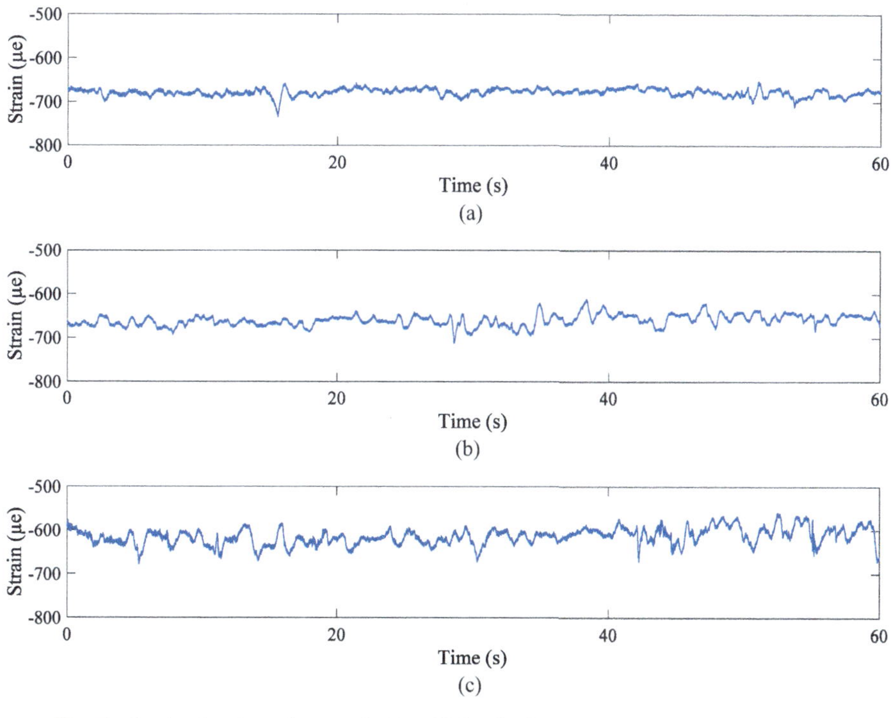

Fig. 1. Strain signals at the speed: (a) <20 km/h, (b) 40–50 km/h, and (c) >70 km/h

Then, the stress σ that was obtained in Eq. 1 was converted into force P using the following equation:

$$\sigma = \frac{P}{A_0} \tag{2}$$

Table 1. Mechanical properties of the SAE 5160 carbon steel [5].

Properties	Values
Ultimate tensile strength (MPa)	1,584
Modulus of Young (GPa)	207
Yield strength (MPa)	1,487
Cyclic strain-hardening exponent	0.05
Cyclic strength coefficient (MPa)	1,940
Poison ratio	0.27

where A_0 is the coil spring cross-sectional area. The forces that were obtained from Eq. 2 were used as the input loadings in the finite element analysis. The coil spring model in this simulation was the same as the coil spring model in the research by Putra et al. [3]. The finite element analysis is carried out by creating partitions on the model into small sizes or also called meshing [6] at a certain level of size and type. The type of mesh element used in this analysis was tetrahedral [7–9] with a size of 0.2 mm so that the total element after the meshing process was 16,437 with 30,917 nodes. Figure 2 shows the meshing and the boundary conditions, where the loads were applied vertically from the bottom (at the yellow area) and the upper was fixed.

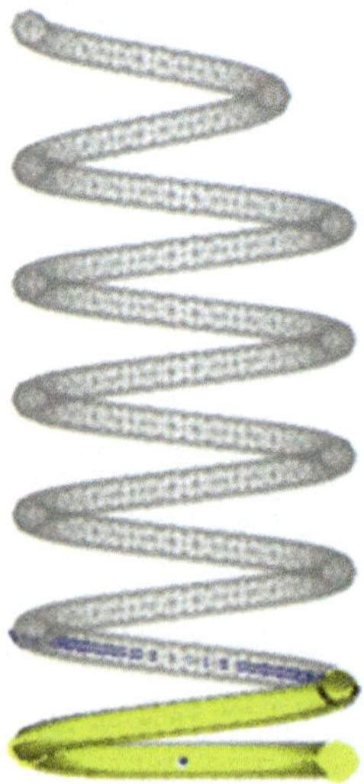

Fig. 2. The mesh shape and boundary conditions

Yield occurs when the equivalent stress σ_e is equal to yield stress, expressed by [10]:

$$\sigma_e = \frac{1}{\sqrt{2}}\sqrt{(\sigma_1 - \sigma_2)^2 + (\sigma_2 - \sigma_3)^2 + (\sigma_3 - \sigma_1)^2} \tag{3}$$

where σ_1, σ_2, σ_3 respectively are the main stresses for the x, y and z directions.

3 Results and Discussion

The results of the stress distribution are shown in Figs. 3, 4 and 5 for the speeds of below 20 km/h, 40–50 km/h, and above 70 km/h, respectively. The critical point occurred on the inside of the second coil from the top, shown in red. This location was similar to the static analysis that was conducted by Putra et al. [3], but the maximum stress values that occurred were different. The highest stress occurred at a speed above 70 km/h, which was 650 kPa, followed by a speed of 40–50 km/h with a stress of 632 kPa, and a speed below 20 km/h with the stress of 615 kPa.

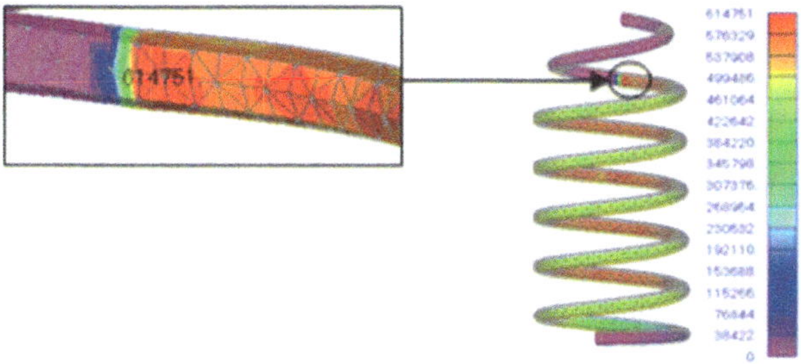

Fig. 3. Stress distributions at speed below 20 km/h

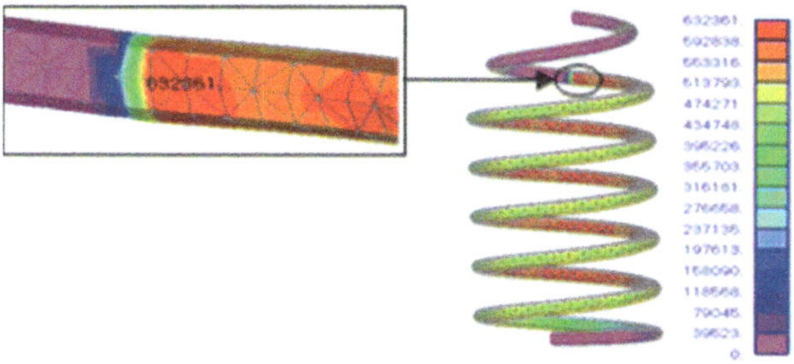

Fig. 4. Stress distributions at speed of 40–50 km/h

These results show that at a high speed, the vibration received by the coil spring was also high, which resulted in high stress. However, the stress that occurred was only 43.7% compared to the yield strength of the SAE 5160 carbon steel, which was 1.487 GPa (Table 1). Even though the stress that occurred was relatively small compared to the yield strength of the material, it can still cause damage to the coil spring. This is because damage can occur to a component that is subjected to continuous loading even though the stress that occurs is still far below the yield strength of the material.

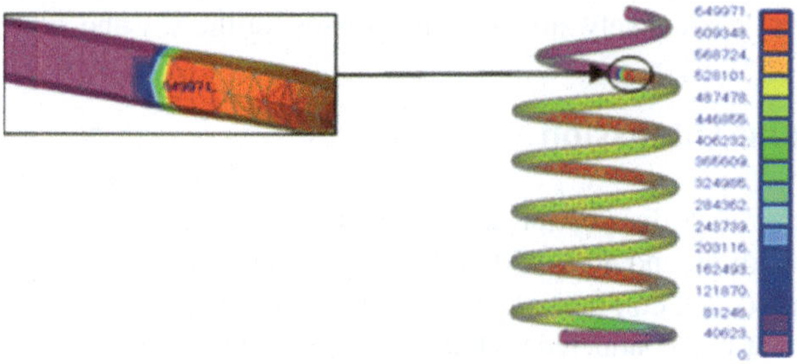

Fig. 5. Stress distributions at speed above 70 km/h

4 Conclusion

The aim of this study is to analyze the stress on the coil spring due to the effect of automotive speed. Strain signals acquired at speeds of below 20 km/h, 40–50 km/h, and above 70 km/h, were used as the loads for the analyses. The results of the dynamic analyses show that each increase in the speed of 10 km/h, was followed by an increase in the stress by almost 10 kPa. It can be concluded that the speed of the car has a direct effect on the coil spring stress, which in turn affects the life of the component. However, the authors strongly recommend validating the results of these numerical simulations by performing a fatigue test using the strain signals that were obtained from the three types of car speeds.

Acknowledgements. The authors would like to express their gratitude to Universitas Syiah Kuala for financial support for this research through grant no. 10/UN11.2.1/PT.01.03/DPRM/2021.

References

1. Tandel, A., Deshpande, A.R., Deshmukh, S.P., Jagtap, K.R.: Modeling, analysis and PID controller implementation on double wishbone suspension using SimMechanics and Simulink. Procedia Eng. **97**, 1274–1281 (2014)
2. Ka'ka, S., Himran, S., Renreng, I., Sutresman, O.: Effects of work on shock absorber and spiral springs against vertical loads of vehicles burdening the road structure. IOP Conf. Ser. Mater. Sci. Eng. **676**(1), 012042 (2019)
3. Putra, T.E., Husaini, Dongoran, M.R.: Effect of vehicle speed to the fatigue life of a coil spring based on strain-life approach. In: AIP Conference Proceedings, vol. 2187, no. 1, p. 040003 (2019)
4. Ramberg, W., Osgood, W.R.: Description of stress-strain curves by three parameters. Natl. Adv. Committee Aeronaut. (1943)
5. nCode: GlyphWorks, nCode International, Ltd., Sheffield (2018)
6. Aykan, M., Celik, M.: Vibration fatigue analysis and multi-axial effect in testing of aerospace structures. Mech. Syst. Sig. Process. **23**(3), 897–907 (2009)
7. Putra, T.E., Abdullah, S., Schramm, D.: A novel way to overcome problems arising in strain signal measurements leading to a fatigue failure characterization. Int. J. Automot. Mech. Eng. (IJAME) **17**(3), 8104–8115 (2020)

8. Putra, T.E., Machmud, M.N.: Predicting the fatigue life of an automotive coil spring considering road surface roughness. Eng. Fail. Anal. **116**, 104722 (2020)
9. Putra, T.E., Husaini, Ikbal, M.: Automotive suspension component behaviors driven on flat and rough road surfaces. Heliyon **7**, e07528 (2021)
10. von Mises, R.: Mechanik der festen Körper im plastisch-deformablen Zustand. Nachrichten von der Gesellschaft der Wissenschaften zu Göttingen, Mathematisch-Physikalische Klasse **1**, 582–592 (1913)

Stress Analysis on the Lower Arm and Coil Spring of an Automotive Due to Dynamic Loadings

Teuku Marjuni, Husaini, and Teuku Edisah Putra(✉)

Department of Mechanical Engineering, Universitas Syiah Kuala, Banda Aceh, Indonesia
marjuni@mhs.unsyiah.ac.id, {husainiftm,edi}@unsyiah.ac.id

Abstract. This study aims to analyze the stresses that occur in the lower arm and coil spring of an automotive subjected to dynamic loads. The strain signals produced from the lower arm and coil spring of a car were used as the loads in these dynamic analyses. Based on the simulation results, the stress that occurs in the lower arm was 244,612 kPa or only 54.35% of the yield strength of the SAE 1513 carbon steel. Meanwhile, the stress on the coil spring was 542,893 kPa or only 36.5% of the yield strength of the SAE 5160 carbon steel. In conclusion, the risk of failure of the lower arm is greater than the coil spring. Although the stresses that occurred were relatively small compared to the yield strength of the material, these stresses can still cause failure because the component receives loads continuously.

Keywords: Lower arm · Coil spring · Dynamic · Strain · Stress

1 Introduction

Suspension is a mechanism of a set of rigid objects that connect the mainframe of the car to the wheels functioning as the support for the weight of the car to achieve stability when the car is running. In general, the suspension system stores kinetic energy in the form of strain and releases the energy to the environment without causing any effect [1]. There are several important components in the suspension system which have different functions according to their respective roles, such as the lower arm and the coil spring. The lower arm functions as a human hand to control the movement of the wheels, such as up-down motions, back-forth motions, as well as when the car turns [2]. While the coil spring serves to minimize the car body from shock loads to the frame and the engine by utilizing the elastic properties of the material, as well as to maximize the friction between the road surface and tires to provide good steering stability [3].

When the car is driving on a road surface and then braking, the existence of the lower arm and the coil spring are expected to be able to withstand vibrations and also be able to keep the load caused by car manoeuvers. It means that both components must withstand, reduce, and absorb impact loads, torsional loads, and cyclic loads. In its application, when the lower arm and the coil spring are subjected to continuous loading,

Akhyar et al. (Eds.): ICECME 2021, LNME, pp. 154–160, 2023.
https://doi.org/10.1007/978-981-19-3629-6_17

the possibility of fatigue failure of these components results in limited service life. If these components fail suddenly, it will cause considerable economic loss and cause accidents which will be very dangerous for the driver and other road users [4].

Previous studies that analyzed the stresses on a car's suspension system were carried out separately. As it is known that the components of the car suspension system, especially the lower arm and coil spring, are an integral part of their application to ensure driver comfort while driving. Therefore, this study aims to analyze the stresses that occur in the lower arm and the coil spring of a car simultaneously. By determining the stresses that occur in the lower arm and the coil spring, it can identify which component fail first.

2 Materials and Methods

Considering that the lower arm and the coil spring receive dynamic loads from the road surface, therefore, this study simulated the stresses on the components by applying dynamic loads as shown in Fig. 1. The dynamic loads were obtained from the works by Putra et al. [5, 6]. In those works, strain gauges were installed on the lower arm and the coil spring of a car steered over a road surface for 60 s. For simulation purposes, the materials used for the lower arm and the coil spring were the SAE 1513 and SAE 5160 carbon steel, respectively, as shown in Table 1.

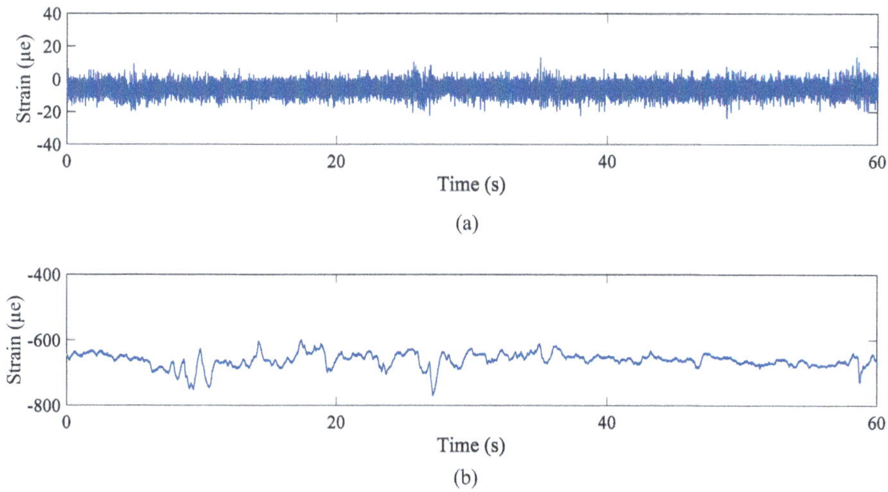

Fig. 1. Strain signals: (a) lower arm, and (b) coil spring

Table 1. Mechanical properties of the SAE 1513 and SAE 5160 Carbon steel [7]

Properties	SAE 1513	SAE 5160
Ultimate tensile strength (MPa)	585	1,584
Material modulus of elasticy (GPa)	210	207
Yield strength (MPa)	450	1,487
Cyclic strain-hardening exponent	0.1169	0.05
Coefficient of cyclic strength (MPa)	978	1,940
Poison ratio	0.27–0.3	0.27

In the dynamic analysis carried out in the current study, the strain signals in Fig. 1 were utilized as the loads. The strain ε was converted into the stress σ utilizing the Ramberg-Osgood equation [8], which is expressed as follows:

$$\varepsilon = \frac{\sigma}{E} + \left(\frac{\sigma}{K'}\right)^{1/n} \tag{1}$$

where E is the material modulus of elasticity, K' is the coefficient of cyclic strength, and n is the exponent of cyclic strain-hardening, listed in Table 1. The stress σ was then converted into force P using the following equation:

$$\sigma = \frac{P}{A_0} \tag{2}$$

where A_0 is the coil spring cross-sectional area.

The forces from Eq. 2 were fed into the finite element analysis. The lower arm in this simulation was the same as the model in [5], while the coil spring is the same as the model in [6]. In the simulation, the model is divided into small elements [9]. The shape of the element used in these simulations was a tetrahedral [10–12] with a size of 0.2 mm. It resulted in 46,822 elements and 85,769 nodes for the lower arm as well as 16,437 elements and 30,917 nodes for the coil spring. The support areas are shown in Fig. 2, while Fig. 3 shows the loading areas applied vertically for both the components.

Failure occurs when the total strain energy of the triaxial stress is equal to the strain energy when yielding begins in the material. The stress that causes failure in the material when the material gets triaxial stress that produces strain energy is called the von Mises stress [13], described by:

$$\sigma_e = \frac{1}{\sqrt{2}}\sqrt{(\sigma_1 - \sigma_2)^2 + (\sigma_2 - \sigma_3)^2 + (\sigma_3 - \sigma_1)^2} \tag{3}$$

where $\sigma_1, \sigma_2, \sigma_3$ respectively are the main stresses for the x, y and z directions.

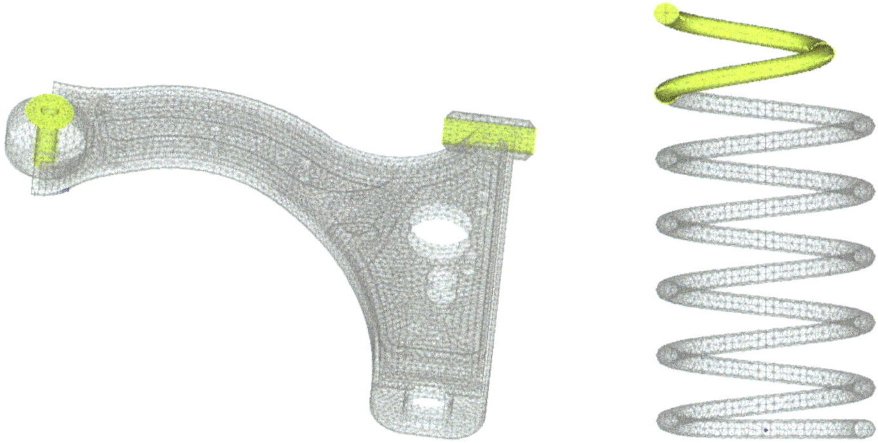

Fig. 2. The support areas

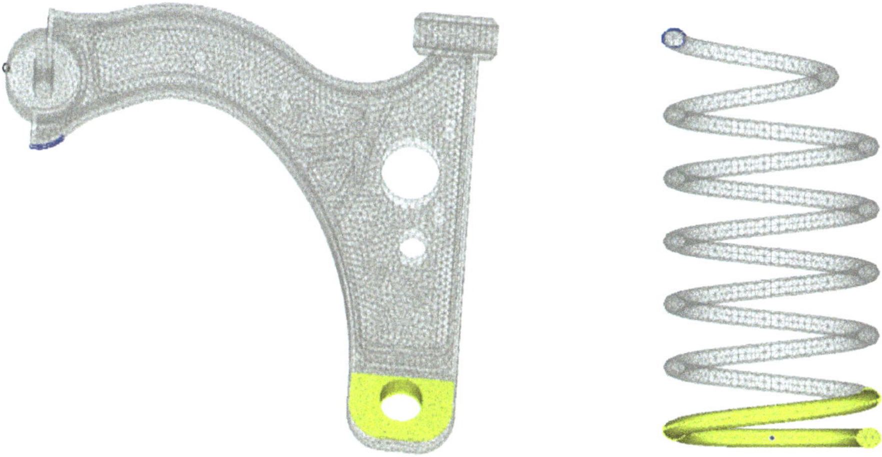

Fig. 3. The loading areas

3 Results and Discussion

The results of stress distributions at the lower arm are shown in Fig. 4. The critical point was located on the inside of the centre near the small hole at the bottom, shown in red. The maximum stress was 244,612 kPa, or only 54.35% of the yield strength of the SAE 1513 carbon steel, which was 450 MPa, as given in Table 1. The location of the maximum stress was similar to the static analysis in [5], which produced a stress of 358,820 kPa. The same location was also generated from the dynamic simulation conducted by Wahyudin et al. [14], which produced a stress of 278,522 kPa.

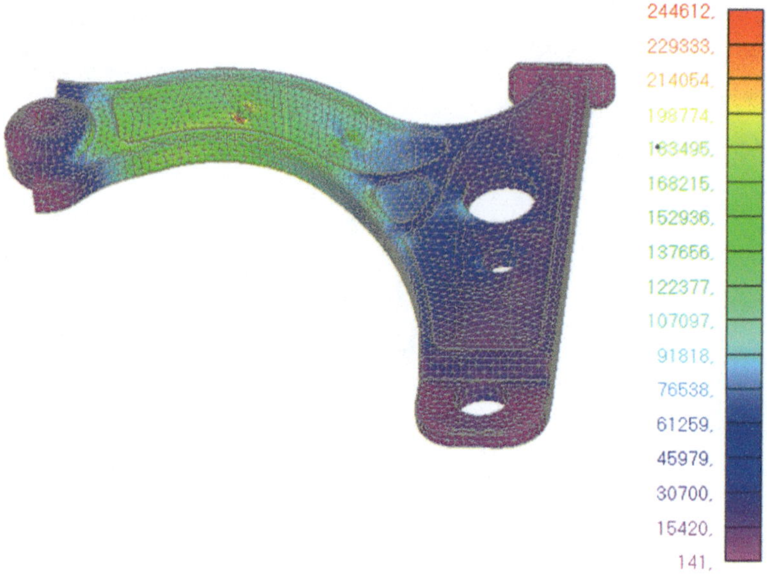

Fig. 4. Stress distribution on the lower arm

The results of the stress distribution at the coil spring are shown in Fig. 5. The critical point was located on the inside of the second coil, shown in red. The maximum stress

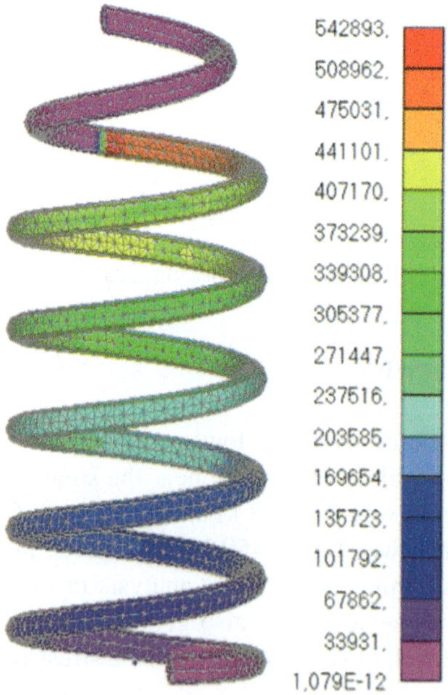

Fig. 5. Stress distribution on the coil spring

was 542,893 kPa, or only 36.5% of the yield strength of the SAE 5160 carbon steel, which was 1,487 MPa, as given in Table 1. The location of the maximum stress was similar to the static analysis in [6], which produced a stress of 997,510 kPa. The same location was also generated from the dynamic simulation conducted by Harahap et al. [15], which produced a stress of 621,967 kPa.

4 Conclusion

This study aims to identify suspension system components that have a greater potential for failure. This study found that the lower arm received a stress of 244,612 kPa. This value was 54.35% of the yield strength of the SAE 1513 carbon steel. Furthermore, the stress that occurred at the coil spring was 542,893 kPa, which was only 36.5% of the yield strength of the SAE 5160 carbon steel. In conclusion, the risk of failure of the lower arm was greater than the coil spring. Although the stresses that occurred were still below the yield strength of the material, these stresses can cause the failure of the lower arm and the coil spring because these components receive loads continuously.

Acknowledgements. The authors would like to express their gratitude to Universitas Syiah Kuala for financial support for this research through grant no. 10/UN11.2.1/PT.01.03/DPRM/2021.

References

1. Savkin, A.N., Klimov, M.A., Doluda, A.O.: The automobile design element fatigue life modeling due to its dynamic model. Procedia Eng. **206**, 416–420 (2017)
2. Kumar, S.A., Balaji, V., Balachandar, K., Kumar, D.P.: Analysis and optimization of lower control arm in front suspension system. Int. J. Chem. Sci. **14**(2), 1092–1098 (2016)
3. Yazar, G.: Design and analysis of helical coil spring forms for independent suspensions of automobiles. Thesis, Middle East Technical University (2015)
4. Prawoto, Y., Ikeda, M., Manville, S.K., Nishikawa, A.: Design and failure modes of automotive suspension springs. Eng. Fail. Anal. **15**(8), 1155–1174 (2008)
5. Putra, T.E., Husaini, M., Ikbal, I., Hasanuddin, M., Rizal, M.N.: Machmud, fatigue life assessments of the AISI 1513 carbon steel subjected to road surface strains. Defect Diffus. Forum **402**, 39–44 (2020)
6. Putra, T.E., Husaini, Ikbal, M.: Effect of road surface contour to the fatigue life of a coil spring based on strain-life approach. IOP Conf. Ser. Mater. Sci. Eng. **739**, 012014 (2020)
7. nCode: GlyphWorks, nCode International, Ltd., Sheffield (2018)
8. Ramberg, W., Osgood, W.R.: Description of stress-strain curves by three parameters. Natl. Adv. Committee Aeronaut. (1943)
9. Aykan, M., Celik, M.: Vibration fatigue analysis and multi-axial effect in testing of aerospace structures. Mech. Syst. Sig. Process. **23**(3), 897–907 (2009)
10. Putra, T.E., Abdullah, S., Schramm, D.: A novel way to overcome problems arising in strain signal measurements leading to a fatigue failure characterization. Int. J. Automot. Mech. Eng. (IJAME) **17**(3), 8104–8115 (2020)
11. Putra, T.E., Machmud, M.N.: Predicting the fatigue life of an automotive coil spring considering road surface roughness. Eng. Fail. Anal. **116**, 104722 (2020)

12. Putra, T.E., Husaini, Ikbal, M.: Automotive suspension component behaviors driven on flat and rough road surfaces. Heliyon **7**, e07528 (2021)
13. von Mises, R.: Mechanik der festen Körper im plastisch-deformablen Zustand. Nachrichten von der Gesellschaft der Wissenschaften zu Göttingen, Mathematisch-Physikalische Klasse **1**, 582–592 (1913)
14. Wahyudin, Husaini, Putra, T.E.: Stress analysis on an automotive lower arm steered on the straight and turning roads. In: Akhyar (ed.) Proceedings of the 2nd International Conference on Experimental and Computational Mechanics in Engineering. LNME, pp. 271–276. Springer, Singapore (2021). https://doi.org/10.1007/978-981-16-0736-3_27
15. Harahap, J., Husaini, T.E., Putra, D.: Schramm, stress analysis on an automotive coil spring driven on flat, uphill, and downhill road surfaces. Key Eng. Mater. **892**, 124–128 (2021)

Planning for Placement of Distributed Generators (DG) in Feeders with the Artificial Intelligent Hybrid GA-ANN Method

Ramdhan Halid Siregar[1,2](✉), Yuwaldi Away[1,2], Tarmizi[1,2], and Ira Devi Sara[1,2]

[1] Doctoral School of Engineering Science, Universitas Syiah Kuala, Banda Aceh 23111, Indonesia
[2] Department of Electrical and Computer Engineering, Universitas Syiah Kuala, Banda Aceh 23111, Indonesia
{ramdhan,yuwaldi,mizi,ira.sara}@unsyiah.ac.id

Abstract. It is planned that in 2025 the target for the utilization of New Renewable Energy (NRE) is 23% of the energy needs in Indonesia. Meanwhile, until now, the achievement of this figure is still very far from expectations. The demand for electrical energy is increasing with population growth and the progress of the times. One solution is to use a distributed generator (DG). DG is easy to construct the generator and improve power flow because it is close to the load. In this case, DG is prioritized from NRE. DG of NRE requires forecasting in the acquisition of output power. Artificial Neural Network (ANN) is used to predict radiation intensity in determining the DG output of renewable energy. The training resulted in a significant regression value (R) close to 1; then, a simulation was carried out. The results obtained were the solar panel area of 1703.86, 1676.98, and 881.49 m^2 needed to get the optimal power output. The optimization results with GA show a decrease in power losses of 81.5709% and an improvement in the voltage profile of 0.032745 p.u. The optimized DG results are placed on buses 9, 22, and 31 and are 317, 312, and 164 kW, respectively.

Keywords: Distributed generation · Voltage profiles · Power losses · Genetic algorithm · Artificial neural network

1 Introduction

Currently, renewable energy sources are a promising alternative in the modern electricity world [1]. A technology called distributed generation is a technology in electric power that prioritizes power generation on a small scale and prioritizes the use and utilization of renewable energy sources that do not significantly affect the environment. One of the issues that prompted the development of DG is the high cost of constructing transmission lines and distribution lines. In the construction process, the transmission line requires a high investment value. In Indonesia, renewable energy sources such as solar and wind have begun to be applied in various areas with considerable potential. Distributed generation is also an electricity distribution model wherein a distribution network can be

Akhyar et al. (Eds.): ICECME 2021, LNME, pp. 161–176, 2023.
https://doi.org/10.1007/978-981-19-3629-6_18

another generator to get better power flow and further distribution. In an electric power system operation, many things must be considered, such as generation and distribution, as for the things that must be regarded as in its process, such as emphasizing operating costs and improving the quality of power flow. In practice, the installation of DG has a very significant effect on the voltage profile and power losses. Therefore, DG is usually placed in a location with high power losses. DG with a small capacity can serve peak loads at certain times, such as at night [2].

The Lingke-Krueng Cut feeder is one of the distribution feeders in Banda Aceh. In their daily life, Lingke-Krueng Cut feeders still need reliability in the distribution of electrical energy. The location of the power plant far from the load center results in significant power losses and a poor voltage profile. One way to reduce the effect of the above problem is by inserting small generators that are placed with the load centre or known as DG. In this study, the author focuses on the effect of DG installation on the voltage profile and power losses in the Lingke-Krueng Cut feeder distribution network to obtain reliable power flow. Power flow studies were validated with ETAP to obtain a voltage profile and initial power losses starting with distribution network modeling. The next step is to optimize the location and capacity of DG with a genetic algorithm. Then look for radiation intensity with Artificial Neural Network simulation to determine the required DG area.

2 Theoretical Consideration and Method

2.1 Distributed Generation

Distributed Generation (DG) is often used to express a small-scale power plant. Therefore the International Council on Large Electricity Systems (CIGRE) defined it as a generating unit with a maximum capacity of 50 MW to 100 MW, connected to a distribution network. However, several studies have defined DG based on its connection and location, excluding its generating capacity [2].

1) Advantages of Using DG: It is a constantly evolving technology that can flexibly adapt to economic changes due to its small size and simpler construction. The following are some of the advantages associated with using DG:

    a) Provides higher reliability in power utilization.
    b) Acts as a local energy source used to save power.
    c) It has a higher efficiency in power distribution, increases system efficiency, and reduces losses.
    d) It is environmentally friendly in producing electrical energy, with low emissions [3]

2) Disadvantages of Using DG: One of the disadvantages of using this system is the increase in short circuit current in the event of a system disturbance. Therefore, several parameters need to be considered before installation, such as the amount of short circuit current, voltage level, and losses in the system. One of the most important factors in determining the optimal location and capacity needed to produce additional

short-circuit currents and minimal power losses. Furthermore, it maintained voltage levels between the minimum and maximum values [4].

2.2 Power Flow in Radial Distribution Network

The voltage is lowered back to 20 kV with a step-down transformer at the distribution substation and transmits electricity to customers using the distribution line with a low voltage system.

1) *Back Forward Sweep*
   The power flow in the distribution network does not use an administration matrix, as opposed to the transmission network. In this case, due to the low connectivity of the channels to each other, the KVL and KCL equations are directly formed in the channel section.
2) *Backward Sweep*
   The procedure for resolving power flows starts with a backward sweep. In the first iteration, all voltages are assumed to be equal to the main source. Therefore, when multiple sources are on the network, the compensating injection current becomes zero in the first iteration. When it is known, the load current is determined by the following Eq. (2.1):

$$I_{ldi} = \left[\frac{P_i + jQ_i}{V_i}\right]^* \tag{2.1}$$

   where,

   I_{ldi}: Load current at point *i*.
   P_i: Active power requirement at point *i*.
   Q_i: Required reactive power at point *i*.
   V_i: Voltage at point *i*.

3) *Forward Sweep*
   At forwarding, sweep starts from the main source point where the voltage value, impedance, and current flowing in each line are known. Equation (2.2) is used to calculate all voltage at point i:

$$V_i = V_{ui} - Z_i I_{Li} i = I, \ldots, N \tag{2.2}$$

   where:

   V_i: Voltage at point *i*.
   V_{ui}: The voltage at the top of the point *i*.
   Z_i: Line impedance *i*.
   I_{Li}: The current flowing in line *i*.

Based on the example of the six bus radial distribution system as shown in Fig. 1, a formation is made using the BBIC, BCBV, and DLF matrix [5] (Fig. 2).

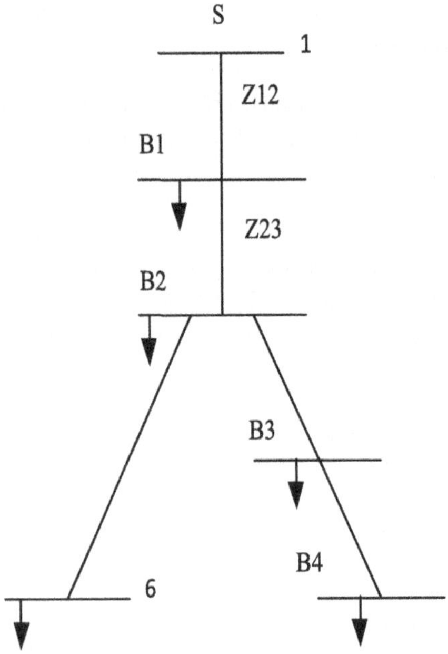

Fig. 1. Example of a six bus radial distribution system

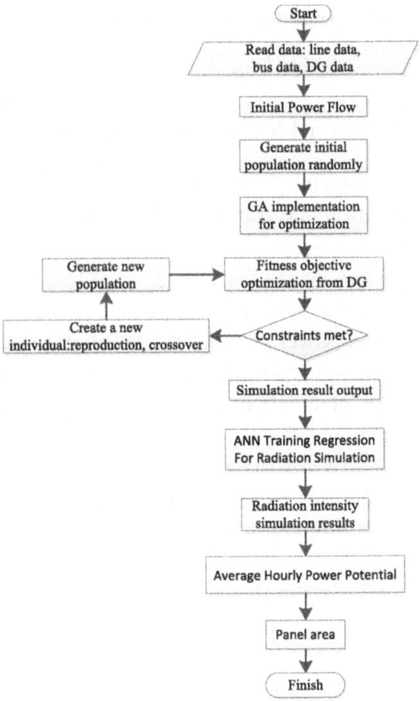

Fig. 2. Flow chart optimization with GA and ANN

a) Forming the BBIC matrix *(Matrix Bus Injection to Branch Current)*

For the distribution system, on the 1st bus, the complex load and to calculate the branch current obtained from the relationship of complex power and the individual voltages Eq. (2.3) is used:

$$I_i^k = conj\left(\frac{P_i + jQ_i}{V_i^k}\right) \tag{2.3}$$

where:

I_i^k = current on the bus I at the k iteration.

V_i^k = voltage on the bus I at the k iteration.

b) Composing the BCBV matrix (Branch Current to Bus Voltage)

The relationship between branch current and bus voltage is expressed as follows:

$$[\Delta V] = [BCBV][BBIC][I] \tag{2.4}$$

$$[\Delta V] = [DLF][I] \tag{2.5}$$

The DLF (Distribution Load Flow) matrix is obtained from the multiplication between the BCBV and the BBIC matrix. The following is the DLF matrix [6]:

2.3 Artificial Neural Network (ANN)

ANN is a computational system in which the architecture and operations are inspired by knowledge of the biological neurons in the brain. This makes ANN very suitable for solving problems of the same type as the human brain. An artificial neural network is defined by three things:

1. Patterns of relationships between neurons are called network architecture.
2. The method of determining the weight of the link is called the training/learning/algorithm method.
3. The activation function used [7].

2.4 Object Function

Fitness is used as a reference in achieving the optimal value of a GA method, which determines the individual value of the lowest fitness [8]. Therefore, the DG placement is determined by randomly generating a population and processing it to find the most insufficient fitness criteria. For optimal fitness selection based on the population raised, fitness chooses the most optimal variables. This function minimizes the total power loss in the radial distribution network. The objective function is as follows:

$$F = Min\left\{Pt = \sum_{i=1}^{n} I_i^2 R_i\right\} \tag{2.6}$$

where:

F = Objective function
Pt = Total active power loss
Ii = i bus line current
Ri = i bus resistance
n = number of buses on the system

Meanwhile, limits or constraints are rules in the value parameters form that must be passed as a selection function. Constraints make the selection process more effective because a condition needs to be met. The limitations used are as follows:

$$Vmin < Vbus < Vmarks \tag{2.7}$$

$$Vmin = (Vbase - 5\%\, x\, Vbase)/Vbase \tag{2.8}$$

$$Vmax = (Vbase - 5\%\, x\, Vbase)/Vbase \tag{2.9}$$

These limits are the standard tolerances for voltages allowed on the bus system, which applies to most large companies, which max and min values of 1.05 p.u. and 0.95 p.u.

2.5 New Individual Form, Generate New Population, Selection, Crossover, and Mutation

In the GA optimization process, a new individual formation sometimes needs to be carried out because the initial population has a less optimal fitness value. The first step in generating a new population is randomly determining the condition of the individuals, which is directly processed into *fitness* for optimal results. The iteration method's fitness value has a solution in the decimal number form in the next stage. It is derived from a binary value, referred to as the population, selected from the crossover and mutated to become a new population.

2.6 Artificial Neural Network Training to Predict the Solar Intensity

After the data supporting the solar intensity training, namely ambient temperature, solar power, humidity, have been obtained. The next step is to design a network that will be used to train the Artificial Neural Network. For that, we need Matlab R2017a software [9]. The neural network architecture that will be used is backpropagation (Fig. 3).

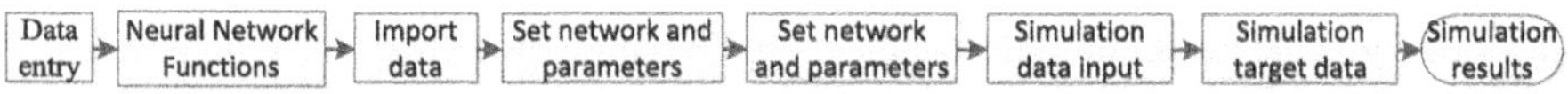

Fig. 3. Flow chart simulation simplified algorithm using ANN

2.7 Simulation of Solar Intensity Using Artificial Neural Network

After the neural network is trained, the data used to predict the sun's power is entered.

2.8 Conversion of Solar Intensity To PV Output Power

After the simulation results from ANN are obtained, the prediction of solar intensity is converted into power according to the PV specifications.

2.9 Calculating the PV Area Needed to Meet the Large Output DG

The sun's intensity that will be used is the sun's intensity on November 21, 2020, at 13.00 WIB. The solar panels used are Canadian Superpower CS6K-MS solar panels with an efficiency of 18%. While the inverter used is Fronius Galvo 3.1-1 with an efficiency of 95.4%. So by using the following formula [4]:

$$P_{out} = I.A.E_p.E_i \quad (2.10)$$

3 Results and Discussions

The data obtained from Electricity Company (PLN) UP3 Banda Aceh for the Lingke-Krueng Cut feeder consists of 37 distribution transformer buses with power data for each transformer, loading, resistance, and reactance as shown in Tables 1 and 2.

Table 1. Load power distribution transformer feeder Lingke-Krueng Cut

No. Bus	Kode Gardu Trafo	Power (kVA)	Pembebanan (%)	Active Power (kW)	Reactive Power (kVAr)
1	GH KRC	0	0	0	0
2	LKE Krc 69-01	100	45.03	53.05125	29.35237
3	LKE Krc 69-00	160	37.67	52.738	29.17906
4	LKE Krc 68-00	160	73.52	94.108	52.06839
5	LKE Krc 64-00	100	28.17	24.64875	13.63774
6	LKE Krc 63-00	160	9.43	70.602	39.06291
7	LKE Krc 61-01	100	27.06	23.6775	13.10036
8	LKE Krc 61-00	100	21.6	18.9	10.45705
9	LKE Krc 62-00	164	77.82	109	62.25
10	LKE Krc 58-00	100	69.05	60.41875	33.42869
11	LKE Krc 57-00	100	68.29	59.75375	33.06075
12	LKE Krc 51-00	160	34.77	48.678	26.93273
13	LKE Krc 49-00	160	48.47	67.858	37.54470
14	LKE Krc 50-00	160	32.5	45.5	26.05
15	LKE Krc 64-01	100	20	17.5	9.68246
16	LKE Krc 65-00	100	38.54	33.7225	18.65810
17	LKE Krc 66-00	100	38.5	33.6875	19.25
18	LKE Krc 59-00	50	61.29	26.814375	14.83594
19	LKE Krc 60-00	100	27.8	24.325	13.9
20	LKE Krc 52-00	100	43.9	40.181	22.22
21	LKE Krc 53-00	100	17.05	14.91875	8.254297
22	LKE Krc 54-00	100	60.97	53.34875	29.51697
23	LKE Krc 55-00	160	60.16	84.224	46.59974
24	LKE Krc 56-0	100	60.03	52.5	29.04738
25	LKE Krc 56-1	100	14.58	12.7575	7.058513
26	LKE Krc 56-2	100	47.4	40.1475	22295
27	LKE Krc 70-00	200	46.48	81.34	45.00407
28	LKE Krc 71-00	100	54.2	50.4875	28.85
29	LKE Krc 70-01	100	46.48	40.67	22.50203
30	LKE Krc 72-00	100	39.39	34.475	19.7
31	LKE Krc 73-00	100	35.39	30.84346	17.65
32	LKE Krc 74-00	50	61.81	25.83875	14.29615
33	LKE Krc 74-01	100	42.73	37.38875	20.68657
34	LKE Krc 74-02	50	61.81	27.041875	14.96182

(continued)

Table 1. (*continued*)

No. Bus	Kode Gardu Trafo	Power (kVA)	Pembebanan (%)	Active Power (kW)	Reactive Power (kVAr)
35	LKE Krc 75-00	100	77.23	67.57625	37.38881
36	LKE Krc 76-00	25	95.7	21.85	12.5
37	LKE Krc 76-1	160	67.97	87.5	50

Table 2. Linke-Krueng Cut feeder distribution channel data

NumBus	From Bus	To Bus	Resistance (Ohm)	Reactance (Ohm)
1	1	2	0.3179865	0.130755
2	2	3	0.130755	0.053587472
3	3	4	0.567573267	0.23338426
4	4	5	0.482642148	0.198460859
5	5	6	0.066943701	0.027527029
6	6	7	0.093365744	0.038391686
7	7	8	0.138514415	0.056956671
8	7	9	0.266319583	0.109509734
9	8	10	0.1216644	0.050028
10	10	11	0.378653275	0.155700978
11	11	12	0.414765	0.17055
12	12	13	0.121426412	0.04993014
13	13	14	0.424388926	0.174507327
14	5	15	0.09575943	0.039375962
15	15	16	0.130159191	0.05352103
16	16	17	0.424708438	0.174638709
17	8	18	0.325781	0.13388865
18	18	19	0.470229715	0.193356908
19	12	20	0.461178648	0.189635139
20	20	21	0.109124051	0.04487145
21	21	22	0.083859639	0.034482807
22	22	23	0.059560446	0.024491059
23	23	24	0.37808468	0.155467173
24	24	25	0.083060511	0.034154208
25	25	26	0.135800369	0.055840664
26	1	27	0.313378	0.12886
27	27	28	0.423709203	0.174227827
28	27	29	0.292819342	0.120406348

(*continued*)

Table 2. *(continued)*

NumBus	From Bus	To Bus	Resistance (Ohm)	Reactance (Ohm)
29	29	30	0.153252811	0.06301705
30	29	31	0.401073872	0.164920253
31	27	32	0.350737088	0.144221934
32	32	33	0.133852588	0.055039743
33	33	34	0.351192285	0.144409109
34	34	35	0.184166032	0.075728465
35	35	36	0.537310511	0.22094031
36	35	37	0.441921708	0.181716749

3.1 Lingke-Krueng Cut Feeder Data

See Fig. 4.

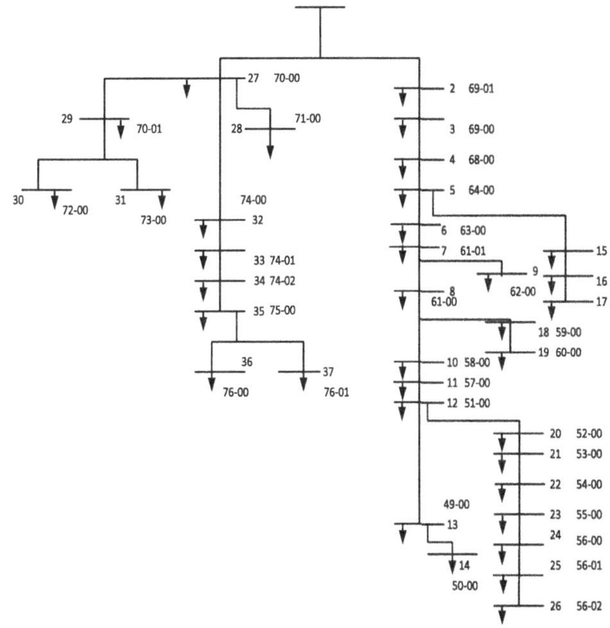

Fig. 4. Single line diagram of Lingke-Krueng Cut's feeder radial distribution system

3.2 Analysis Before DG Placement

At the experiment before placing DG using the *software ETAP* power station, the value of active and reactive losses was 54,673 kW and 13,768 kVAR. In this case, adding DG using the GA method is necessary to reduce the total losses and improve the voltage profile.

3.3 After DG Placement

In the 37 bus system, the optimization process uses the GA (Genetic Algorithm) process in Matlab to obtain the correct position and size on the DG to reduce power losses and improve the voltage profile. The experiment is limited to 3 placements. The total value results of power losses are likely to affect the obtained voltage profile because each power loss is considered very influential. Therefore, the smaller the real values of the power losses, the higher the existing voltage profile (Table 3).

The place and capacity are generated randomly by GA, with a maximum capacity of 500 kW. The population raised and experienced the Genetic Algorithm process from the optimization results, with the placement of the DG on buses 9, 22, and 31 of 317 kW, 312 kW, and 164 kW, respectively. Table 4 shows the results of location determination and DG capacity. The Maximum Power Potential That Can Be Generated From Renewable Energy Sources is as follows:

Table 3. Results of determining the location and capacity of DG

No	Location (No Bus)	DG Capacity (kW)	PV area(m^2)
1	Bus 9	317	1703.86
2	Bus 22	312	1676.98
3	Bus 31	164	881.49

Table 4. Average hourly power potential that PV can produce

Hours	Average hourly radiation value (w/m^2)	Rated power conversion (kW)
7	79.84	191.94
8	131.68	316.57
9	250.27	601.67
10	404.68	972.88
11	572.82	1377.11
12	625.91	1504.74
13	589.32	1416.77
14	469.76	1129.34
15	354.16	851.43
16	237.6	571.21

(continued)

Table 4. (*continued*)

Hours	Average hourly radiation value (w/m^2)	Rated power conversion (kW)
17	147.64	354.94
18	88.49	212.74

After determining the placement and size of the DG, using the *Backward Forward Sweep* power flow method, the power losses value and voltage profiles are obtained, as shown in Table 5.

Table 5. DG optimization power losses and voltage profile

Dari Bus	Ke Bus	P loss (kW)	Q loss (kVAR)	No Bus	Voltage Frofile (p.u)
1	**2**	**3**	**4**	**5**	**6**
1	2	1.2829	0.3873	Bus 1	1.0000
2	3	0.6669	0.1335	Bus 2	0.9976
3	4	0.2324	0.4653	Bus 3	0.9962
4	5	1.2864	0.2575	Bus 4	0.9958
5	6	0.1097	0.0220	Bus 5	0.9925
6	7	0.1254	0.0251	Bus 6	0.9922
7	8	0.3588	0.0718	Bus 7	0.9919
7	9	0.6364	0.1274	Bus 8	0.9895
8	10	0.1944	0.0389	Bus 9	0.9910
10	11	0.3673	0.0735	Bus 10	0.9903
11	12	0.2306	0.0462	Bus 11	0.9886
12	13	0.1739	0.0348	Bus 12	0.9873
13	14	0.0261	0.0052	Bus 13	0.9860
5	15	0.0200	0.0040	Bus 14	0.9856
15	16	0.0172	0.0034	Bus 15	0.9923
16	17	0.0141	0.0028	Bus 16	0.9920
8	18	0.0245	0.0047	Bus 17	0.9917
18	19	0.0082	0.0016	Bus 18	0.9906
1	**2**	**3**	**4**	**5**	**6**
12	20	0.2396	0.0445	Bus 19	0.9903
20	21	0.1236	0.0153	Bus 20	0.9878

(*continued*)

Table 5. (*continued*)

1	2	3	4	5	6
21	22	0.3063	0.0135	Bus 21	0.9881
22	23	0.1446	0.0125	Bus 22	0.9888
23	24	0.0399	0.0245	Bus 23	0.9881
24	25	0.0304	0.0014	Bus 24	0.9878
25	26	0.0031	0.0029	Bus 25	0.9873
1	27	1.2910	0.1881	Bus 26	0.9872
27	28	0.0093	0.0064	Bus 27	0.9837
27	29	0.0586	0.0186	Bus 28	0.9836
29	30	0.0057	0.0011	Bus 29	0.9841
29	31	0.3507	0.0702	Bus 30	0.9840
27	32	0.7459	0.1493	Bus 31	0.9856
32	33	0.2326	0.0466	Bus 32	0.9814
33	34	0.4367	0.0874	Bus 33	0.9806
34	35	0.1726	0.0346	Bus 34	0.9788
35	36	0.0077	0.0015	Bus 35	0.9780
35	37	0.1020	0.0204	Bus 36	0.9777
Total		**10.0758**	**2.4438**	Bus 37	0.9771

3.4 Comparison Before and After DG Placement

Furthermore, a comparison will be made between active power loss, reactive power loss, and voltage profile before and after the placement of DG on the system, as shown below (Figs. 5, 6 and Table 6):

Table 6. Comparison of power losses and voltage profiles before and after optimization of laying DG

From the bus	To the bus	Before		After	
		PLoss (kW)	*Voltage Profile (pu)*	*PLoss (kW)*	*Voltage Provile (pu)*
1	**2**	**3**	**4**	**5**	**6**
1	2	8.4524	0.99	1.2829	1.0000
2	3	4.9406	0.9837712	0.6669	0.9976
3	4	1.9566	0.9801505	0.2324	0.9962
4	5	13.852	0.9772589	1.2864	0.9958

(*continued*)

Table 6. *(continued)*

From the bus	To the bus	Before		After	
		PLoss (kW)	*Voltage Profile (pu)*	*PLoss (kW)*	*Voltage Provile (pu)*
1	**2**	**3**	**4**	**5**	**6**
5	6	1.5111	0.9656002	0.1097	0.9925
6	7	1.775	0.9641674	0.1254	0.9922
7	8	1.821	0.9623341	0.3588	0.9919
7	9	0.0971	0.9676032	0.6364	0.9895
8	10	1.276	0.9668812	0.1944	0.9910
10	11	3.1993	0.965107	0.3673	0.9903
11	12	2.7521	0.9601583	0.2306	0.9886
12	13	0.1853	0.9553579	0.1739	0.9873
13	14	0.0277	0.9540462	0.0261	0.9860
5	15	0.0211	0.9535607	0.0200	0.9856
15	16	0.0181	0.9653971	0.0172	0.9923
16	17	0.0149	0.9651779	0.0141	0.9920
8	18	0.0256	0.9648198	0.0245	0.9917
1	**2**	**3**	**4**	**5**	**6**
18	19	0.0085	0.9664744	0.0082	0.9906
12	20	1.388	0.9661897	0.2396	0.9903
20	21	0.3702	0.9516542	0.1236	0.9878
21	22	0.7089	0.9505518	0.3063	0.9881
22	23	0.1567	0.9484112	0.1446	0.9888
23	24	0.0432	0.9477902	0.0399	0.9881
24	25	0.0329	0.9473905	0.0304	0.9878
25	26	0.0033	0.9469340	0.0031	0.9873
1	27	3.6005	0.9468412	1.2910	0.9872
27	28	0.0101	0.9414030	0.0093	0.9837
27	29	0.0670	0.9412035	0.0586	0.9836
29	30	0.0062	0.9408739	0.0057	0.9841
29	31	0.0123	0.9407338	0.3507	0.9840
27	32	0.8128	0.90021	0.7459	0.9856
32	33	0.2534	0.9390190	0.2326	0.9814
33	34	0.4760	0.9381970	0.4367	0.9806

(continued)

Table 6. *(continued)*

1	2	3	4	5	6
34	35	0.1881	0.9363736	0.1726	0.9788
35	36	0.00841	0.9355440	0.0077	0.9780
35	37	0.1112	0.9352449	0.1020	0.9777
Total		**50.187**	0.934558	**10.0758**	0.9771

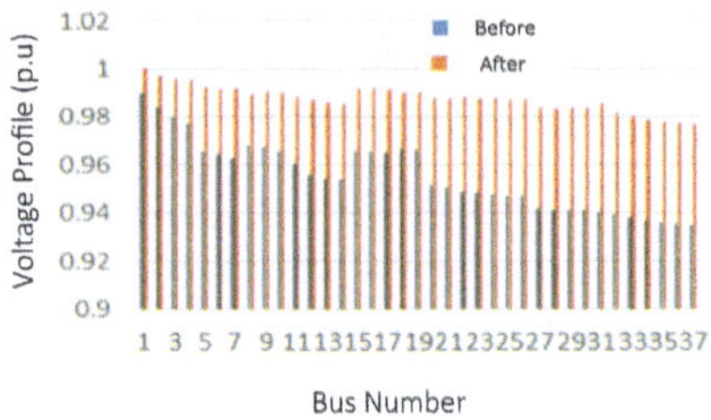

Fig. 5. Improved voltage profile on each bus

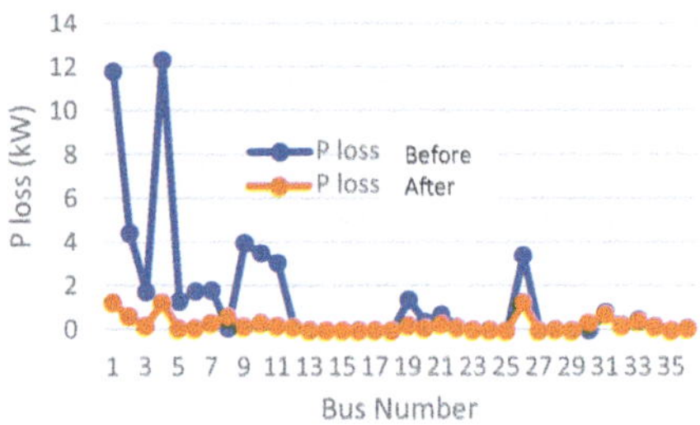

Fig. 6. Reduction of active power losses after laying DG

3.5 Required DG elder

The following are the Artificial Neural Network training results for simulating the intensity of solar radiation, followed by comparing the simulation results with the actual data (Figs. 7 and 8).

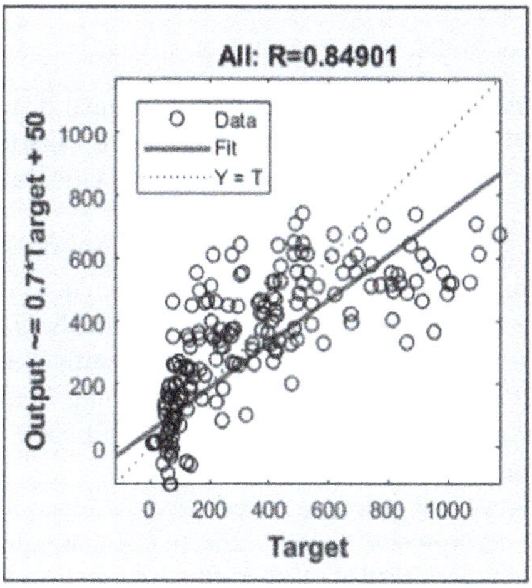

Fig. 7. ANN training regression results for simulating solar radiation intensity

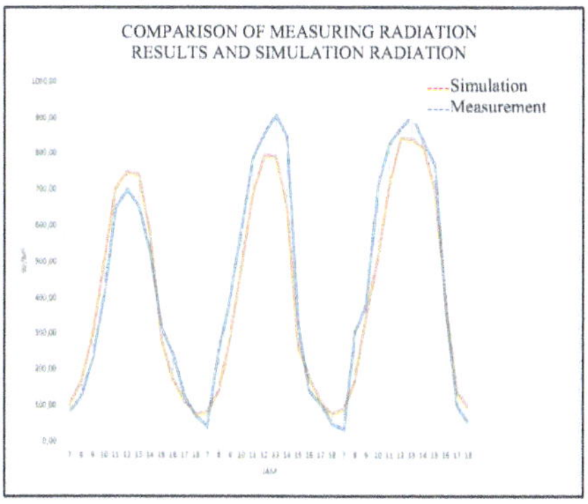

Fig. 8. Comparison of simulation data and measurement results

4 Conclusions

The genetic algorithm can determine the layout and distributed generating capacity (DG) required for power flow optimization, namely 317, 316, and 164 kW on buses 9, 22, and 31, respectively.

Furthermore, to realize the large capacity, it is carried out using ANN through training and simulations. Then the solar radiation intensity data obtained is used to calculate the power generated by the solar panels every hour. Furthermore, the PV area is calculated according to the energy required to meet the optimized DG capacity. The benefits of optimization are reduced power losses and improved voltage profile, as shown in the graph.

References

1. Bahramara, S., Moghaddam, M.P., Haghifam, M.R.: Optimal planning of hybrid renewable energy systems using HOMER: a review. Renew. Sustain. Energy Rev. **62**, 609–620 (2016)
2. Purchala, K., Belmans, R.: Distributed Generation and Grid Integration Issues. Imperial College, London (2003)
3. Sari, P.D., Kumara, I.N.S.: The development of Jatiluwih micro-hydro power plants to support tourism destinations. In: 2018 International Student Conference on Electrical and Computer Engineering (ISCECE), Bali, Indonesia, pp. 9–14 (2018). ISBN 978-602-294-346-4
4. Vasilki, V.: The impact of distributed generation in the distribution networks voltage profile and energy losses. In: IEEE European Modelling Symposium (2015)
5. Darji, A.D.R.J.B., Pandya, M.: Backward/Forward Sweep Load Flow Algorithm for Radial Distribution System, vol. 2, no. 1, pp. 398–400 (2014)
6. Prakash, K.M.: Topological and primitive impedance based load flow method for radial and weekly meshed distribution systems. Iranian J. Electr. Comput. Eng. **10**(1) (2011)
7. Hamid, N.A., Rozaida Ghazali, N.M.N., Salleh, M.N.M.: Accelerating learning performance of back propagation algorithm by using adaptive gain together with adaptive momentum and adaptive learning rate on classification problems. Int. J. Softw. Eng. Appl. **5**(4) (2011)
8. Bahramara, S., Moghaddamn, M.P., Haghifam, M.R.: "Optimal planning of hybrid renewable energy systems using HOMER": a review. Renew. Sustain. Energy Rev. **62**, 609–620 (2016)
9. Al Shamisi, M.H., Assi, A.H., Hejase, H.A.N.: Using MATLAB to develop artificial neural network models for predicting global solar radiation in Al Ain City – UAE. United Arab Emirates University United Arab Emirates (2011)

Structural Analysis of Mobile Robot Frame for Spinach Water Seed Planting Using Finite Element Method

Arhami(✉), Aulia Nanda Rizki, and Kurniawan Rudi

Department of Mechanical Engineering and Industry, Syiah Kuala University, Banda Aceh, Aceh 23111, Indonesia
{arhami,kurniawan}@unsyiah.ac.id

Abstract. This paper discusses the analysis of mobile robot frames for planting water spinach seeds. This analysis aims to determine the displacement and the stresses distribution caused by the load acting on the frame. The method used in this research is simulation-based using the finite element analysis with Autodesk Inventor 2020. The material used for the frame is Galvanized Steel. The load location is carried out with a magnitude of 261.9 N on the left and right beam frame, and 284.09 N in the middle of the beam frame. This load is given from the calculation of the total weight of the storage volume and the total weight of the mobile robot components. The results showed that with the given loads, the maximum Von misses stress was 38.03 MPa and the maximum displacement was 1.49 mm. From these results, it can be concluded that the structural frame used for a mobile robot is safe at a safety factor of up to 5.4. It is proven that the yield stress, of course, occurs 38.03 MPa less than 207 MPa of the material.

Keywords: Mobile robot · Frame structure · Stress analysis · Displacement · Finite element analysis

1 Introduction

Each vehicle has a frame to support all components used, including the frame on the mobile robot for water spinach seed planting. But in design, it is necessary to calculate the received load and see the reaction on the frame. The purpose of this study is to analyze the reactions that occur due to loading and to analyze the strength of the construction using stress distribution parameters, so that the frame is suitable or not for use in mobile robots for water spinach seed planting. The software was used to simulate the Autodesk Inventor Professional 2020. According to [1], the first thing done in developing a design is carrying out the design process. Design is to formulate new concepts and ideas or change existing concepts. ideas in new ways to meet human needs. So, this method begins with the design process, material selection, load location determination, analysis process, and results. According to [2], the design stages that will be carried out in solving this case are collecting field data and technical drawings, carrying out the design transport mechanism, implementing the conceptual design, performing dimensional optimization, and analyzing the design results aided by the computer simulation software.

Akhyar et al. (Eds.): ICECME 2021, LNME, pp. 177–186, 2023.
https://doi.org/10.1007/978-981-19-3629-6_19

2 Literature Review

The simulation has the basic equation to determine the amount that occurs when the frame is given a load, while the related equations are force, a moment of inertia, deflection, and stress.

Force

Force can be described intuitively as the action of an object to another and generally determined by its working point, its size, and direction [3]. Force has magnitude and direction is represented by the symbol F. The basic formula of the force has shown in Eq. 1, where m and g are the mass and acceleration of gravity respectively.

$$F = m \cdot g(N) \tag{1}$$

Moment of Inertia

The moment of inertia, *I*, Eq. (2), is a measure of the magnitude of the rotational tendency which is determined by the state of the object or the constituent particles.

$$I = \frac{(2 \cdot b^2)h^2 t_1 t_2}{b \cdot t_1 + h \cdot t_2} (\text{mm}^4) \tag{2}$$

where b = width of stalbush cross section (mm), h = height of stalbush cross section (mm), t_1 = thickness of cross section 1 (mm), t_2 = thickness of cross section 2 (mm).

Deflection

Deflection, δ, is a change in the shape of the beam in the y-direction due to the vertical loading applied to the beam or rod, as shown in Eq. (3).

$$\delta = \frac{F \cdot a(3L^2 - 4a^2)}{48 \cdot E \cdot I} (\text{mm}) \tag{3}$$

where F = force (N), L = length of bar (mm), E = modulus of elasticity (MPa), I = moment of inertia (mm^4), and a = cross-sectional area divided by 2 (mm^2).

Stresses

Stress, σ, (Eq. 4) shows the strength of the force that causes the object to change the shape, while a strain is defined as the ratio between the increase in the length of the stem and the initial length.

$$\sigma = \frac{F}{A} (\text{MPa}) \tag{4}$$

Strength is the ability of a material to accept stress without causing the material to break. There are several kinds of this strength depending on the type of load that is working, namely tensile strength, shear strength, compressive strength, punter strength/torsion, and bending strength, increasing the load that reaches the strength of the material which causes the material to be failure.

Finite Element Analysis

Finite element analysis is one of the numerical methods that can be used to obtain solutions for various engineering problems. This method is widely used to calculate the structure due to the loading or other effects on the structure. The main objective is to obtain the value of the stress and strain approach that occurred in the structure. This approach is an analysis based on stress or strain assumptions [4].

3 Methodology

Frame Design Method

In developing the design, the water spinach plant mobile robot is designed using the Autodesk Inventor 2020 software, each component is designed to get the right shape, and each component has a special load so that the load is totalled and converted into a force equation. The stages in the design process include size determination, initial sketches, component design according to needs, and usage analysis process [5].

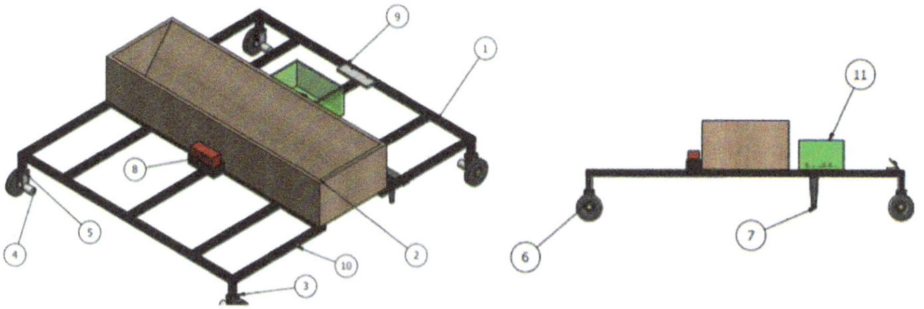

Fig. 1. Design of the mobile robot spinach water seed planting

The frame design model of the mobile robot is shown in Fig. 1 along with the component list shown in Table 1.

Table 1. Component list

Number	Name
1	*Frame*
2	Seedlings
3	Hook 1

(*continued*)

Table 1. *(continued)*

Number	Name
4	Motor DC
5	Hook 2
6	Wheel
7	Scratching the soil
8	Battery
9	Switch
10	Camera
11	Controller

The purpose of this design is to get a complete mobile robot shape. After knowing the component list on the tables, the location of the points on the frame must be known.

Location of the External Load and Assessment Location

In each part of the frame, there are several points to support the robot car planter plant seeds of spinach. The frame size used for the mobile robot water spinach seed planting is 1500 × 1600 × 250 mm using Galvanized Steel material. To see the location on the frame can be seen in Fig. 2.

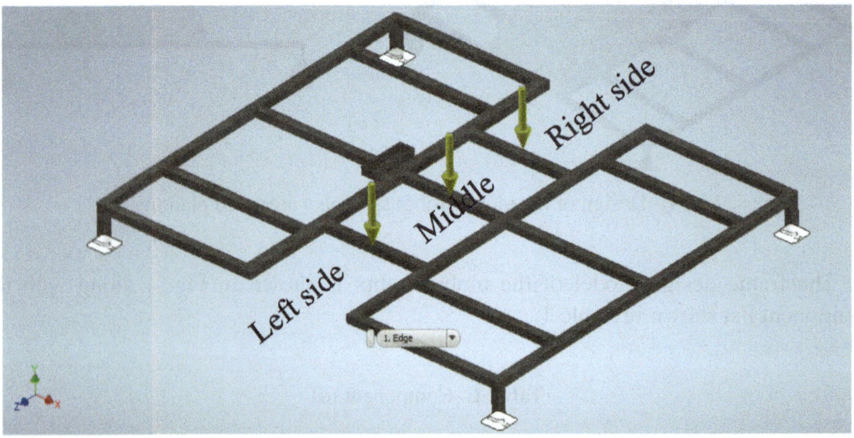

Fig. 2. Location of the pointed load

From Fig. 2, the load location is carried out with a magnitude of 261.9 N on the left and right beam frame, and 284.09 N in the middle of the beam frame. This load is given from the calculation of the total weight of the storage volume and the total weight of the mobile robot components. After that, a fixed constraint as the boundary condition is attached to the four legs of the frame. In Fig. 3 that the reaction laying point occurs on the frame.

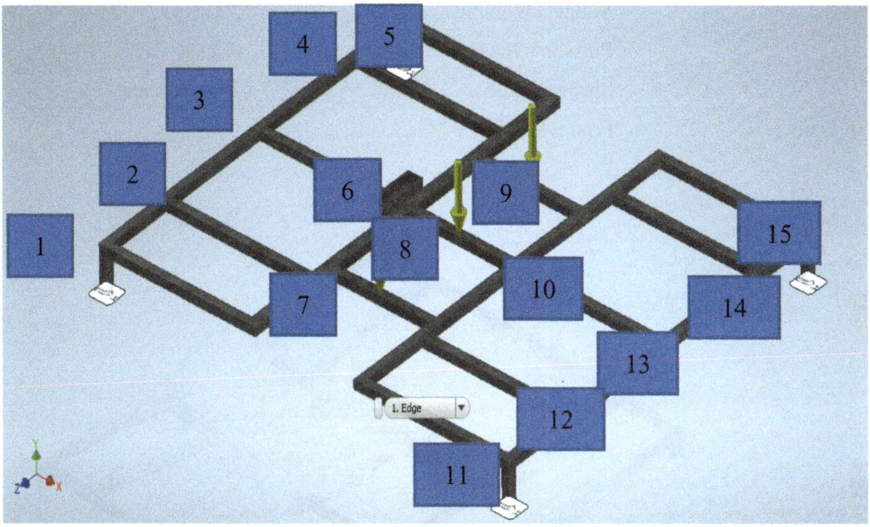

Fig. 3. Location of the displacement and stress distribution

The determination of the points from Fig. 3, explains that these points are to determine the area where the reaction occurs. The data can be displayed at a predetermined point if the reaction has been given to the mobile robot frame. After determining the point, then the results and conclusions of the research are taken. There is consist of 15 points, each point has a value to display reactions such as stress, strain, displacement, safety factor, and principal Stress.

4 Result and Discussion

First, it starts by determining the material that will be used in this simulation software, for material on this robot car frame can be seen in Table 2.

Table 2. Material specification

Name	Steel, galvanized	
General	Mass density	7,85 g/cm^3
	Yield strength	207 Mpa
	Ultimate tensile strength	345 Mpa

(continued)

Table 2. (*continued*)

Name	Steel, galvanized	
Stress	Young's modulus	200 Gpa
	Poisson's RATIO	0,3 ul
	Shear modulus	76,9231 Gpa
Part Name(s)	Frame VS.ipt	

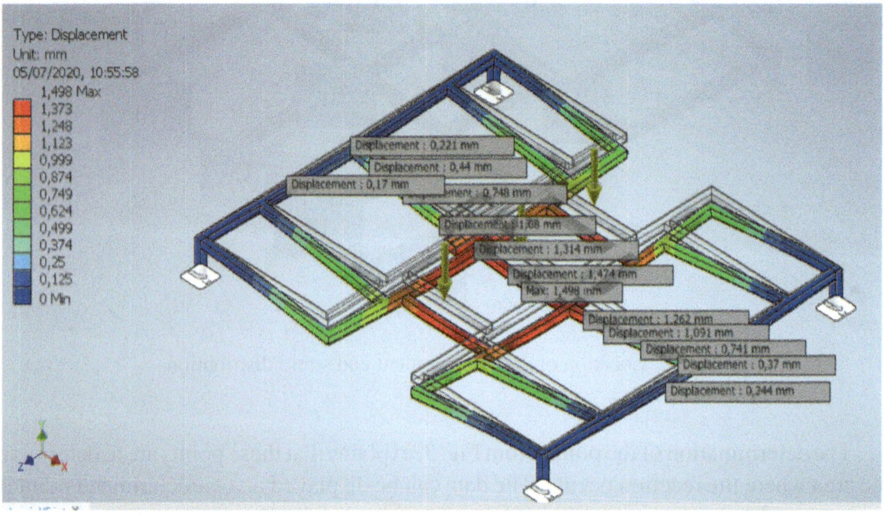

Fig. 4. Simulation displacement

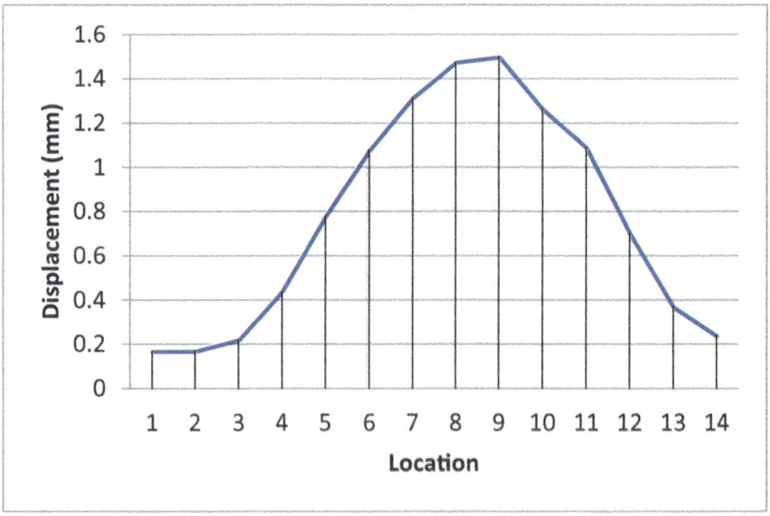

Fig. 5. The displacement result of frame in milimeter

Simulations carried out to see the results of von mises stress, displacement, safety factor and feasibility. Figures 4 and 5 show the high displacement section.

From Fig. 4 shows that the biggest displacement point is at location 4–10, it happens because the biggest loading is at that point. The effect of displacement also shows the effect of stress on the frame, for its effect can be seen in Figs. 6 and 7.

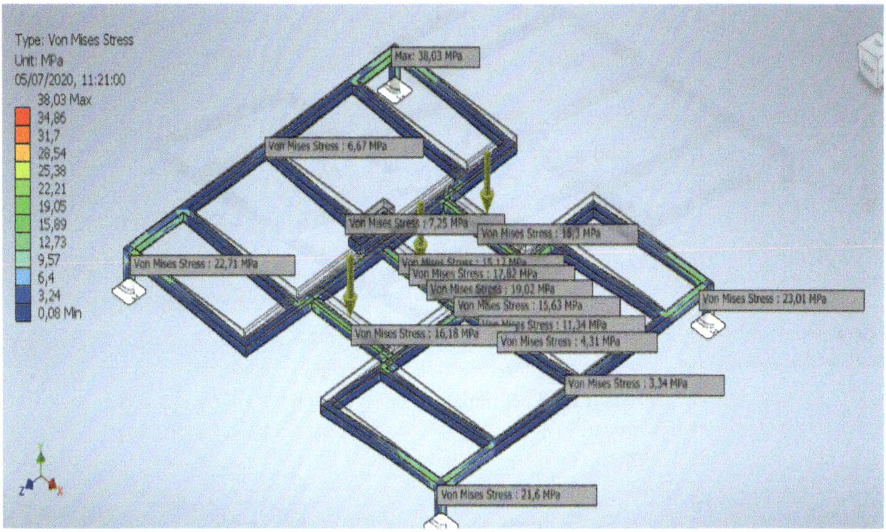

Fig. 6. Simulation Von Mises stress

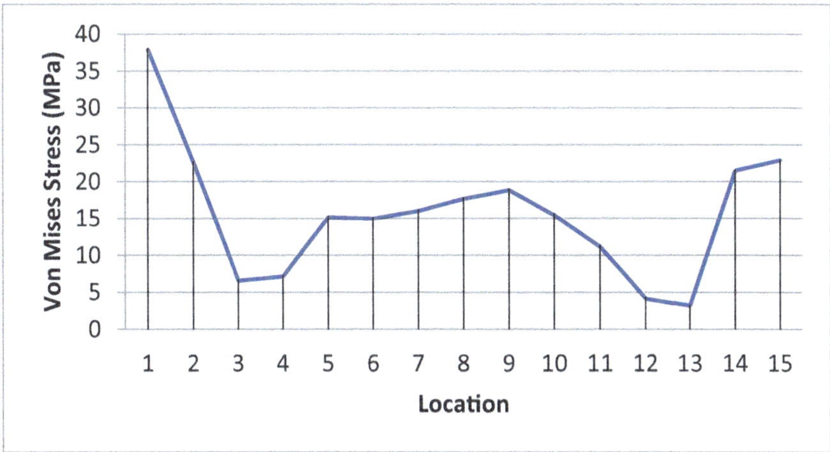

Fig. 7. Von mises stress v.s location

From Fig. 6 it can be seen that the effect of loading makes the frame experience stress. The highest stress can be seen at the location 1,2,4,5,6,7,8,9,10,14,15. The reason for the stress is high because the displacement reaction is in the location of 5,6,7,8,9,10. thus affecting the robot car frame legs. So from that it can be seen the level of safety that occurs in the frame when loading. In Figs. 8 and 9 can be seen the location of the safety factor on the frame.

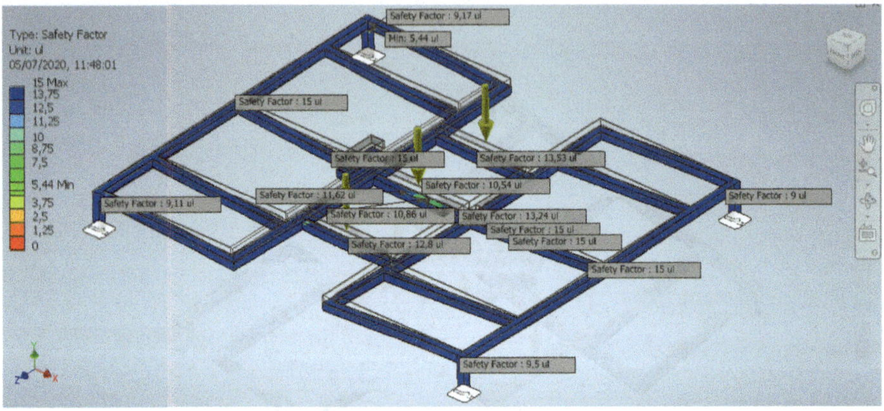

Fig. 8. Simulation safety factor

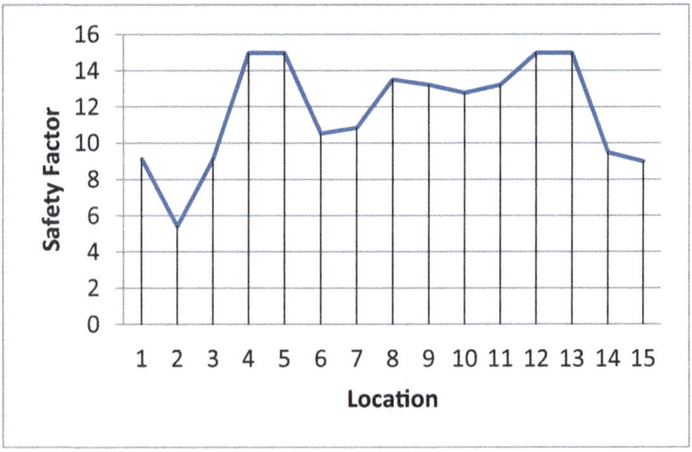

Fig. 9. Result of safety factor

In Fig. 9 can be seen that the level of safety is less at the same point in the location of von mises stress. So it can be explained that the safety factors can be reduced if the stress also increases. In Table 3 is a recap of the results of the analysis using the Autodesk Inventor Software.

Reports from this frame analysis are shown in Table 3. It's can be explained that the maximum Von Mises of 38.0258 MPa produces a maximum displacement of 1.49 mm with a maximum stretch value of 0.0001764, a safety factor or feasibility of this frame

Table 3. Total result simulation

Name	Minimum	Maximum
Volume	3748750 mm^3	
Mass	29,4277 kg	
Von Mises stress	0,0786106 Mpa	38,0258 MPa
1st principal stress	−7,84853 Mpa	47,467 MPa
3rd principal stress	−28,4155 Mpa	10,1702 MPa
Displacement	0 mm	1,49778 mm
Safety factor	5,44367 ul	15 ul
Equivalent strain	0,000000350243 ul	0,000176434 ul

is found, Based on [6] (machine element) that the static load numbers safety factor is: 1.25–2; and for the dynamic load: 2–3; shock load 3–5 [7]. This frame is a dynamic object so the minimum safety factor value is 2, the results of this analysis produce a minimum of 5.4 safety factor, then it can be concluded with a magnitude of force of 284.09 N in the center of the frame and 261.9 N on the left and right. The frame is said to be suitable for use. Figure 10 a mobile robot spinach water on tester.

Fig. 10. The real model of a mobile robot for spinach water seed planting

Conclusion

The serial simulation to analyze the displacement and stresses on the mobile robot frame using finite element analysis was successfully done. The design process is carried out of the frame model. Caused by the load of 261.9 N on the left and the side, also the load of 284.09 N on the middle, the results of the frame analysis is safe to use at maximum load until 284.09 N. In the displacement study, it was shown that the displacement occurred maximally at points 4–10 with a maximum value of 1.5 mm. The von misses stress occurs at a maximum point 1,2,4,5,6,7,8,9,10,14,15 with a value of 38.08 MPa. The safety factor shows that the highest safe point is at 3,4,5 and 11,12,13 and the lowest point is at points 2 and 15 which is a part that must be considered in mobile robot safety.

References

1. Salimin, S., Adha, I.: Design and analysis of tourist bike frame loading simulation for two passengers using software autodesk inventor 2017. Enthalpy-Sci. J. Mech. Eng. Students **3**(3) (2018). ISSN 2502-8944
2. Eder, W.E., Hosnedl, S.: Design Engineering, a Manual for Enhanced Creativity. CRC Press, Taylor & Francis Group, New York (2008)
3. Raudin, P.M.: Design and manufacturing a pedal drive system car frame. J. Mech. Eng. Educ. (2016). (Department of Mechanical Engineering, Faculty of Engineering, University of Padang)
4. Moaveni, S.: Finite Element Analysis Theory and Application with Ansys. Prentice Hall, Inc., Hoboken (1999)
5. Setyono, B., Mrihrenaningtyas, H.A.: Design and strength analysis of the "TRISONA" hybrid bicycle frame using autodesk inventor software. J. IPTEK **20**(2) (2016)
6. Dobrovolsky: A Text Book Machine Elements. Peace Publishers, Moscow (1989)
7. Caesar, B.P.P., Hazimi, H., Sukanto, H., Prabowo, A.R.: Development of novel design and frame structural assessment on Mitutoyo's auto checking hardness machine using reverse engineering approach: series HR-522 hardness tester. J. Eng. Sci. Technol. **15**(2), 1296–1318 (2020). School of Engineering, Taylor's University

Halal Supply Chain Performance Measurement Model in Food Industry Using SCOR Model, AHP Method and OMAX

Ratna Wulandari[1](✉), Ari Yanuar Ridwan[1](✉), and Suksessanno Muttaqin[2](✉)

[1] Industrial Engineering Department, Telkom University, Bandung, Indonesia
ratnawullandari@student.telkomuniversity.ac.id,
ariyanuar@telkomuniversity.ac.id

[2] Logistics Engineering Department, Faculty of Industrial Engineering, Telkom University, Bandung, Indonesia
prafajars@telkomuniversity.ac.id

Abstract. Indonesia is a country that has the largest Muslim population in the world, which makes Indonesia a large market share for Sharia products and services, especially Indonesia has potential in the halal food industry. There are several standards for the halal management system that must be considered, such as in SNI 99001: 2016 Clause 3.4.1 which contains criteria, methods, including measurements and related performance indicators needed to ensure effective operation, and inter-process control. Thus, the organization or company is needed to be able to establish company criteria, methods and measurement of performance indicators in order to ensure effectiveness in the process. The purpose of this study is to determine the criteria for measuring the performance of the halal supply chain in the planning process, to design a system and interface for measuring the performance of the halal supply chain in the planning process based on SNI 99001: 2016. In the process of completing this research using the SCOR model and the AHP, OMAX, and TLS method. The results of the performance measurement for the criteria and sub-criteria are obtained. In this study, three criteria were produced for each part of the planning, namely reliability, responsiveness, and cost. For planning the procurement of raw materials there are 6 sub-criteria reliability, 1 sub-criteria responsiveness, and 1 sub-criteria cost. For production planning, there are 5 reliability sub-criteria, 3 responsiveness sub-criteria, and 1 cost sub-criteria. For raw material procurement planning, there are 7 reliability sub-criteria, 3 responsiveness sub-criteria, and 1 cost sub-criteria.

Keywords: Halal · Performance measurement model · SNI 99001:2016 · SCOR · AHP · OMAX · TLS

1 Introduction

Indonesia currently is developing in several industrial sectors, one of which is the food industry. The halal food industry needs to be developed with support in the form of halal supply chain management which is commonly called halal supply chain management

Akhyar et al. (Eds.): ICECME 2021, LNME, pp. 187–197, 2023.
https://doi.org/10.1007/978-981-19-3629-6_20

(HSCM). Halal supply chain management (HSCM) is a development of the supply chain concept with guaranteed halal integrity starting from raw materials to products purchased by consumers until they reach the hands of consumers by maintaining their halal status so that they are safe for consumption [1].

Vannisa Brownies is one of the SMEs located in Bandung, which produces brownies with various variants, as well as other cakes. Vannisa Brownies has implemented a halal system and has been certified by MUI. MUI halal certification is a fatwa that certifies a halal product in accordance with Islamic law written by the Indonesian Ulema Council.However, based on the results of interviews with Mr. Nasrul always the owner of Vannisa Brownies explained that in the planning section in terms of resources, product manufacturing, and delivery there are no halal activities that aim to maintain product halalness in accordance with SNI 99001:2016, therefore, it is necessary the business process of the Vannisa Brownies production planning section was developed to ensure the halalness of a product.

In accordance with the current condition of the company, which has not implemented a halal management system due to the company's lack of understanding regarding halal management. In addition, the company has never measured supply chain performance, so it is difficult for the company to determine KPIs and the method that will be used to measure supply chain performance [2]. Based on these problems, it is necessary to design a halal supply chain performance measurement system in the planning section that is integrated with the activity process based on halal standards to meet halal management system standards [3, 4]. The purpose of this study is to create a performance measurement system, namely to create a halal Key Performance Indicator (KPI) using the SCOR (Supply Chain Operation Reference) model [5]. SCOR is a framework used to describe business activities between components in the supply chain from upstream (suppliers) to downstream (customers) [6, 7]. Furthermore, the determination of halal criteria is based on SNI 99001:2016. The Indonesian National Standard (SNI) 99001:2006 is a standard in the halal management system that has been established by the National Standardization Agency (BSN) as a reference and principle in implementing a company's halal management system. SNI 99001:2016 is regulated in the Regulation of the Minister of Industry Number 17 of 2020.

Based on BSN, SNI 99001:2016 is a standardization of halal management based on HAS 2300 and ISO 9001:2015. This standard regulates organization, risk-based management, bridges fatwas into positive law in certain fields which aims to make organizations that produce halal products apply SNI 99001:2016 as a guarantee of product halal quality [8, 9].

The methods used in this research are AHP (Analytical Hierarchy Process), OMAX (Objective Matrix), and TLS (Traffic Light System). The AHP method is used for weighting in each attribute

After weighting, the next step is the calculation of the scoring system using the OMAX method which aims to determine the performance achievement of each performance indicator [10, 11]. Then the OMAX method will be integrated with TLS which aims to make it easier to categorize the results of achieving company performance.

2 Methodology

Systematic problem solving is an explanation of the steps taken in solving problems in research.

Systematics of problem solving can be seen in Fig. 1.

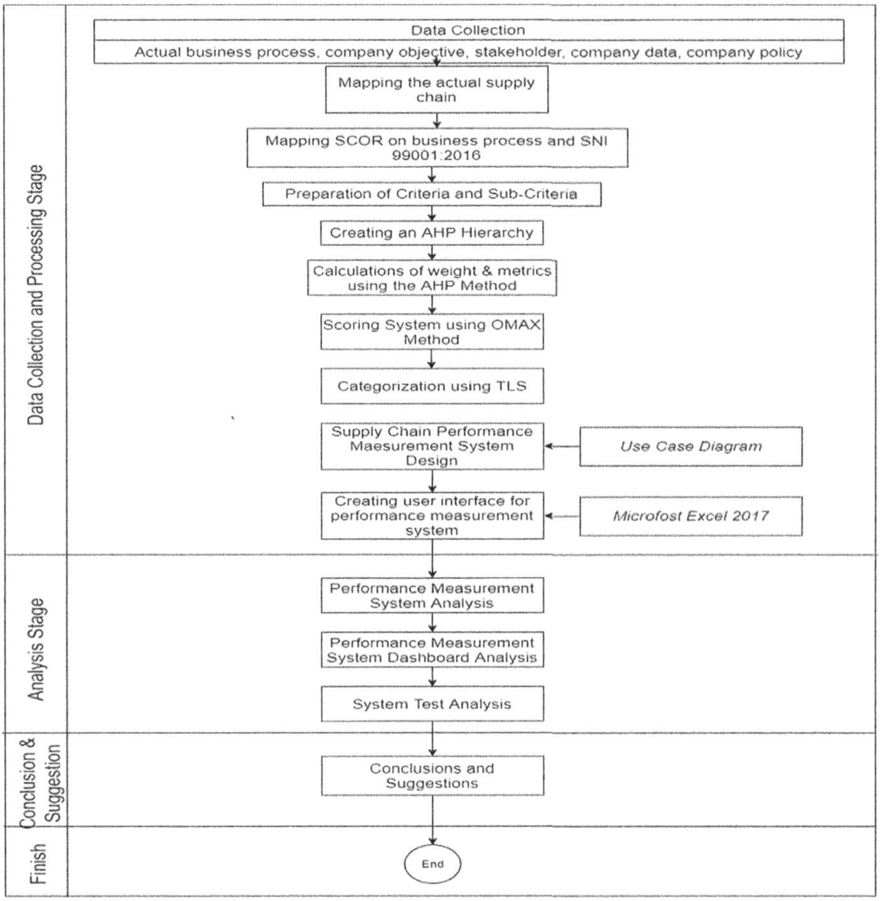

Fig. 1. Methodology

3 Data Collection

Data collection by direct observation to the company to obtain the company's actual business processes and interviews with relevant stakeholders in the planning process for raw material procurement, production planning, and distribution planning consisting of company owners, production managers, heads of production divisions, and heads of distribution divisions. As for some of the data needed in the research including actual business processes, and company goals. The company's objectives can be seen in Table 1.

The company's actual business processes are then modeled into the SCOR model. This stage aims to identify several processes contained in SCOR which are adapted to the company's actual processes in the planning section which consists of planning for the procurement of raw materials, planning for the production process, and planning for distribution.

Table 1. Company objectives

Objectives	Criteria	References
Meet the target of material procurement, production planning per day, and distribution in terms of quality and quantity that have been determined	Reliability	Company interview
Meet procurement targets, customer requests, and distribution within the specified time	Responsiveness	Company interview
Meet procurement, production, and distribution targets at an efficient cost	Cost	Company interview

4 Determine the Criteria

The preparation of the sub-criteria is done by identifying the sub-criteria of the planning performance that is adjusted to the company's goals and mapped into the SCOR Model. Furthermore, the sub-criteria will be verified by the company. The verified sub-criteria can be seen in Table 2.

5 Determine the Weighting Criteria Using AHP Method

Next is the stage of weighting the criteria and sub-criteria using the AHP method. The decision model in the AHP method is carried out by describing complex multi-factors or multi-criteria to become one hierarchy. In the AHP model, if the data processing is in the form of qualitative data, then in equalizing the processing weights, you can use pairwise comparisons [6, 9].

- Calculation of Consistency Index (CI), with the following equation:

$$CI = \frac{\lambda maxs - n}{n - 1}$$

Table 2. Verification of criteria and sub-criteria

Code	Sub-criteria	Code	Sub-criteria
SKS1	Raw material forecasting accuracy [9]	SKM7	Speed in identifying production process needs [10]
SKS2	Scheduling accuracy rate [12]	SKM8	Speed set production planning [10]
SKS3	Accuracy in making reports on raw material requirements [13]	SKM9	Costs required in the production process [10]
SKS4	The raw materials needed are in accordance with quality and quality standards [13]	SKD1	Delivery quantity accuracy [13]
SKS5	Planning for the procurement of raw materials in accordance with halal requirements [14]	SKD2	Delivery type accuracy [12]
SKS6	Guaranteed raw materials in accordance with halal requirements [13]	SKD3	Adjusting the amount of transportation to send products to retail [Company Interview]
SKS7	The speed of determining the raw materials needed and which will be purchased [9]	SKD4	Products are delivered according to quality standards and quality [13]
SKS8	Costs required in the procurement of raw	SKD5	The accuracy of the documents related to the product to be sent [13]
SKM1	Production forecasting accuracy [10]	SKD6	Distribution planning in accordance with halal requirements [13]
SKM2	Products made in accordance with quality and quality standards [13]	SKD7	Guaranteed delivery transportation in accordance with halal requirements [13]
SKM3	Accuracy of production requirement reports [13]	SKD8	Time required in the delivery schedule [13]
SKM4	Production Planning in accordance with halal requirements [14]	SKD9	The time it takes to create a delivery route [13]
SKM5	Guaranteed production process equipment in accordance with halal requirements [13]	SKD10	Speed setting distribution planning [Company Interview]
SKM6	Speed in scheduling the production planning process [13]	SKD11	Costs required in the delivery process

Notation: λmaxs = maximum eigenvalue matrix A

'n = Ordo of matrix A

- Calculation of Consistency Ratio (CR) using the following equation:

$$RI = \frac{1.98(n-2)}{n}$$

$$CR = \frac{CI}{RI}$$

6 Determine the Scoring System Using OMAX Method

The next step is the calculation of the scoring system using the OMAX method which aims to determine the performance achievement of each performance indicator. Then the OMAX method will be integrated with TLS which aims to make it easier to categorize the results of achieving company performance [8]. Calculation of interval in OMAX method with the following equation [15]:

- The interval between levels 10–4:

$$\text{Level n} = \text{Value of level n} + 1 - \left(\frac{valueoflevel(n+1) - valueoflevel3}{(n+1)-3}\right)$$

- The interval between levels 3–0:

$$\text{Level n} = \text{Value of level n} - \left(\frac{valueoflevel(n) - valueoflevel0}{(n)-0}\right)$$

7 Result and Discussion

Measurement of productivity in each sub-criteria is carried out using the OMAX method which is then classified into the Traffic Light System category using three color indicators, such as red, yellow, and green. Categorization of performance productivity can be seen in Tables 3, 4 and 5. And then, the last step is the design of the performance measurement system described by the Fig. 2.

In the visualization of productivity performance categorization using TLS, one example is taken from the planning aspect, that is the production planning aspect. Scoring system calculation results of performance measurement on reliability criteria in the aspect of plan make using OMAX can be seen in the Table 3.

The results of the Omax calculation are obtained from the perspective of reliability criteria on the aspect of the plan make with the highest level, were SKM4 and SKM5 with a level value of 10. While the lowest level was SKM1 with a level value of 3.82. Therefore, the index value was 3.30. In this perspective urgent improvements was needed to the SKM1 sub-criteria.

Scoring system calculation results of performance measurement on responsiveness and cost criteria in the aspect of plan make using OMAX can be seen in the Tables 4 and 5.

The results of the Omax calculation are obtained from the perspective of responsiveness criteria on the aspect of the plan deliver at SKM6 with a level value of 5.75, at SKM7 with a level value of 7.20, at SKM8 with a level value of 6.50 and the index value was 2.08. And then, The results of the Omax calculation are obtained from the perspective of cost criteria on the aspect of the plan make at SKM9 with a level value of 4.62 and the index value was 1.20. So, In this perspective no urgent repairs are needed.

Table 3. Realibitily (plan make)

KPI		SKM1	SKM2	SKM3	SKM4	SKM5
Reliability Performance		85.00	99	94	100.00	100.00
Unit		%	%	%	-	-
	10	100.00	100.00	100.00	100.00	100.00
	9	97.57	99.71	98.86	100.00	100.00
	8	95.14	99.43	97.71	100.00	100.00
	7	92.71	99.14	96.57	100.00	100.00
	6	90.29	98.86	95.43	100.00	100.00
	5	87.86	98.57	94.29	100.00	100.00
	4	85.43	98.29	93.14	100.00	100.00
	3	83.00	98.00	92.00	100.00	100.00
	2	80.33	97.00	90.67	100.00	100.00
	1	77.67	96.00	89.33	100.00	100.00
	0	75.00	95.00	88.00	100.00	100.00
Level		3.82	6.50	4.75	10.00	10.00
Weight		0.04	0.13	0.02	0.14	0.08
Value		0.16	0.87	0.09	1.40	0.77
Index	3.30 2.08					

Scoring system calculation results of performance measurement on responsiveness and cost criteria in the aspect of plan make using OMAX can be seen in the Tables 7 and 8

The last step performance measurement system design. The design of this integration system is carried out starting from the identification of criteria and sub-criteria related to company activities which will then be integrated with the integrated system. The purpose of this integration system is to find out what aspects are related to the system to be made. This performance measurement system consists of several menus or displays consisting of a dashboard, specifications, input data, and performance productivity which can be seen in the Fig. 2.

Table 4. Responsiveness (plan make)

KPI		SKM6	SKM7	SKM8
Responsiveness Performance		3.00	2.00	3.00
Unit		Day	Day	Day
	10	1.00	1.00	1.00
	9	1.57	1.36	1.57
	8	2.14	1.71	2.14
	7	2.71	2.07	2.71
	6	3.29	2.43	3.29
	5	3.86	2.79	3.86
	4	4.43	3.14	4.43
	3	5.00	3.50	5.00
	2	5.67	4.67	5.67
	1	6.33	5.83	6.33
	0	7.00	7.00	7.00
Level		5.75	7.20	6.50
Weight		0.15	0.09	0.08
Value		0.86	0.68	0.55
Index	2.08			

Table 5. Cost (plan make)

KPI		SKM9
Cost Performance		500000000
Unit		Rp/Month
	10	400000000
	9	418571429
	8	437142857
	7	455714286
	6	474285714
	5	492857143
	4	511428571
	3	530000000
	2	553333333
	1	576666667
	0	600000000
Level		4.62
Weight		0.26
Value		1.20
Index	1.20	

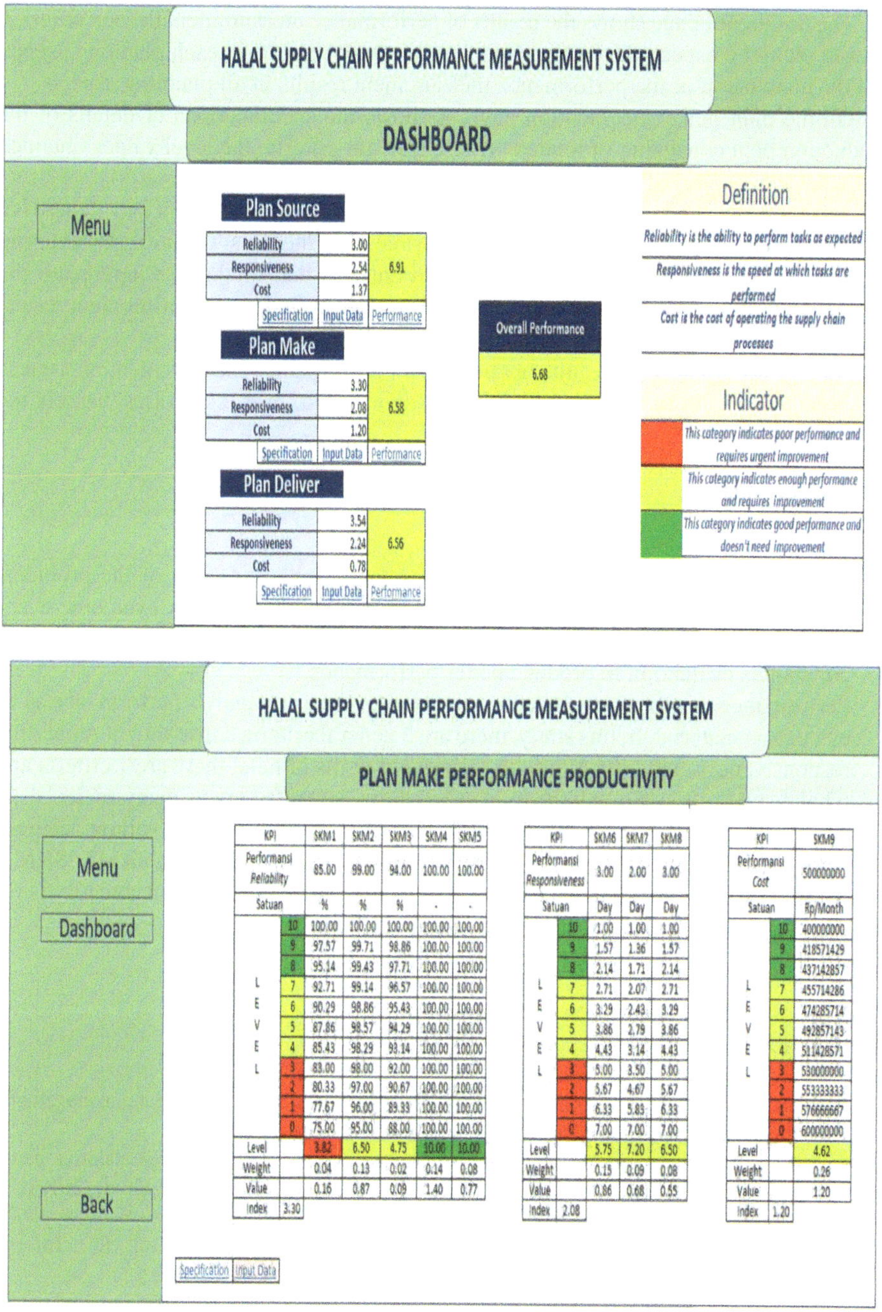

KPI		SKM1	SKM2	SKM3	SKM4	SKM5
Performansi Reliability		85.00	99.00	94.00	100.00	100.00
Satuan		%	%	%	-	-
LEVEL	10	100.00	100.00	100.00	100.00	100.00
	9	97.57	99.71	98.86	100.00	100.00
	8	95.14	99.43	97.71	100.00	100.00
	7	92.71	99.14	96.57	100.00	100.00
	6	90.29	98.86	95.43	100.00	100.00
	5	87.86	98.57	94.29	100.00	100.00
	4	85.43	98.29	93.14	100.00	100.00
	3	83.00	98.00	92.00	100.00	100.00
	2	80.33	97.00	90.67	100.00	100.00
	1	77.67	96.00	89.33	100.00	100.00
	0	75.00	95.00	88.00	100.00	100.00
Level		3.82	6.50	4.75	10.00	10.00
Weight		0.04	0.13	0.02	0.14	0.08
Value		0.16	0.87	0.09	1.40	0.77
Index	3.30					

KPI		SKM6	SKM7	SKM8
Performansi Responsiveness		3.00	2.00	3.00
Satuan		Day	Day	Day
LEVEL	10	1.00	1.00	1.00
	9	1.57	1.36	1.57
	8	2.14	1.71	2.14
	7	2.71	2.07	2.71
	6	3.29	2.43	3.29
	5	3.86	2.79	3.86
	4	4.43	3.14	4.43
	3	5.00	3.50	5.00
	2	5.67	4.67	5.67
	1	6.33	5.83	6.33
	0	7.00	7.00	7.00
Level		5.75	7.20	6.50
Weight		0.15	0.09	0.08
Value		0.86	0.68	0.55
Index	2.08			

KPI		SKM9
Performansi Cost		500000000
Satuan		Rp/Month
LEVEL	10	400000000
	9	418571429
	8	437142857
	7	455714286
	6	474285714
	5	492857143
	4	511428571
	3	530000000
	2	553333333
	1	576666667
	0	600000000
Level		4.62
Weight		0.26
Value		1.20
Index	1.20	

Fig. 2. Dashboard and plan make specification page

The dashboard page shows the results of performance measurement in each criterion in each planning aspect, the performance measurement results in each planning aspect, and the conclusion of the performance measurement results in all planning aspects.

On the plan make specification page, it shows the specifications or details of the production plan consisting of related criteria and sub-criteria, the level values obtained along with the values of each sub-criteria or KPI.

The make input data page is a page that can be accessed to enter or edit data if there is a data update and you want to measure the performance of the latest planning. By entering related data contained in the data input page, by entering data on the data input page, the final result will automatically come out in the form of performance measurement results on the plan make performance.

On this page shows the results obtained from performance measurement on aspects of production planning. The categorization is divided into three categories, namely the red, yellow, and green categories.

8 Conclusion

1. The design of this system, can produce performance to analysis of the problems that occur, so that by implementing this system the company can overcome every risk with a strategy to overcome the adverse effects that occur as material for the company's evaluation to produce better performance.
2. Performance measurement is needed to determine the company's performance as an evaluation material. In this study, there are 3 general criteria consisting of reliability, responsiveness, and cost. While the criteria in terms of halal there are 1 criteria and 6 halal sub criteria. So that the total sub-criteria generated are as many as 28.
3. There is a red category for KPI SKS3, SKM1, and SKD11. The yellow category is for SKS1, SKS2, SKS4, SKS7, SKS8, SKM2, SKM3, SKM6, SKM7, SKM8, SKM9, SKD1, SKD2, SKD4, SKD5, SKD8, and SKD10. The green category is for SKS5, SKS6, SKM4, SKM5, SKD3, SKD6, SKD7, and SKD9.

References

1. Tieman, M., Van der Vorst, J.G., Ghazali, M.C.: Principles in halal supply chain management. J. Islamic Mark. (2012)
2. Fradinata, E., Kesuma, Z.M., Rusdiana, S.: Support vector regression and adaptive neuro fuzzy to measure the Bullwhip effect in supply chain. J. Phys. Conf. Ser. 022010 (2018)
3. Mefid, K.N., Ridwan, A.Y., Sari, W.P.: Global industry perspective of halal cosmetics applying sales and distribution process based on enterprise resources planning. Bull. Soc. Inform. Theory Appl. **3**, 61–68 (2019)
4. Zahra, D.K., Ridwan, A.Y., Saputra, M.: Halal SCM: how to monitor halal performance of procurement process for food industry in Indonesia by using ERP system. In: ICORE, vol. 5 (2020)
5. Fauziyah, I., Ridwan, A., Muttaqin, P.: Food production performance measurement system using halal supply chain operation reference (SCOR) model and analytical hierarchy process (AHP). In: IOP Conference Series: Materials Science and Engineering, p. 012074 (2020)

6. Saaty, T.L.: Decision making with the analytic hierarchy process. Int. J. Serv. Sci. **1**, 83–98 (2008)
7. Fradinata, E., Kesuma, Z.M., Rusdiana, S.: Optimal lot-sizing algorithms on stochastic demand at the retailer. In: Proceedings of MICoMS 2017. Emerald Publishing Limited (2018)
8. Moharamkhani, A., Bozorgi-Amiri, A., Mina, H.: Supply chain performance measurement using SCOR model based on interval-valued fuzzy TOPSIS. Int. J. Logist. Syst. Manag. **27**, 115–132 (2017)
9. Fradinata, E., Kesuma, Z.M.: Numeric model to predict the location of market demand and economic order quantity for retailers of supply chain. In: IOP Conference Series: Materials Science and Engineering, p. 012010 (2018)
10. Anugrah, S., Anggraini, Absor, W., Fauzi, S.: Integrated analytical hierarchy process and objective matrix in balanced scorecard dashboard model for performance measurement. Telkomnika 16 (2018)
11. Fradinata, E., Suthummanon, S., Sunthiamorntut, W.: Comparison of hybrid ANN models: a case study of instant noodle industry in Indonesia. Int. J. Adv. Appl. Sci **4**, 19–28 (2017)
12. Wahyuniardi, R., Syarwani, M., Anggani, R.: Pengukuran Kinerja supply chain Dengan Pendekatan supply chain operation references (SCOR). Jurnal Ilmiah Teknik Industri **16**, 123–132 (2017)
13. Nusran, M., Triana, D., Raehana, S., Damis, H., Syarifudin, A., Wekke, I.S.: Policy on halal slaughtering availability for halal chicken needs Makassar City Indonesia. Int. J. Eng. Technol. **7**, 75–81 (2018)
14. Nurhidayanti, S., Rusdi, N., Abdul, S., Yusuf, B.: Analysing and modelling gripper arm using shape optimization of fusion 360 and 3d printing of PLA (2020)

Experimental Study of Modular Brushless Permanent Magnet Axial Flux Motor Performance on Micro Electric Car

Muhammad Tadjuddin[1](✉), Zahrul Fuadi[1], Nazaruddin[2], and Muhammad Tulus[1]

[1] Department of Mechanical and Industrial Engineering, Syiah Kuala University, Banda Aceh, Indonesia
m.tadjuddin@unsyiah.ac.id

[2] Department of Mathematics and Science, Syiah Kuala University, Banda Aceh, Indonesia

Abstract. Electric vehicles are now a trend. Various types of electric vehicles have been produced. The development of electric vehicle propulsion motors is also continuously carried out to get the best performance. One type of electric motor that has a high efficiency is a brushless permanent magnet axial flux motor. This type of motor has a relatively large torque at low speeds. Therefore, this paper presents the results of testing the micro electric car propulsion system using a brushless permanent magnet axial flux motor with a modular system. Tests are carried out to obtain the value of torque, electric power, mechanical power, speed, and temperature rise of the motor. The driving motor is mounted on the rear axle of the micro electric car, then tested using a chassis dynamometer with disk brake loading. It tests with maximum torque of 37.8 Nm, at 550 rpm, with an electric power input of 1197 Watts. The highest efficiency is 76% obtained at 83rpm with 37,8Nm Torque. From the test results, the total efficiency of the drive system is considered low, it was due to mechanical losses on transmission and gear of the axle.

Keywords: Micro electric car · Axial flux motor · Modular system · High efficiency

1 Introduction

Electric vehicles (EVs) are a trending mode of transportation. In addition to the limitations of fossil fuel energy in the future, it is also a kind of effort to reduce air pollution. Today's electric vehicles have sophisticated technology and are comfortable to drive. Unlike petroleum-fueled vehicles, it has instant power so that it feels new in the driving experience. The instant power and comfort of this vehicle are not a debate. Still, the limitation of electric vehicles is limited power storage, so this vehicle is not yet practical to replace vehicles with long distances operations.

The mileage of this vehicle is not only associated with power storage, but there are also other contributing factors, including electric motor efficiency, inverter components

Akhyar et al. (Eds.): ICECME 2021, LNME, pp. 198–204, 2023.
https://doi.org/10.1007/978-981-19-3629-6_21

(motor drive) and energy management algorithms, and power loss in mechanical transmissions. Many studies and simulations on energy consumption management in electric vehicles have been carried out either by software simulation or by field testing [1, 2].

The main prime-mover of electric vehicles is the electric motor. Various types of electric motors have been used in electric cars with their characteristics. Some electric vehicles use switch reluctance motors (SRM) with the advantage that they do not use permanent magnets [3]. IPM motors are also widely used in electric vehicles or hybrid vehicles, where these motors have good efficiency [4, 5]. There are also electric vehicles that use axial flux electric motors [6]; axial electric motors have good efficiency and relatively large torque due to their dimensions which tend to be thin like discs. This motor is a slotless stator that has no teeth on the stator. The slotless system was chosen because of its advantages in that there is no cogging force between the stator and rotor [7].

The axial flux permanent magnet motor (AFPM) has a disk-like stator and rotor. It can be assembly with multi stator and rotor modules (Fig. 1). Previous research has made multiple stator and rotor modules; any changes in the power specification are also easier to configure [8]—motor with multi modules as shown in the following figure.

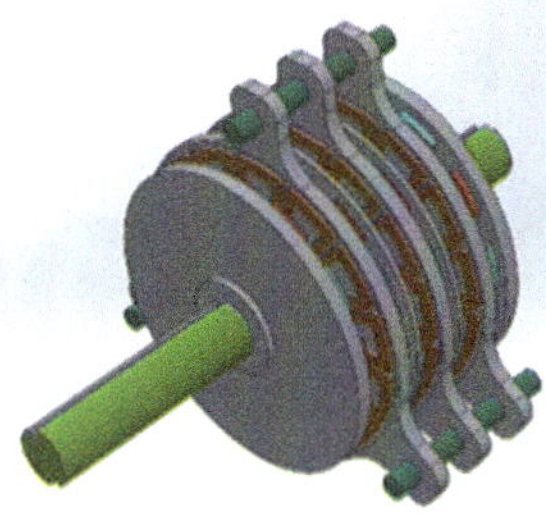

Fig. 1. Multi-module permanent magnet axial flux motor

The working principle of an electric motor, in general, is to convert electrical energy into mechanical power. Among the types of motors, permanent magnet motors have high efficiency when compared to induction motors. One type of permanent magnet motor is an axial flux permanent magnet motor [9]. This motor is thin or disc. The diameter of the motor is relatively large compared to its length. Therefore, this motor has a large specific torque at low revs. This type of motor has the best efficiency, power density, and torque. A new design by combining an axial flux motor with a radial motor has been made, and it is hoped that the advantages of both will be found [10]. Comparison between axial flux permanent magnet motor (AFPM) and radial flux permanent magnet motor (RFPM) has also been carried out using the 3D Finite Element Method (FEM). The result is that AFPM has advantages, especially for coreless ones with low ripple torque at low speeds. With a relatively large torque, this motor can be used as a direct drive [11]. Large torque is more accessible to obtain due to the somewhat wider motor diameter. Its slim shape also allows it to be hidden inside for in-wheel drives.

2 Micro Electric Vehcle

Electric vehicle technology continues to develop. Ongoing research is an attempt to increase the overall efficiency of the vehicle. Another area of interest recently is micro electric vehicles [12]. Micro electric vehicles are small vehicles for sufficient mileage in the city. Maximum vehicle for two people, one driver and one passenger, or one driver without passengers. This type of electric vehicle is predicted to be a vehicle that has excellent business prospects for at least the next decade. These electric vehicles include micro electric cars with one or two passengers, e-bikes, scooters, hoverboards. In general, electric vehicles are expected to reduce the contribution of transportation equipment to environmental degradation [13].

This paper has designed and tested a micro electric motor whose driving motor uses a multi-stack axial flux permanent magnet motor. The motor is installed by connecting the main shaft to the differential gear on the Micro EV. The motor used is an axial flux magnet with two stators and three rotor disks containing a permanent magnet arrangement. The design can be seen in Fig. 2 below.

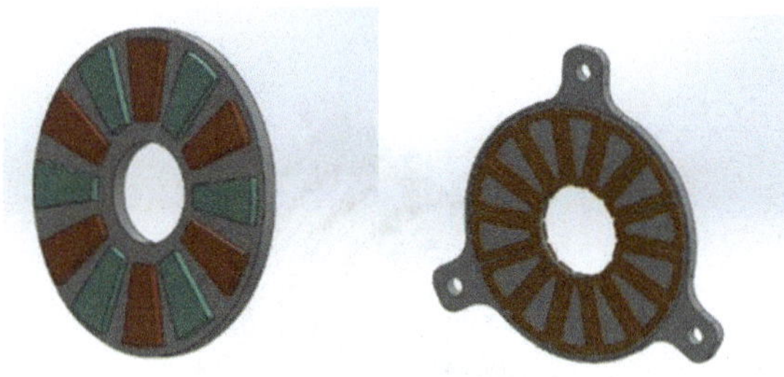

.(a) (b)

Fig. 2. a) Rotor with permanent magnet array, b) Stator with windings

The specifications of the multi-module axial flux motor used are as follows:

Parameters	Amount	Unit
Magnet dimension P × L1 × L2 × T	50 × 30 × 20 × 8	mm
Type of magnet	Trapezoid	NdFeB-N 52
Magnet per module	3 × 16	Magnet
Windings	2 × 12	pcs
Modul	2	pcs
Number of turns	91	turns
Air gap	1	mm
Inner diameter	120	mm
Outer diameter	220	mm

This motor is paired with the rear axle of a micro EV (Fig. 3). A universal joint connects the motor to the rear axle. It has a gear ratio of 1:3. The motor uses a DC voltage of 48 V while the tire size is 165/55-R13.

Fig. 3. Multi-module axial flux motor mounted on the rear axle

The figure above shows the rear axle of the micro EV on a testbed. The chassis dynamometer has two rolls that will provide a torsion load. So that several driving scenarios can be simulated, both speed steps and torque. The value taken in the test with a dynamometer is the mechanical output power. While the input power P_i is the electric power calculated from the battery terminal voltage (V_B) and Battery terminal current (I_B)

$$P_i = V_B.I_B$$

While the output mechanical power is calculated as follows

$$P_m = 2\pi f.T_T$$

where:

Pm= mechanical power (Watts)

f = rotation (hertz)

T_T = torque on the tire (Nm)

The torque value is obtained by measuring with a dynamometer. The torque measured is on the wheels, so the results of the mechanical power measurement will show how much power the drive system will push the micro electric vehicle.

While the system efficiency:

$$\eta = \frac{P_m}{P_i} \times 100\%$$

3 Results and Discussion

A performance test of AFPM multi-module with chassis dynamometer was done. Measurement parameters were 3-speed steps 250, 400, 550 rpm and three different torque loads 4.2, 8.4, and 12.6 Nm. The input power was recorded using a data logger. Speeds and loads are measured directly on the electric motor. The measurement results are as follows (Fig. 4) the graph above shows that the greater load and speed will cause the high input power. The change in value also looks linear, with the maximum power in this experiment was 1197 watts at 550 rpm motor speed.

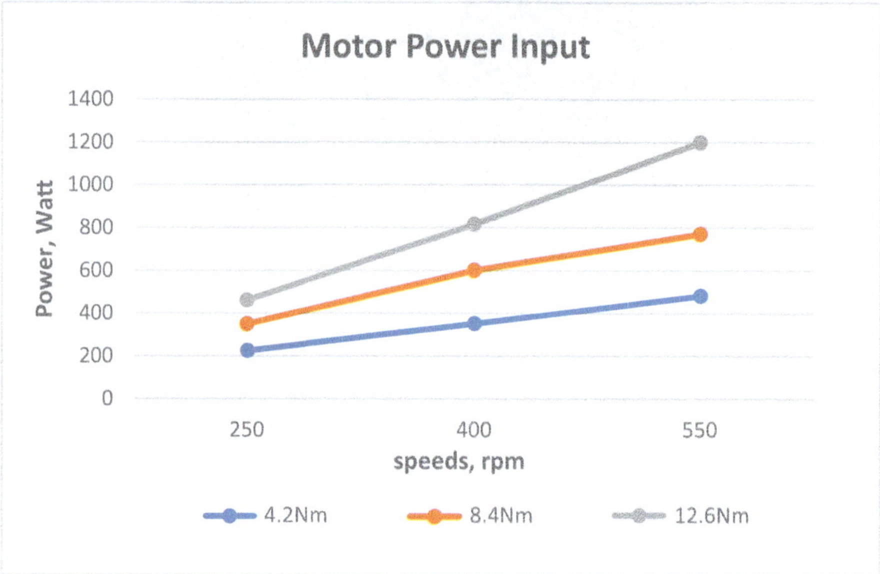

Fig. 4. Power input of AFPM

The output power, which is mechanical power, is measured at the wheel with three-speed steps and three variations of load then the results are calculated. The graph of the measurement results can be seen in Fig. 5, as follow:

The most significant output power is also found in extensive loading and rotation, where the maximum load is 775 W. The figure shows a linear relationship at speed, loading, and output power. The efficiency of the drive system can be seen in Fig. 6 below.

The overall efficiency value was considered low. Efficiency at maximum load and low speed has the highest efficiency value of 76%, while the lowest is low load and speed.

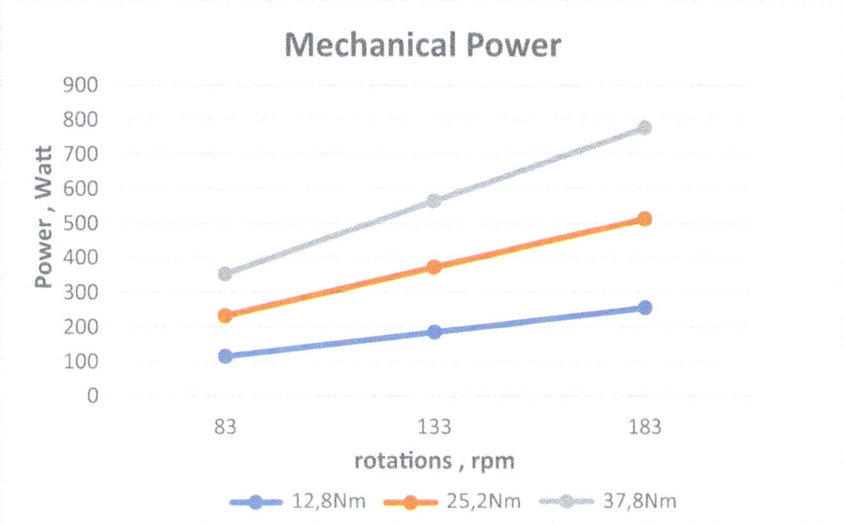

Fig. 5. The mechanical power of micro EV propulsion

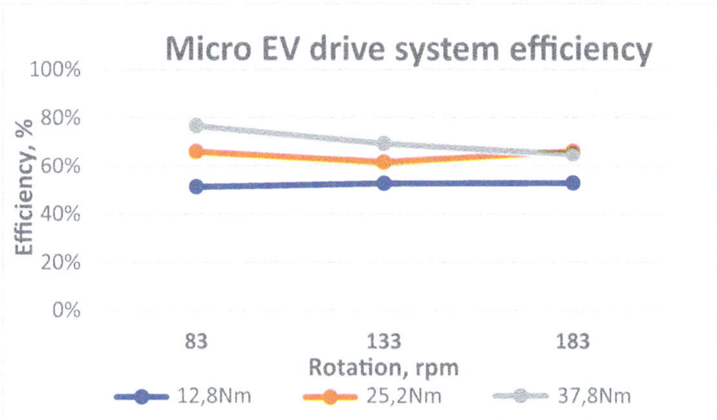

Fig. 6. The efficiency of the micro-EV drive system

4 Conclusion

Tests have been carried out, and the results have also been displayed. The most significant efficiency is 76%, at 83 rpm (the lowest), and the torque is 37.8Nm. Some points that must be considered in this Micro-EV drive system are mechanical components that contribute to power loss. In this test, the total efficiency is relatively low due to high power loss on the poor quality of the mechanical components manufacturers such as differential gears, bearings, universal joints, and shaft alignment.

Acknowledgment. DRPM Kemenristek/BRIN/Kemdikbud supported this research under Applied Research Grant (Penelitian Terapan) 2021 with LPPM Universitas Syiah Kuala. We thank our colleagues from LDM-USK, who provided insight and expertise that greatly assisted the research.

References

1. Guo, J., Jiang, Y., Yu, Y., Liu, W.: A novel energy consumption prediction model with combination of road information and driving style of BEVs. Sustain. Energy Technol. Assess. **42**(October), 100826 (2020)
2. Miri, I., Fotouhi, A., Ewin, N.: Electric vehicle energy consumption modelling and estimation—a case study. Int. J. Energy Res. **45**(1), 501–520 (2021)
3. Belhadi, M., Krebs, G., Marchand, C., Hannoun, H., Mininger, X.: Evaluation of axial SRM for electric vehicle application. Electr. Power Syst. Res. **148**, 155–161 (2017)
4. Hew, W.P., Jamaludin, J., Tadjuddin, M., Nor, K.M.: Fabrication and testing of linear electric generator for use with a free-piston engine. In: National Power Engineering Conference, PECon 2003 – Proceedings, pp. 277–282 (2003)
5. Aslan, M., Özpolat, A.B., İşçi, C., Eroğlu, F., Vural, A.M.: Design and modelling of internal permanent magnet motor. Int. J. Energy Eng. Sci. **5**(2), 80–104 (2020)
6. Tadjuddin, M., Hasanuddin, I., Fuadi, Z., Mukminin, A., Arief, M.R.: Power density comparison of multi-air gap axial flux permanent magnet motor for electric vehicle. In: IOP Conference Series: Materials Science and Engineering, vol. 931, no. 1 (2020)
7. Tadjuddin, M., Hasanuddin, N. I., Mukminin, A.: Thrust force analysis of slotless permanent magnet linear generator for wave energy converter. In: IOP Conference Series: Materials Science and Engineering, vol. 931, no. 1 (2020)
8. Djebarri, S., Charpentier, J.F., Scuiller, F., Benbouzid, M.: Design and performance analysis of double stator axial flux PM generator for rim driven marine current turbines. IEEE J. Ocean. Eng. **41**(1), 50–66 (2016)
9. Gieras, J.F., Wang, R.-J., Kamper, M.J.: Axial Flux Permanent Magnet Brushless Machines, no. May 2014. Springer, Dordrecht (2008). https://doi.org/10.1007/978-1-4020-8227-6
10. Zhou, L., Guo, F., Wang, H., Wang, B.: High-torque direct-drive machine with combined axial-and radial-flux out-runner vernier permanent magnet motor. In: 2021 IEEE Internatinal Electric Machines & Drives Conference. IEMDC 2021 (2021)
11. Rahim, N.A., Ping, H.W., Tadjuddin, M.: Design of an in-wheel axial flux brushless DC motor for electric vehicle. In: 2006 1st International Forum on Strategic Technology, pp. 16–19 (2006)
12. Loustric, I., Matyas, M.: Exploring city propensity for the market success of micro-electric vehicles. Eur. Transp. Res. Rev. **12**(1) (2020)
13. Rosa-Santos, P., et al.: Experimental study of a hybrid wave energy converter integrated in a harbor breakwater. J. Mar. Sci. Eng. **7**(2), 1–18 (2019)

Prediction of Shear Capacity of Reinforced Concrete Column Using Non-linear Finite Element Method

Mochammad Afifuddin[1(✉)], Muttaqin[1], and Iin Shabrina Hilal[2]

[1] Jln. Syech Abdul Rauf No: 7, Kopelma Darussalam, Banda Aceh 23111, Indonesia
{m.afifuddin,muttaqin}@unsyiah.ac.id

[2] Jl. Tgk. Tanoh Abee No. 1 Kopelma Unsyiah Darussalam, Banda Aceh 23111, Indonesia

Abstract. Shear failure in reinforced concrete columns in buildings often found after an earthquake. This type of failure occurs due to the column does not have sufficient shear reinforcement to support earthquake loads. Many studies have been carried out on shear failure in reinforced concrete. Munawir et al. (2012) have conducted research with several different placements of shear reinforcement. It has been found that the placement of shear reinforcement affects the confinement of reinforced concrete columns. Predicting the failure pattern of reinforced concrete columns using the finite element method is a challenge today. With the development of finite element method software this has become possible. This type of research has been developed recently, because it can be more efficient than research in the laboratory. The objectives of this research are to study the behavior of columns that carried axial loads and lateral loads and to predict crack pattern using software non-linear finite element method called ATENA V5. There were two specimens which have been tested experimentally by Munawir, et.al, 2012. There were two variations, namely three arms and crossties of stirrup reinforcement with axial load $0.3P_0$. Reinforced concrete column specimen with a square cross section of size 20×20 cm^2 with 58 cm height. Main reinforcement were 12D11.6 mm, and stirrup reinforcement was D5.4 mm. Concrete compression strength (f_c') was 24.19 MPa. Reinforcement tensile strength was 356 MPa. In three arms of stirrup reinforcement specimens, It has been found that based on ATENA V.5 result, the shear load was 150 Mpa. The difference with experimental result was 2.08%. Based on ATENA V.5 result for deflection was 19.36 mm. The difference with experimental result was 12.43%. On the other hand, in crossties of stirrup specimen, it has been found that based on ATENA V.5 result, the shear load was 150 Mpa. The difference with experimental result was 3.88%. Based on ATENA V.5 result for deflection was 23.21 mm. The difference with experimental result was 23.21%. Meanwhile, prediction of crack *pattern* by non-linear finite element show the same pattern with experimental results for both three arms and crossties of stirrup reinforcement. The failure mode which has been found from numerical analysis showed that all columns had shear failure.

Keywords: Shear capacity · Reinforced concrete · Column · ATENA · Finite element method

Akhyar et al. (Eds.): ICECME 2021, LNME, pp. 205–214, 2023.
https://doi.org/10.1007/978-981-19-3629-6_22

1 Introduction

Different methods have been utilized to study the response of structural components. Experimental based testing has been widely used as a means to analyze individual elements and the effects of concrete strength under axial and lateral loading. While this is a method that produces real life response, it is extremely time consuming, and the use of materials can be quite costly. In recent years, however, the use of finite element analysis has increased due to progressing knowledge and capabilities of computer software and hardware. The use of finite element analysis to study these components has also been used widely. Unfortunately, early attempts to accomplish this were also very time consuming and infeasible using existing software and hardware. It has now become the choice method to analyze concrete structural components.

The use of computer software to model concrete structural components is much faster, and extremely cost-effective. To fully understand the capabilities of finite element computer software, data obtained from a finite element analysis package is not useful unless the necessary steps are taken to understand what is happening within the model that was created using the software. Also, executing the necessary checks along the process was the key to make sure that what is being output by the computer software is valid. By understanding the use of finite element packages, more efficient and better analyses can be made to fully understand the response of individual structural components and their contribution to a structure as a whole [2].

The objectives of this research are to study the behavior of columns that carried axial loads and lateral loads and to predict crack pattern using non-linear finite element method called ATENA V5. There were two specimens, which have been tested experimentally by Munawir, et.al, 2012. There were two variations, namely three arms and crossties of stirrup reinforcement with axial load $0.3P_0$ where P_0 was lateral load. A shear behaviour of reinforced concrete columns predicted using a commercial finite element analysis package ATENA V.5. Analytical result compared with experimental result.

2 Methodology

This research used finite element software, namely ATENA V.5 which has license number No. 250–2016 for WR 556 specifically for analyzing static loads and assisted by other software such as Autocad 2007 for drawing needs and for obtaining points as input in the ATENA V5 and Microsoft Excel 2013 software for processing and computation of data.

2.1 Detail Specimens

The object that was used as a model was the research specimen by Munawir, 2012 [6], namely with a reinforced concrete column specimen with a square cross section of 20×20 cm^2 height 58 cm using 12D11.6 mm reinforcement. Concrete compression strength f'c was 24.19 MPa. The stirrup reinforcement used is D5.4 mm. Detail Column specimens showed in Fig. 1. The configuration of the stirrup reinforcement used were three-arm stirrups and crossties as shown in Fig. 2.

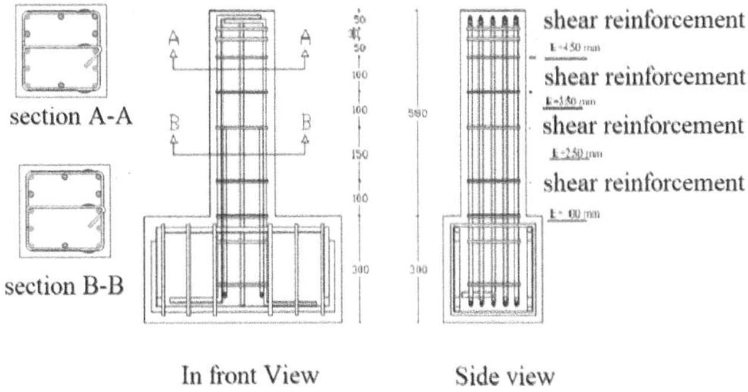

Fig. 1. Detail column specimens

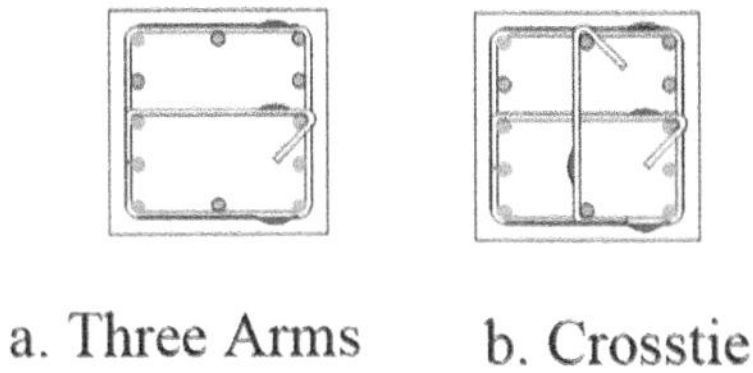

Fig. 2. Shear reinforcement configuration

2.2 Constitutive Model for Concrete

Concrete is a non-linear material even under service load conditions. The constitutive concrete model in ATENA V.5 was taken from the SBeta constitutive model developed by program maker Al Faridi [4]. According to Cervenka et al. [5] the non-linear behaviour of concrete at biaxial stress is described by the effective stress σ_{cef} and the equivalent uniaxial strain ε_{eq}. Equivelent uniaxial strain was introduced to eliminate the Poisson effect in the stress statement. The equivalent uniaxial strain can be thought of as the strain that will be produced by the stress σ_{ci} in the uniaxial test with the modulus E_{ci} in the i direction. The equivalent uniaxial stress-strain diagram for concrete can be seen in Fig. 3. The default values for concrete material parameters can be seen in Table 1.

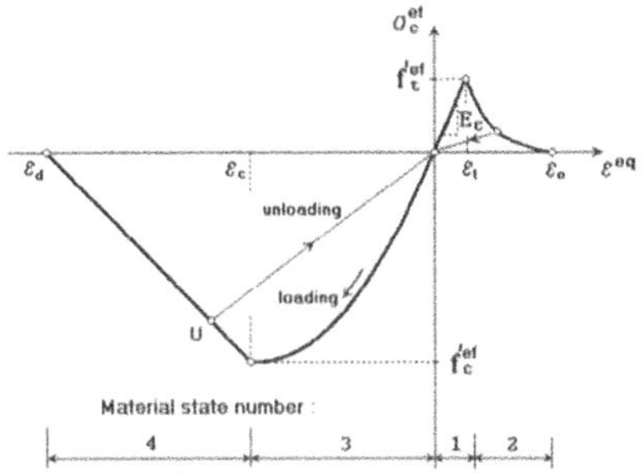

Fig. 3. The law of uniaxial stress-strain in concrete [5]

Table 1. Parameters of sbeta constitutive model [5].

Parameters	Formula
Concrete cylinder compression strength	$f_c' = -0,85 f_{cu}'$
Tensile strength	$f_t' = -0,24 f_{cu}'^{\frac{2}{3}}$
Initial modulus elasticity	$EC = (6000 - 15,5 f_{cu}')\sqrt{f_{cu}}$
Poisson ratio	$\upsilon = 0,2$
Softening compression	$w_d = 0,0005$ mm
Tension softening type	1–eksponensial, based on G_f
Compressive strength of cracked concrete	$c = 0,8$
Tension stiffening stress	$\sigma_{st} = 0$
Shear retention factor	Variable
Tension-compression failure function type	Linier
Fracture energy G_f based on Vos (1983)	$G_f = 0,000025 f_t'^{ef}$ [MN/m]
Strain localization orientation factor	$\gamma_{max} = 1,5$
The unit used was MPa	

2.3 Constitutive Model for Reinforcement

Reinforcement can be modeled in two ways, namely discrete and smeared. Discrete reinforcements were in the form of reinforcing bars and were modeled with truss elements. Smeared reinforcement was a composite material and can be considered as a single material (only one constituent) in an element or one or more constituents. Cervenka et al. [5]. Faridi [4] stated that the stress-strain relationship for reinforcing steel can be modeled in two ways, namely the law of bilinier and multi-linear law.

2.4 Bilinear Law

The biliary stress-strain relationship can be seen in Fig. 4 (a). According to Cervenka et al. [5] the elastic part has the modulus of elasticity of E_s, while the next part was the plastic condition with the slope of the curve is the hardening modulus E_{sh}. For perfectly plastic conditions, the value of $E_{sh} = 0$.

2.5 Multi Linear Law

According to Cervenka et al. [5] the multi-linear law consists of four lines, as shown in Fig. 4 (b). This law represents four parts of the steel's behavior, namely elastic, yield, strain hardening, and fracture. The four lines are defined through four predetermined points.

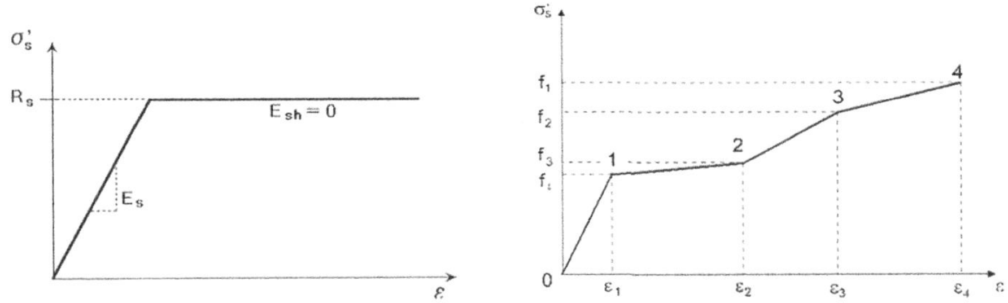

Bilinear (b) Multi-Linear

Fig. 4. Stress-strain relationship for reinforcement [5]

2.6 Newton-Raphson Iteration

In ATENA V.5, the concept of a solution to non-linear equations was using the Newton-Rhapson Method. Figure 5 below shows the method.

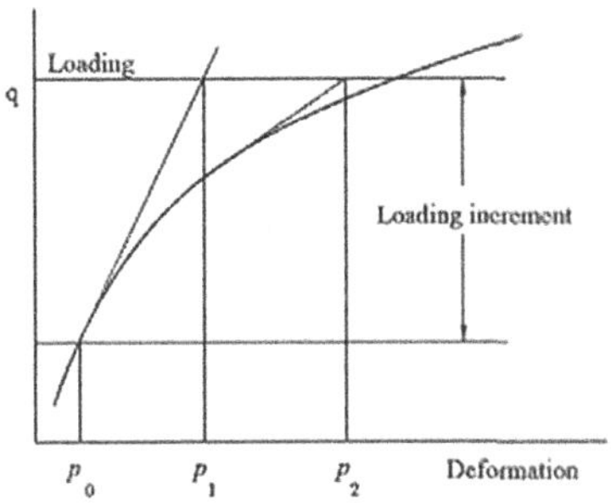

Fig. 5. Newton Rhaphson method

2.7 Running Program

After the column has been modeled, the next step was running the program (non-linear analysis). This analysis was carried out with the ATENA 3D (RUN) program through ATENA 3D which can be accessed by selecting the button on the menu bar which was located in the upper right corner of the screen.

2.8 Analysis Results

The results of the non-linear analysis of the ATENA program consist of numerical data and graphical data. Numerical data was in the form of structural response results monitored by the monitoring point, while graphic data was in the form of crack patterns, stresses, strains, and so on. To read data from the analysis of ATENA 3D (RUN), it can be opened it by selecting the button on the menu bar located in the upper right corner of the screen.

2.9 Constitutive Model for Concrete Confined by Stirrups

Confined and unfinished concrete have different strengths, Concrete with good confinement, the higher lateral stress that was transmitted to the concrete withstands axial loads, so that the strength of the concrete increases. Increasing the compressive strength of the concrete for each confinement requires analysis by calculating f'$_{cc}$ on confined concrete which refers to the analysis of (Mander et al., 1988) [2]. Figure 6 shows constitutive model for Concrete under Compression load [3]. Based on research conducted by (Saidi et al., 2018) [7] regarding the calculation of f'$_{cc}$ in a column with a three-arm configuration and crossties, the compressive strength f'$_{cc}$ was produced as shown in Table 2.

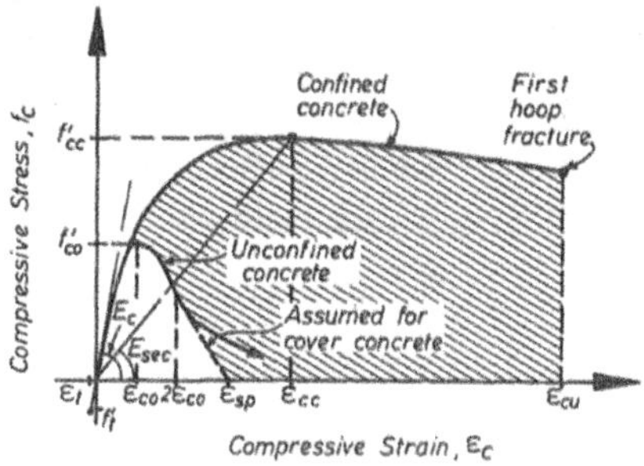

Fig. 6. Constitutive model for concrete under compression load [3]

Table 2. Strength of concrete with restraint.

Configuration	f_c'[Mpa]	f_{cc} [Mpa]	Ratio f_{cc}/f_c'
S_1P_2	24.19	30.24	1.25
S_2P_2	24.19	31.21	1.29

Note:
S_1 = Three arm shear reinforcement
S_2 = Crosstie shear reinforcement

3 Result and Discussion

3.1 Load and Deflection Relationship

Based on numerical analysis of ATENA V5, the maximum deflection of columns with a configuration of three-arm stirrups and crossties loaded with constant axial pressure can be seen in Table 3. Figures 7 and 8 shows the load and deflection relationship for specimen S_1 and S_2.

Table 3. Comparisson ultimate load and deflectiont.

Specimen	Ultimate load [kN]		Difference [%]	Ultimate deflection [mm]		Difference [%]
	ATENA V5	Experiment		ATENA V5	Experiment	
S_1P_2	150	146.95	2.08	19.36	17.22	12.43
S_2P_2	150	144.40	3.88	12.14	15.81	23.21

Note:
S_1 = Three arm shear reinforcement
S_2 = Crosstie shear reinforcement

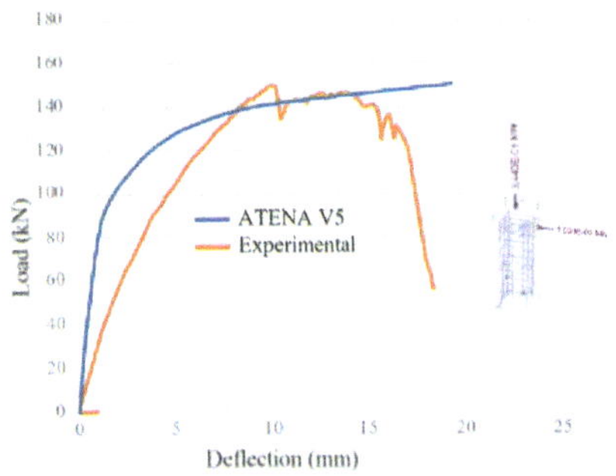

Fig. 7. Load deflection relationship for three arms shear reinforcement

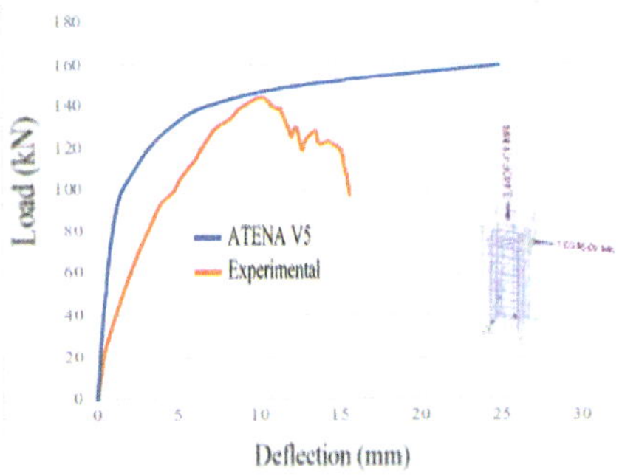

Fig. 8. Load deflection relationship for Crosstie shear reinforcement

3.2 The Initial Crack

Based on numerical analysis of ATENA V5, the first crack of the column which was subjected to constant axial pressure had cracked especially after the shear load exceeded 51 kN. The results of numerical analysis of ATENA V5 show a value close to the experimental results where the first crack of the column under constant axial load has cracked, especially after the shear load exceeds 68.67 kN. Complete results regarding the initial cracks that occur in the column can be seen in Table 4.

Table 4. Comparison of shear and deflection forces in the initial crack.

Specimen	Shear force (kN)		Deflection (mm)	
	ATENA V5	Experiment	ATENA V5	Experiment
S_1P_2	46	58,87	0,59	2,23
S_2P_2	40	49,05	0,42	1,62

Note:
S_1 = Three arm shear reinforcement
S_2 = Crosstie shear reinforcement

3.3 Failure Pattern

Cracks in the column will continue to appear as the loading increases. The numerical model crack pattern has similarities with the experimental result. The experimental results showed that the column was failed in shear, which is indicated by the increase in the constant axial load acting as the shear angle that was formed decreases with a more ductile column collapse pattern which is characterized by more crack growth on the side of the column that experiences tension. In Table 4, it can be seen that the results of the numerical analysis of the column experienced collapse to shear where the ratio of the shear capacity of the analysis results to the theoretical capacity were all greater than one. Comparison of experimental column failure patterns and numerical analysis can be seen in Figs. 9 and 10.

Table 5. Comparison of the results of theoretical calculations with the results of the ATENA V5

Specimen	Comparison shear strength results [kN]			
	Theoretical	ATENA V5	ATENA V5/theoretical	The type of collapse
S_1P_2	110,09	150	1,36	Shear
S_2P_2	110,09	150	1,36	Shear

Note:
S_1 = Three arm shear reinforcement
S_2 = Crosstie shear reinforcement

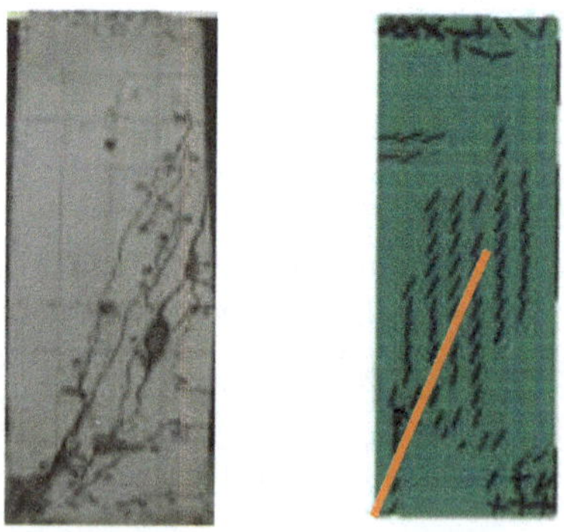

EXPERIMENT ATENA V5

Fig. 9. Comparison of experimental and numerical crack pattern with three arm

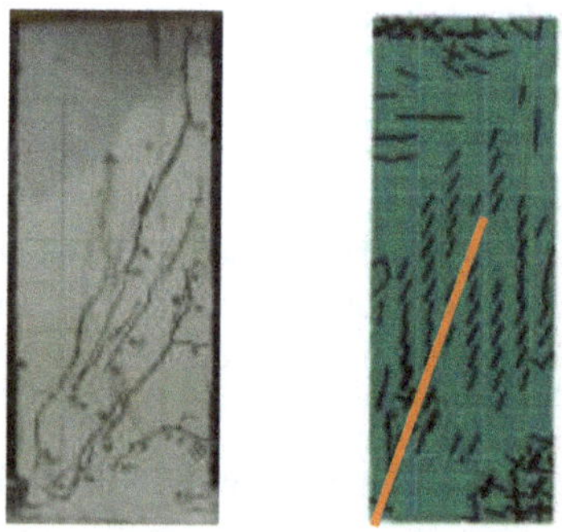

EXPERIMENT ATENA V5

Fig. 10. Comparison of experimental and numerical crack pattern with crosstie

Figures 11 and 12 show stress in Three arm shear reinforcement and Crosstie shear reinforcement, respectively. In the Figure, the circled area receives the greatest shear force so that the maximum stress in that area. It can be seen that stress maximum for Three arm was 628.1 MPa, and 663.2 MPa for Crosstie, respectively.

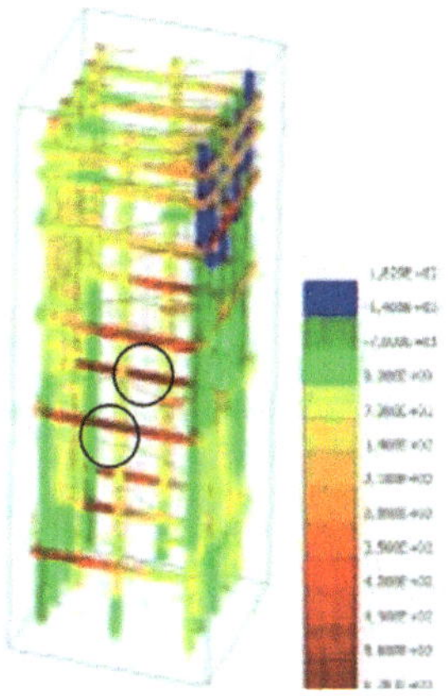

Fig. 11. Stress reinforcement with three arm shear reinforcement

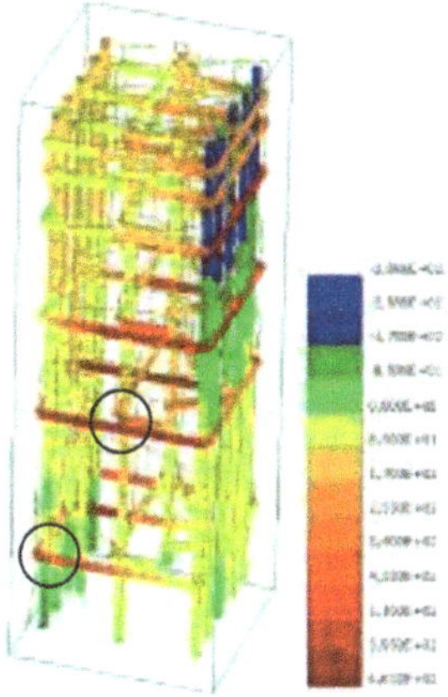

Fig. 12. Stress reinforcement with crosstie shear reinforcement

4 Discussion

Numerical analysis of ATENA V5 gives quite good results. Comparison of the results of numerical analysis with experimental shows a considerable difference between the two types of analysis, namely the initial crack and the load-deflection curve. This is due to differences in constitutive models and laboratory behavior with the modeling of materials used in the analysis both in modeling concrete and steel materials. At the time of loading, the results of the numerical analysis of ATENA V5 stop after the maximum crack occurs, while in the laboratory the ability to support loads will decrease because the concrete elements have experienced cracks.

5 Conclusions

1. Ultimate load for Three arm shear reinforcement and Crosstie shear reinforcement based on experimental result were 146.95 kN and 144.40 kN, respectively. On the other hand, analytical results got 150 kN for both three arm and crosstie. The difference were 2.08% and 3.88% for three arm and cross tie, respectively.
2. Ultimate deflection, experimental result give 17.22 mm and 15.81 mm for three arm and crosstie, repectively. Analytical result give 19.36 mm and 12.14 mm for three arm and crosstie, respectively. The difference were 12.43% and 23.21% for three arm and cross tie, respectively.
3. Maximum stress in Three arm was 628.1 MPa, and 663.2 MPa for Crosstie, respectively.
4. Crack pattern and Failure pattern predicted by finite element give same result as an experimental result. Failure pattern was shear failure.

References

1. ACI C 318: Building code requirements for structural concrete(ACI 318S-05) and commentary (ACI 318SR-05). ACI Struct. J. (2005)
2. Mander, J.B., Priestley, M.J., Park, R.: Theoretical stress-strain model for confined concrete. J. Struct. Eng. (United State) (1988)
3. Paulay, T., Priestly, M.J.N.: Seismic Design of Reinforced Concrete and Masonry Buildings (1992)
4. Al Faridi, S., Analisis, M.P.: Non-Linier ElemenHingga dengan ATENA 2.1.8 Demo. Universitas Gadjah Mada, Yogyakarta (2010)
5. Cervenka, V., Jendele, L., Cervenka, J.: ATENA Program Documentation Part 1-Theory, Cervenka Consulting s.r.o. Prague (2016)
6. Munawir, M., Saidi, T., Putra, R.: Pengaruh Konfigurasi Tulangan Sengkang Terhadap Kapasitas Geser dan Pola Kehancuran Kolom yang dibebani Tekanan Aksial (2012)
7. Saidi, T., Putra, R., Amalia, Z., Munawir, M.: Contribution of transverse reinforcement configuration on concrete shear capacity of RC column. Aceh. Int. J. Sci. Technol. **7**(2), 77–84 (2018)

Coal Ash Utilization as a Filler in Flexible Pavement Construction

Fitrika Mita Suryani(✉), Irvandi Yusuf, Hanif Aida, and Afnan Farhan

Civil Engineering Department, Engineering Faculty, Universitas Syiah Kuala, Darussalam, Banda Aceh 23111, Indonesia
fitrika_mitasuryani@unsyiah.ac.id

Abstract. Coal ash, a fine, rounded and pozzolanic aggregate, is a waste product of coal combustion from steam power plant. The ash is an inorganic material undergoing mineral changes from the combustion process. It is a type of fly ash material with fine grain size and greyish color. The utilization of coal ash as a filler is expected to improve the quality of asphalt concrete mixture. The purpose of this study was to evaluate the performance of asphalt-concrete mixture using coal ash in dense-graded and gap-graded gradations. These gradations were used due to their higher amount of filler and fine particles than uniform and open-graded gradations. The study results revealed that the performance of the mixture using coal ash filler met the specifications of asphalt concrete mixture. In addition, this result indicated that the mixture would be able to withstand heavy traffic loads. Thus, the utilization of coal ash was recommended as a filler in asphalt-concrete mixtures.

Keywords: Coal ash · Filler · Flexible pavement construction

1 Introduction

A steady road infrastructure is required to maintain traffic ease, comfort and safety. Asphalt concrete mixture consists of coarse aggregate, fine aggregate, filler and binder in the form of hot mix asphalt. The relatively high asphalt content in the mixture aims to increase flexibility, durability, and resistance to heat. Pavement using dense-graded and gap-graded gradation is suitable to apply in tropical climates area with relatively high temperature due to its high flexibility and resistance to heat.

Most of the materials in asphalt composition are natural products, such as rock ash or fly ash, which is limited in number, thus an alternative substitute material is necessary. One of the substitute materials is in the form of waste such as coal ash. Several studies have been carried out in an order to obtain more economical substitute materials, and it is found that the quality of lightweight concrete with a substitution of fly ash as a filler has fulfilled the compressive strength requirements for the general specification criteria. A study by Suryani in 2021 revealed that fly ash utilization in the subbase layer of lightweight concrete construction does not significantly affect the mixture strength [1]. The results of previous studies suggest a cheaper and more efficient alternative material in road pavement construction.

Akhyar et al. (Eds.): ICECME 2021, LNME, pp. 215–223, 2023.
https://doi.org/10.1007/978-981-19-3629-6_23

The process of coal combustion produces ash in the form of fly ash and bottom ash. Fly ash is an inorganic, fine grain size material undergoing mineral changes from the combustion process. Its amount increases over time, and may be a potential risk to cause problems for the environment as a pollutant [2]. Accordingly, the utilization of this material as a road construction material can reduce the negative impact on the environment. Furthermore, the coal ash utilization is an alternative way in handling the waste material resulting from the combustion of industrial works that requires high cost.

Coal ash is suitable for use in the mixture of foundation layer of road pavement construction, as the gradation and fineness of coal ash particle meet the grading requirements for mineral fillers [2]. Compared to using Portland cement filler, the use of coal ash filler in the AC-WC surface layer does not significantly affect the mixture strength [3]. Based on this background, this research was conducted to evaluate the performance of asphalt-concrete mixture using coal ash in dense-graded and gap-graded gradation.

2 Literature Review

Filler. According to Tahir, the filler added consists of stone dust that must be dry, free from lumps, non-plastic, and pass the sieve No. 200 [4] and not less than 75% of which pass the sieve No. 30 [5]. The material used as filler in this study is coal ash.

Fly Ash. Aggregate, based on the processing technique, is divided into natural aggregates, first-processed aggregates and artificial aggregates. Artificial aggregate is a mineral filler (particles with sized less than 0.075 mm) obtained from the by-products of cement factories and stone crushing machines. Fly ash is a small greyish material with fine grains and can be obtained from the residue of coal burning [6]. Qualified fly ash can increase the long-term concrete strength [7]. The fine grains of coal ash pass the sieve No. 325 (45 millimicrons) 5–7% in specific gravity between 2.15–2.8 gr/cm3 [6]. According to the ASTM C 618–05, fly ash is classified into:

- Class C (CaO component > 10%) is produced from coal burning with a carbon content of ±60%.
- Class F (CaO component < 10%) is produced from burning anthracite coal at 1560 °C.
- Class N (natural pozolan or combustion products classified into diatomic soil, opaline chertz and shales, tuff and volcanic ash), is processed with or without burning [8].

Coal Ash. According to SNI 03-6414-2002, the coal ash is a fine, rounded and pozzolanic aggregate which is a waste resulting from coal burning in a steam power plant furnace [9]. It is a material that has a fine grain size, grey in color and is obtained from the coal combustion. In addition, it contains high concentration of silica (SiO2), alumina (Al2O3), phosphorus oxide (P2O5) and ferrous oxide (Fe2O3) and calcium oxide (CaO), magnesium oxide (MgO), titanium oxide (TiO2), alkaline (Na2O and K2O), sulfur trioxide (SO3), and carbon [10].

Asphalt Concrete Layer. Saodang (2005) stated that the asphalt concrete layer is also known as AC (Asphaltic Concrete) is a surface layer consisting of a mixture of hard asphalt and aggregate which is continuously graded, mixed, spread and compacted, in a hot state at a certain temperature. The characteristics of laston are waterproof, durable, structural valuable with an asphalt content of 4–7% by weight and can be used for light, medium to heavy traffic [11]. Pavements are expected to be adequately strong and durable for their design life. They consist of different layers, more so in the case of asphalt pavements than concrete ones. From the bottom up, these layers are known as the subgrade, subbase, base and binder and/or surface (Fig. 1) [12].

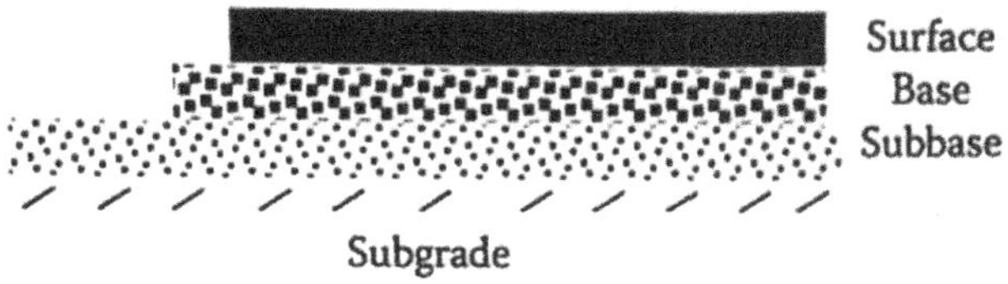

Fig. 1. Flexible pavement structural design [12].

Design Gradation. Aggregate gradation gives the percentage of each of the sizes in a blend. It is typically expressed as the percentage of the aggregate blend passing sieves with standard openings. The size distribution of aggregate particles is directly related to the performance of the pavement layers. In general, aggregate size distributions are classified as gap-graded, uniform, dense-graded or open-graded [13].

Dense-Graded Gradation. Aggregate with this gradation has particle size distributed evenly in each size range. It is also known as well-graded gradation. Dense-graded gradation is accessible, has minimum void and high stability. The stability is determined based on the biggest particle size contained [14].

Gap-Graded Gradation. Aggregate with this gradation has particles distributed unevenly. It may not contain one of the particle size range, leaving some parts with gaps unfilled [14]. Moreover, this gradation has poor interparticle contact, which leads to low density and stability [11].

Asphalt Concrete Mixture Parameters. According to Sukirman [14], the most widely used mixture design in Indonesia is the mixed design method based on empirical testing, using the Marshall test. This test consists of several parameters, which are defined below.

Stability. Stability is the ability of the road pavement to accept traffic loads without permanent changes in shape such as waves, grooves, or bleeding. The need for stability is proportional to the function of the road, and the traffic load to be served. Roads serving high traffic volumes which predominantly consist of heavy vehicles require pavement with high stability. On the other hand, pavements serving light vehicle traffic certainly do not need to have high stability [14].

Flow. Flow is a state of change in the shape of an asphalt mixture that occurs due to a load to the point of collapse [14].

Density. Density is the comparison of the dry weight of a specimen to the weight of water at the same volume [14].

Marshall Quotient. The Marshall quotient (MQ) is the quotient of stability and flow. The higher the MQ is, the higher the stiffness of a mixture is and the more susceptible the mixture is to cracking [14].

Voids in Mix (VIM). VIM is assessed when the coating becomes watertight to prevent air leak into the mixture causing the asphalt to become brittle [14].

Voids in Mineral Aggregate (VMA). VMA is an assessment whether the asphalt blanket can be made thick [14].

Void Filled by Asphalt (VFA). VFA is the ratio between the voids filled by asphalt and the specimen volume [14].

3 Materials and Methods

Materials. The materials used in this research are asphalt, aggregate and coal ash. The asphalt used is Shell Bitumen 60/70 penetration. The aggregate used is crushed stone obtained from Aceh Besar area. The coal ash (Fig. 2) used was coal ash passing sieve No. 200 and was obtained from the steam power plant (Fig. 3).

Fig. 2. Coal ash.

Fig. 3. Steam power plant.

The equipment used is those required to assess the physical properties of aggregates, asphalt and Marshall test. Equipment used in specimen manufacture for Marshall test consisted of a compacting mold with its base plate and a compactor hammer, while the equipment required in performing Marshall test involved a flow meter, a heating bath equipped with temperature regulation, and a Marshall test apparatus.

Methods. The first step in this research is to prepare the materials and equipment needed, as well as to examine the physical properties of aggregates, asphalt and sieve analysis. Gradations used in this study included dense-graded and gap-graded gradation. The gradations used are explained in Table 1 and Table 2. Specimen manufacture is proceeded after obtaining the results of physical properties analysis in accordance with the specifications.

Table 1. Dense-graded gradation.

Sieve size		Asphalt concrete – wearing course				
		% Passing			% Retained	
Sieve no	Size [mm]	Lower limit	Upper limit	Aggregate gradation	Retained	Cumulative
¾″	19	100	100	100.0		
½″	12.5	90	100	95.0	5	5
3/8″	9.5	77	90	83.5	11.5	16.5
No. 4	4.75	53	69	61.0	22.5	39
No. 8	2.36	33	53	43.0	18	57
No. 16	1.18	21	40	30.5	12.5	69.5
No. 30	0.6	14	30	22	8.5	78
No. 50	0.3	9	22	15.5	6.5	84.5
No. 150	0.15	6	15	10.5	5	89.5
No. 200	0.075	4	9	6.5	4.0	93.5
Filler	0	0	0	0	6.5	100

Table 2. Gap-graded gradation.

Sieve size		Asphalt concrete – wearing course				
		% Passing			% Retained	
Sieve no	Size [mm]	Lower limit	Upper limit	Aggregate gradation	Retained	Cumulative
¾″	19	100	100	100.0		
½″	12.5	90	100	94.0	6	6
3/8″	9.5	65	90	70.5	23.5	29.5
No. 8	2.36	35	55	58.5	12.0	41.5
No. 30	0.6	15	35	30.0	28.5	70.0
No. 200	0.075	2	9	8.0	22.0	92.0
Filler	0	0	0	0	8	100

Determination of Asphalt Content Variation. The variation is determined based on the initial estimated value of the middle asphalt content (Pb). The calculation of the estimated value of the middle asphalt content (Pb) used the following equation was based on design gradation producing composition of each fraction presented in the Table 3 below.

Table 3. Composition percentage of aggregate fraction in each gradation tested.

Composition	Percentage [%]	
	Dense-graded	Gap-graded
Coarse Aggregate (CA)	38.8	44.0
Fine Aggregate (CA)	54.7	48.0
Filler	6.5	8.0

The calculation of Pb used the following equation [14].

$$Pb = 0.035(\% \text{ CA}) + 0.045(\% \text{ FA}) + 0.18(\% \text{ Filler}) + k \tag{1}$$

Pb: middle asphalt content, percentage of total mixture weight.
CA: % aggregate retained from sieve No. 8
FA: % aggregate passing sieve No. 8 and retained from sieve No. 200.
Filler: % aggregate (not less than 75%) passing sieve No. 200.
Using the percentage given in Table 3, Pb calculation is as follows.

a) Dense-graded gradation
$Pb = 0.035\ (\%\ CA) + 0.045\ (\%\ FA) + 0.18\ (\%\ Filler) + k$
$Pb = (0.035 \times 38.8\%) + (0.045 \times 54.7\%) + (0.18 \times 6.5\%) + 0.5$
$Pb = 5.5\%$

b) Gap-graded gradation
$Pb = 0.035\ (\%\ CA) + 0.045\ (\%\ FA) + 0.18\ (\%\ Filler) + k$
$Pb = (0.035 \times 44\%) + (0.045 \times 48\%) + (0.18\ x\ 8\%) + 0.85$
$Pb = 5{,}99 \approx 6{,}0\%$

Mix design was then carried out based on the results of the mixture gradation distribution using asphalt content of Pb – 1%, Pb – 0.5%, Pb%, Pb + 0.5% and Pb + 1% of the mixture total weight (asphalt and constituent aggregate). Thus, the asphalt content variation of the specimens for each gradation is presented in Table 4.

Table 4. Mixture gradation distribution for mix design.

Mixture gradation distribution	Asphalt content [%]	
	Dense-graded	Gap-graded
Pb – 1%	4.5	5
Pb – 0.5%	5	5.5
Pb	5.5	6
Pb + 0.5%	6	6.5
Pb + 1%	6.5	7

Table 5. The lists of specimen manufacture to determine optimum asphalt content between gradation variations in Marshall test.

Dense-graded		Gap-graded	
Asphalt Content [%]	Quantity [pcs]	Asphalt Content [%]	Quantity [pcs]
4.5	3	5	3
5	3	5.5	3
5.5	3	6	3
6	3	6.5	3
6.5	3	7	3

Table 6. Marshall test results of mixture using dense-graded and gap-graded gradation.

Mixture characteristic	Asphalt content with variation of gradations		General specifications (2014)
	Dense-graded gradation	Gap-graded gradation	
Density [t/m^3]	2.43	2.45	–
VIM [%]	4.5	5.3	4–6
VMA [%]	18.3	18.1	>18
VFA [%]	75	72	>68
Stability [kg]	1090	1790	>800
Flow [mm]	3.40	5.7	>3
MQ [kg/mm]	340	300	>250

Specimen Manufacture. After assessing the material physical properties, three specimens of each asphalt content category were made using Portland cement filler. The specimens obtained were subsequently tested using the Marshall test to obtain the optimum asphalt content. For more details, see the following Table 5.

Optimum Asphalt Content Determination. After conducting Marshall test, the parameter values were obtained. Next, the relationship between asphalt content variation and each parameter was described in a graphic. After that, we determined the best asphalt content corridor from the parameter using range overlapping charts. In this study, the optimum asphalt content was the middle value of the asphalt content corridor.

Determining Asphalt Concrete Mixture Characteristic on Optimum asphalt Content. The minimum asphalt content obtained was used to assess asphalt concrete mixture characteristic. Each parameter value of the mixture using optimum asphalt content was obtained from the graphic describing the relationship asphalt content variations and each parameter. The value was then referred to the general specifications, to determine whether it met the specifications or not.

4 Results

The Marshall test results of the specimens were as follows (Table 6).

Table 4 shows the parameter value of each gradation tested. Based on the table, it was revealed that all parameters from both gradations met the specifications value which were presented at the table. This result indicated that coal ash, in both gradations, did not affect the mixture characteristics. The presence of silica (SiO2) in coal ash is important in producing a strong bond between aggregates and asphalt when being mixed with water.

The result was considered noteworthy because the gradations used included dense-graded and gap-graded gradations. Gap-graded gradation, as mentioned in the literature review, has particles distributed unevenly and some gaps unfilled [14]. Moreover, this gradation has poor interparticle contact, which leads to low density and stability [11]. In spite of this, the substitution of coal ash as filler in the gap-graded gradation did not decrease the stability of the mixture.

Therefore, it revealed that the method of utilizing coal ash as a substitution in the mixture for flexible pavement construction was applicable. This practice also indicated a potential way in reducing the cost required in flexible pavement construction.

5 Conclusions

Based on the results, all values of parameters from both gradations met the specifications, thus the utilization of coal ash was recommended as a filler in asphalt-concrete mixtures due to its effectivity in filling the voids of the mixture. The stability of mixture using coal ash filler in dense-graded and gap-graded gradation met the specifications. Therefore, this study revealed that the utilization of coal ash as a filler in flexible pavement construction was applicable. This practice also indicated that the mixture would be able to withstand heavy traffic loads and became a potential way in reducing the cost required in flexible pavement construction.

References

1. Suryani, F.M., Rizal, S., Abdullah, M.I.: Fly ash utilization in rigid pavement construction. In: 2nd International Conference on Experimental and Computational Mechanics in Engineering (ICECME) 2020, Banda Aceh, 13–14 Oct 2020. Springer (2021). ISSN 2195-4356
2. Indriyati, T.S., Malik, A., Alwinda, Y.: Kajian Pengaruh Pemanfaatan Limbah FABA (Fly Ash dan Bottom Ash) pada Konstruksi Lapisan Base Perkerasan Jalan. J. Tek. Unilak. **13**, 112–119 (2019)
3. Sugeha, A.L., Sulandari, E., Suyono, R.S.: Pemanfaatan Limbah Abu Batu Bara sebagai Filler pada Campuran Laston, JeLAST: J. PWK, Laut, Sipil Tambang **5**, 1–11 (2018)
4. Tahir, A.: Karakteristik Campuran Beton Aspal (AC-WC) dengan Menggunakan Variasi Kadar Filler Abu Terbang Batu Bara. J. SMARTek. **7**, 256–278 (2009)
5. Direktorat Pekerjaan Umum, Metode Pengujian Jumlah Bahan Dalam Agregat Yang Lolos Saringan No. 200 (0,075 mm), Badan Standarisasi Nasional, Jakarta (1996)
6. Wardani, S., Prabandiyani, R.: Pemanfaatan Limbah Batubara (Fly Ash) untuk Stabilisasi Tanah Maupun Keperluan Teknik Sipil Lainnya dalam Mengurangi Pencemaran Lingkungan, J. Fak. Tek. Univ. Diponegoro, 1 (2008)
7. Nugraha, P.: Antoni, Teknologi Beton, Penerbit Andi, Yogyakarta (2008)
8. ASTM InternationaL, ASTM C 618-05: Standard Specification for Coal Fly Ash and Raw or Calcined Natural Pozzolan for Use in Concrete, ASTM, West Conshohocken (2005)
9. Direktorat Pekerjaan Umum, Pengertian dan Manfaat Fly Ash (SNI 03-6414-2002), Badan Standarisasi Nasional, Jakarta (2000)
10. Caroles, J.D.: Ekstraksi Silika yang Terkandung dalam Limbah Abu Terbang Batu Bara. Fullerene J. Chem. **4**, 5–7 (2019)
11. Saodang, H.: Konstruksi Jalan Raya: Perancangan Perkerasan Jalan Raya, Penerbit Nova, Bandung (2005)
12. Mallick, R.B., El-Korchi, T.: Pavement Engineering: Principles and Practice. CRC Press, United States (2017)
13. Papagiannakis, A.T., Masad, E.A.: Pavement Design and Materials. Wiley, New Jersey (2007)
14. Sukirman, S.: Beton Aspal Campuran Panas, Grafika Yuana Marga, Bandung (2003)

Open Circuit Type Wind Tunnel Fabrication for Testing Wind Turbine Blade Aerodynamics

Amar Yazid, Ahmad Syuhada(✉), and Razali Thaib

Magister of Mechanical Engineering Program, Department of Mechanical and Industrial Engineering, Faculty of Engineering, Universitas Syiah Kuala, Banda Aceh, Indonesia
amar.y@mhs.unsyiah.ac.id, {ahmadsyuhada, razalithaib}@unsyiah.ac.id

Abstract. Experimental studies of wind turbines require a long time and other obstacles such as weather. There is a need for an assessment system on the effect of the aerodynamic system used in wind turbine blades that is more practical and not affected by natural conditions. The purpose of this study is wind tunnel fabrication to facilitate studies of the characteristics of the turbine blade model that will be developed in the future. Wind tunnels are used with an open system type. The NACA 4412 horizontal axis wind turbine prototype with blade lengths of 70, 90, and 110 mm were tested for the number of blades 3, 4, and 5 at wind speeds between 3–8 m/s, the power generated by the rotor experienced a large change starting at wind speed. 6 m/s. By looking at the resulting blade rotation based on wind potential and speed, getting optimal results at a blade length of 110 mm with 5 blades. At a wind speed of 6 m/s, the rotation is 279.2 rpm, the rotor power is 0.39 W. The wind speed is 8 m/s with a rotation of 526.7 rpm, and the rotor power is 0.64 W.

Keyword: Aerodynamics · Wind tunnel · Open circuit · Wind turbine prototype · Naca 4412 · Blade length

1 Introduction

Aerodynamics, as a branch of science, is growing rapidly like other branches of science. Aerodynamics can be defined as a branch of science that deals with the interaction between fluid flow and solid objects that move relatively in the flow [1]. A wind tunnel is a tool used in aerodynamic research to study the effects of air moving through solid objects. A Wind Tunnel is used to simulate the actual state of an object that is under the influence of aerodynamic forces in the aeronautical field [2]. Wind tunnels are used in various fields, one of which is in the field of engineering. Wind tunnels play a significant role in human life. For example, objects that are often tested in wind tunnels are cars, planes, windmills, and wind turbines. The tests of these objects are from the aerodynamic side [3].

Akhyar et al. (Eds.): ICECME 2021, LNME, pp. 224–236, 2023.
https://doi.org/10.1007/978-981-19-3629-6_24

A wind turbine is a device that converts wind energy into mechanical energy and then converts it into electrical energy through a generator. Wind turbines can be increased inefficient to find the optimal energy coefficient. The ability of the turbine blades to absorb kinetic energy from wind speed is influenced by the layout of the turbine blade installation in the direction of wind speed (angle of attack) [4]. To test the effect of turbine blade angle of attack on craftsmanship Energy absorption has been tried by many researchers such as the Effect of Angle of Attack on the Performance of the Gorlov Helical Wind Turbine with the Addition of Curveplate [3], The Effect of Variation in Angle of Attack and Wind Speed Against Turbine Performance Horizontal Axis Wind [5], Effect Performance Coefficient of Horizontal Axis Wind Turbine Micro Scale Naca 4412 [6], Determination of the Angle of Attack on a Research Wind Turbine Rotor Blade Using Surface Pressure Measurements [7], A numerical study on the angle of attack to the blade of a horizontal-axis offshore floating wind turbine under static and dynamic yawed conditions [8].

From this experimental study, it takes a long time and other obstacles such as weather. For this reason, it is necessary to study a system about the effect of the aerodynamic system used on wind turbine blades that is more practical and not affected by natural conditions. Many studies have been conducted to examine the aerodynamic pattern of the turbine blades' ability to absorb wind energy using a system wind tunnel [9]. Systems Wind tunnel has been carried out, including Development of Simple Wind Tunnel Design [3], Design and Fabrication of Low-Speed Wind Tunnel and Flow Analysis [10], Initial Tunnels for Automotive Testing [11].

From the above background, the researcher will design a system wind tunnel to test the characteristics of the turbine blades on the ability of the wind turbine blades to absorb wind energy. This system wind tunnel is to facilitate studies of the characteristics of the turbine blade model that will be developed in the future.

2 Literature Review

Wind Energy. Is the air that moves because of the difference in pressure on the earth's surface. The wind will move from an area of high pressure to an area of lower pressure.

Moving air has mass, density, and velocity. So that with these factors the wind has kinetic energy and potential energy. However, the velocity aspect dominates the mass position with respect to the earth's surface. So kinetic energy is more dominant than potential energy. Thus, wind energy is kinetic energy or energy caused by wind speed to be used to rotate the wind turbine blades. To utilize wind energy into electrical energy, the first step that must be done is is to calculate wind energy with the formula [12]:

$$E = \frac{1}{2} m.V^2 \tag{1}$$

Where,

- E: Kinetic energy (Joule)
- m: Air mass (kg)

- v: Wind speed (m/s)

To get an air mass, suppose an air block has a cross-section with an area of A (m^2) and moves with a velocity v (m/s), then the mass of air that passes through a place is:

$$\dot{m} = A.v.\rho \tag{2}$$

Where,

- $\dot{m}$: Mass of flowing air (kg/s)
- A: Cross-sectional area (m^2)
- v: Wind speed (m/s)
- ρ: Air density (kg/m^3)

With Eqs. (1) and (2) can be calculated the amount of power generated from wind energy, namely:

$$P = \frac{1}{2}.A.\rho.v^3 \tag{3}$$

Where,

- P: Power is energy per unit time (watt)
- A: Cross-sectional area (m^2)
- v: Wind speed (m/s)
- ρ: Air density (kg/m^3)

Wind Turbin. The wind turbine is a windmill used to generate electricity. Wind turbines are mostly used to accommodate people's electricity needs, using the principle of energy conversion and using renewable natural resources, namely wind [13].

The main component of a wind turbine is an energy converter which converts the kinetic energy contained in the moving air into mechanical energy. Basically, the extract properties of the wind energy converter are irrelevant. Extraction of mechanical energy from the moving air stream with the help of blades or blades, rotating according to the surrounding wind conditions [14].

Airfoil. Airfoil or often referred to as airfoil is an important thing to study in aerodynamics because most of the analysis will revolve around the airfoil. Airfoil is a geometric shape which when placed in a fluid flow will produce lift more than the drag force (drag). Currently, the highly varied geometric shapes already have certain naming standards. One of the most frequently used in the international world is the airfoil standard according to the *National Advisory Committee for Aerodynamics* (NACA). The name or writing code of the airfoil can directly provide the main information about the geometric shape of the airfoil [15] (Fig. 1).

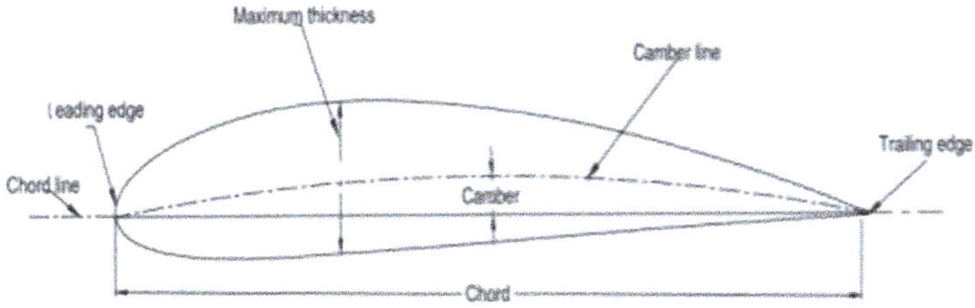

Fig. 1. The geometry of the NACA airfoil series [15]

Description:

- Chord line is a straight line connecting the center of curvature of the leading and trailing edges.
- The leading edge is the front, the shape is in the form of a curve so that the air will flow smoothly.
- Maximum thickness is the maximum thickness measured perpendicular to the chord line. The thickness of the airfoil is expressed as the ratio of the thickness to the chord.
- Camber line, this line is obtained by connecting the midpoints between the top and bottom of the airfoil.
- Camber is the distance maximum camber line as measured from the line chord. Usually expressed as a ratio to chord.
- Chord is the distance from a point on of the curve leading edge to a point on the trailing edge.
- The trailing edge is the point behind the camber line.

The following is an example of a NACA 4412 airfoil (four digits) and an explanation (Fig. 2):

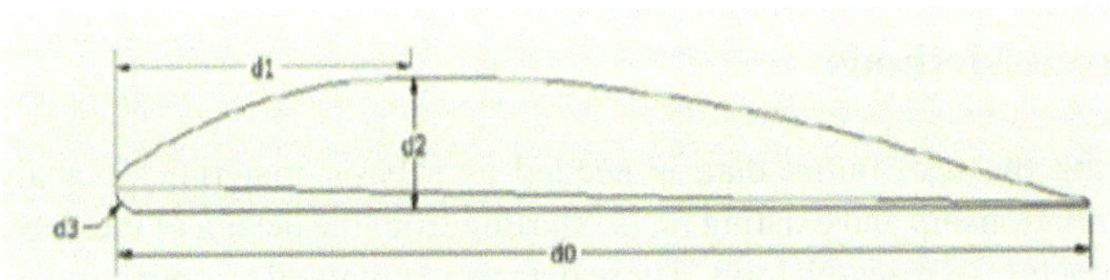

Fig. 2. Airfoil NACA 4412

4 = Shows the maximum camber value in percentage of the *chord* (4% × length *chord*).

4 = Indicates the location of the maximum camber in tenths of the chord (4% × length *chord*).

12 = Number indicating the maximum thickness of the airfoil which is 12% × length chord.

Wind Tunnel. Wind tunnel designed is an open circuit type (Open Return Wind Tunnel), so the main components of wind tunnel this type consists of 5 main components: Settling chamber, Contraction Cone, Test section, Diffuser and drive mechanism (airflow generating system namely blower) (Fig. 3).

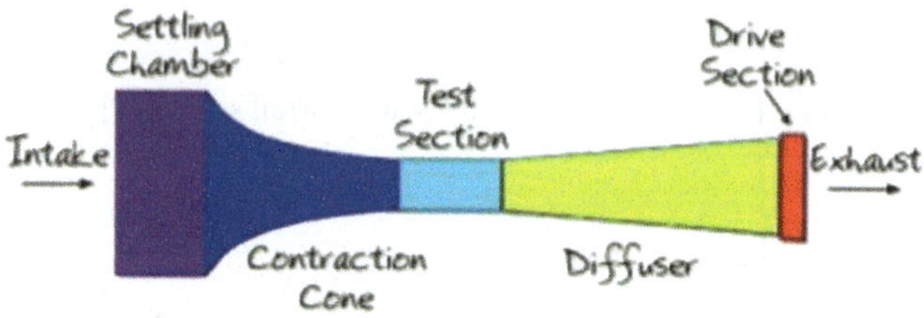

Fig. 3. Open circuit tunnel

- The Settling Chamber is at the very front of the wind tunnel and consists of a filter and honeycomb. The design of the settling chamber aims to obtain a laminar flow type, so for this purpose, the outside air entering the tunnel is passed to the honeycomb system [12].
- The contraction cone serves to take air that has a large volume of low velocity and reduce it to air that has a high velocity of small volume. The smaller the size, contraction cone the higher the air velocity.
- The test section serves as a place to place the model of an object that we want to test. The size of the test chamber must be large enough so that the flow disturbance due to the contraction cone is sufficiently damped before it reaches the object.
- Diffuser serves to slow down the rate of air coming out of the test section before heading out. The slowing down of the air velocity is due to the shape of the diffuser [6].

3 Research Methods

In completing the test, initial data is needed as a basic material for analysis of the theoretical relationship and existing facts. Starting from the design of the drawing design to the manufacture of research tools. These data were obtained by collecting data through literature studies and field studies related to research aspects.

In the literature study, the data were analyzed by conducting a theoretical study related to the test to be carried out. These data are obtained, among others, from research that has been done previously, renewable energy guidebooks, journals, wind turbine articles, and all forms of learning and understanding of related fields, both from print media and electronic media.

The study method using the system wind tunnel is a characteristic test of several elements, namely, data collection to determine the parameters to be tested such as the rotational speed of the wind turbine shaft using NACA 4412 and observing the work performance produced by the wind turbine in the study.

Research Place. Research and data processing were carried out at the Thermal Engineering Laboratory, Mechanical and Industrial Engineering Study Program, Faculty of Engineering, Syiah Kuala University. The research time for the final project is 6 months.

Tool Making. There are several main components that need to be made for this research, including a wind tunnel and a prototype wind turbine blade.

- Blade
  The material used for the prototyping of the blade in this research is PLA (*Polylactic Acid*), with a length of 70, 90, and 110 mm. PLA filament type is the best filament for the manufacture of objects with a small size. To make the shape of the blade using a 3Dmachine printing. Figure 6 is a blade design (Fig. 4).

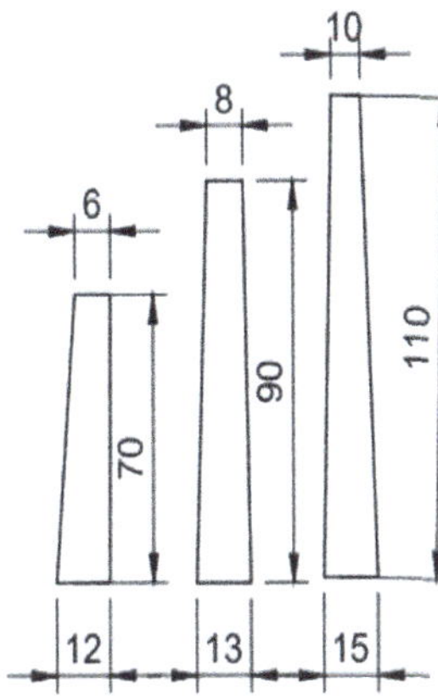

Fig. 4. Blades

The shape of the airfoil is made according to the provisions of NACA 4412, at the base of the blades with chord line lengths of 12, 13, and 15 mm, while at the ends it is 6, 8, and 10 mm (Fig. 5).

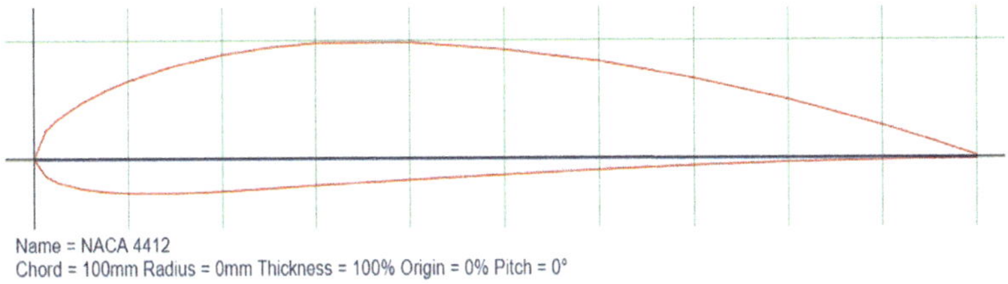

Fig. 5. Base airfoil [16]

At the base the airfoil has a maximum thickness of 2 mm with a chord line of 70 mm, 90 mm, and 110 mm. The size and shape follow the provisions of NACA 4412 which have been scaled according to the standard.
The tip has a maximum thickness of 1.1 mm with the length of chord lines 6, 8, and 10 mm, on the airfoil of the blade from base to tip, the blade is tapered which continues to decrease along the blade from the base to the tip.

Fig. 6. Results of making blades

- Hub

Manufacture of rotor hubs for the number of blades 3, 4, and 5 of PLA material with a diameter of 13 and 7 mm and a distance between blades of 6, 10, and 13 mm. Making holes for the shaft using a hand drill with a diameter of 2 mm (Fig. 7).

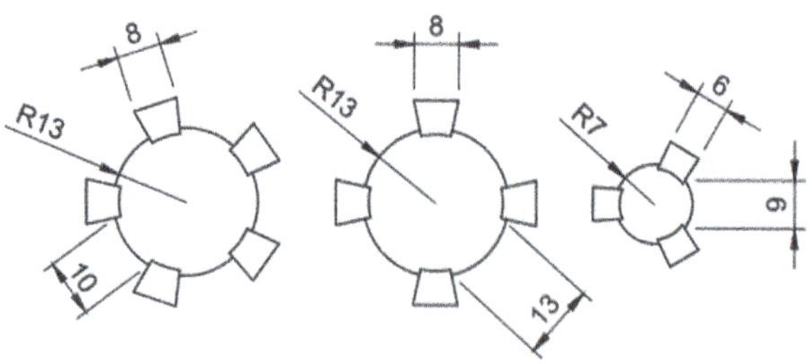

Fig. 7. Hub

Wind Tunnel. In this research wind tunnel, which was designed with an open circuit type (Open Return Wind Tunnel) with 3 mm thick acrylic material, wind tunnel this type of consists of 5 main components, namely, settling chamber, contraction cone, test section, diffuser, and honeycomb. In addition, there is also a driving mechanism (airflow generation system, namely a blower). For more details, can be seen in Fig. 10 below (Fig. 8):

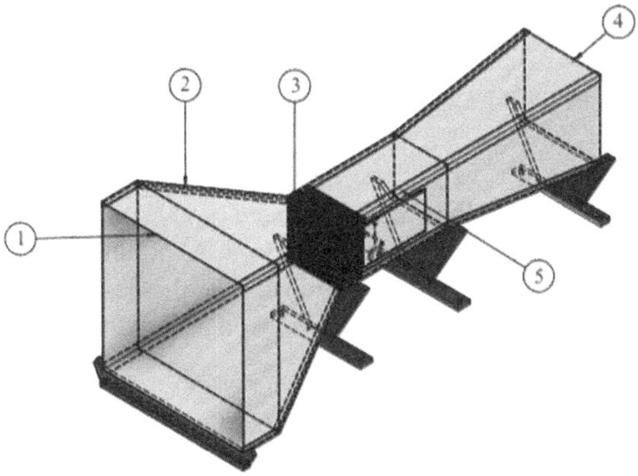

No	Nama
1	Settling Chamber
2	Contraction Cone
3	Honeycomb
4	Diffuser
5	Test Section

Fig. 8. Planning wind tunnel

4 Result and Discussion

In this study, a wind tunnel with an open circuit type system was used. There are 4 main parts of the wind tunnel, namely the settling chamber, contraction cone, test section, and diffuser. Tests were carried out on this tool by taking several points on each part of the wind tunnel, then measuring the wind speed at a predetermined point.

- Point 1 is taken at the settling chamber with dimensions of 800 × 800 mm and a distance of 2200 mm from the blower.
- Point 2 is taken at the contraction cone with dimensions of 400 × 400 mm and a distance of 1700 mm from the blower.
- Point 3 is taken in the test section with a space dimension of 300 × 300 mm and a distance of 1100 from the blower.

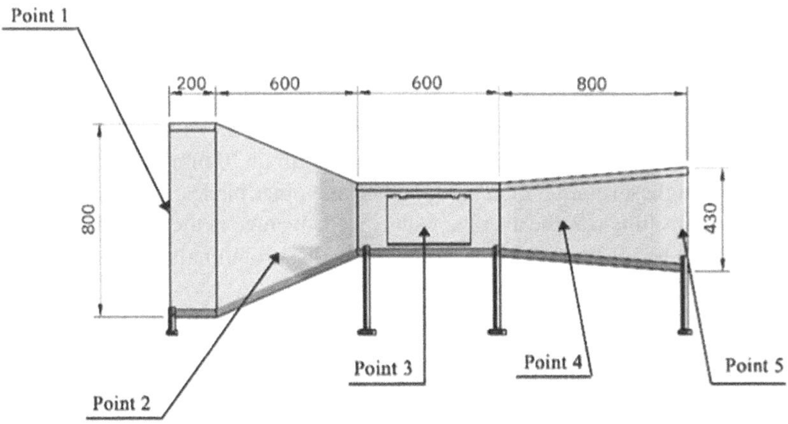

Fig. 9. Wind speed point

- Point 4 is taken at the diffuser with dimensions of 370 × 370 mm and a distance of 500 mm from the blower.
- Point 5 is taken at the diffuser with room dimensions of 430 × 430 mm or the fluid outlet (Fig. 9).

At point 1 the settling chamber, the wind enters at a speed of 1.02 m/s, then enters the section, the contraction cone wind begins to increase in speed. At point 1 of the contraction cone, the wind speed reaches 4.23 m/s. In the test section, the desired wind speed is 3–8 m/s. Then in the section, the diffuser wind experienced a slowdown with a speed of 5.17 m/s (at point 1) and 3.81 m/s (at point 2).

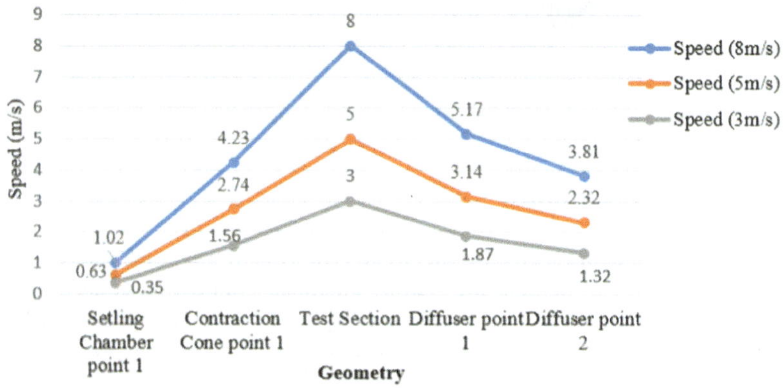

Fig. 10. Graph of velocity against geometry.

Furthermore, in this study used the number of blades 3, 4 and 5 blades. With a length of 70, 90 and 110 mm. Tests carried out on a wind turbine prototype, the parameters measured are wind speed, blade rotation speed and power generated. To measure wind speed, a digital anemometer is used, by pointing the anemometer in the direction of the wind. To measure the rotational speed, a tachometer is used by directing the laser beam contained in the measuring instrument to the wind turbine blades which have been affixed with sticker markers so that the rotation is read by the sensor. To measure current and voltage by determining the direction of the selector switch to be measured then attaching the jack cable to the lamp by paying attention to the (+) and (−) signs (Fig. 11).

In blade 3 it was found that the blade with a length of 70 mm with a wind speed of 4 m/s, got the highest rotation than the length of the other blades, where when the wind speed was 4–6 m/s this was the average wind speed. average in the coastal area of Banda Aceh. When the wind speed reaches 6 m/s, the highest rotation is obtained by 5 blades with a length of 110 mm.

In the 4th blade rotation with a length of 90 mm, it gets the highest initial rotation up to a wind speed of 5 m/s among other blade lengths. However, when the wind speed reaches 6 m/s and above, the highest blade rotation is 5 blades with a length of 110 mm (Tables 1 and 2).

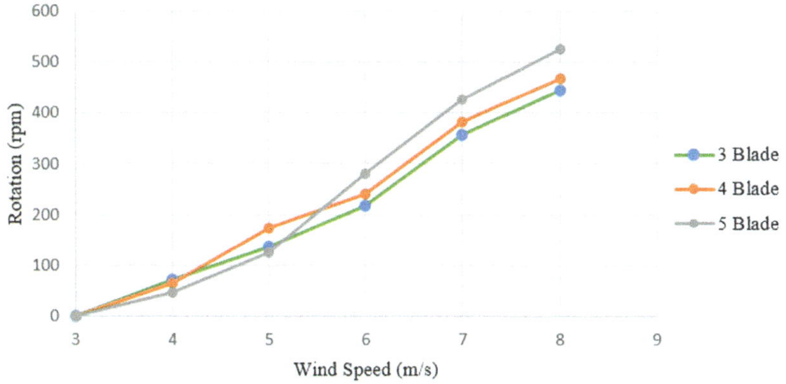

Fig. 11. Graph of wind speed against rotation

Table 1. Data results using 3 blades (NACA 4412)

Wind Speed [V] [m/s]	Blade rotation [n] [rpm]	Voltage [V] [Volt]	Current [I] [A]	Power [P] [W]	Description
3	0	0	0	0	Not on
4	71.6	0	0	0	Not on
5	137.6	1.21	0.2	0.24	Dim
6	218.2	1.38	0.21	0.29	On
7	356.6	1.63	0.23	0.37	On
8	445.4	1.75	0.24	0.42	On

Table 2. Data Results Using 4 Blades (NACA4412)

Wind Speed [V] [m/s]	Blade rotation [n] [rpm]	Voltage [V] [Volt]	Current [I] [A]	Power [P] [W]	Description
3	0	0	0	0	Not on
4	65.7	0	0	0	Not on
5	173.5	1.27	0.21	0.27	Dim
6	240.7	1.42	0.23	0.33	On
7	381.3	1.74	0.26	0.45	On
8	467.2	1.89	0.27	0.51	On (Bright)

Table 3. Data results using 5 blades (NACA4412)

Wind Speed [V]	Blade rotation [n]	Voltage [V]	Current [I]	Power [P]	Description
[m/s]	[rpm]	[Volt]	[A]	[W]	
3	0	0	0	0	Not on
4	45.3	0	0	0	Not on
5	126	1.2	0.19	0.23	Dim
6	279.2	1.56	0.25	0.39	On
7	425	1.87	0.28	0.52	On
8	526.7	2.21	0.29	0.64	On (Bright)

- Turbine Rotor Power

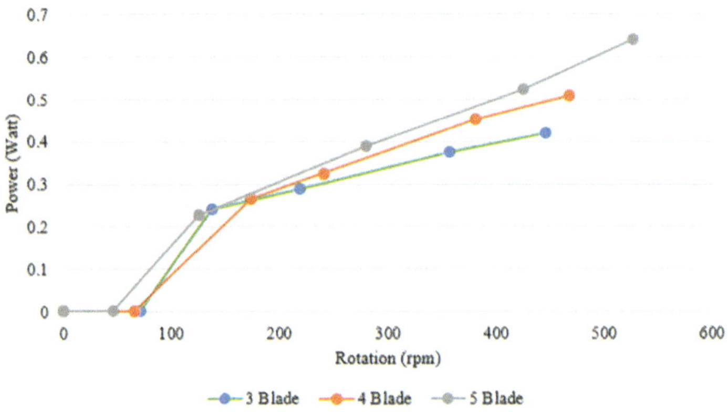

Fig. 12. Rotor power graph of wind turbine prototype

Power generated by the rotor begins to experience a large change at a wind speed of 5 m/s and continues to enlarge the gap in the power difference generated by the rotor from wind energy which is getting higher and higher. It can be seen that 5 blades with a length of 110 mm produce more power than the length and number of other blades with wind speeds above 6 m/s (Table 3).

By using 3 blades with a length of 70 mm, it produces more stable power considering the wind speed in the Banda Aceh area is around 3–6 m/s, but the power produced by the rotor is not too high compared to using 5 blades with a length of 110 mm (Fig. 12).

- Analysis of Result

Results of the research that has been done from collecting sufficient data and selecting data to be analyzed to get the final result that can be taken data on each component of the wind tunnel has a different speed starting from the incoming flow through the steering chamber to the exit through the diffuser. The velocity of the air flowing in each component corresponds to the geometry of the component itself. This relationship is found in the continuity equation which states that the cross-sectional area of 1 times the speed of 1 equals the cross-sectional area of 2 times the speed of 2, so speed 2 is equal to the cross-sectional area of 1 times the speed of 1 divided by the cross-sectional area of 2.

Wind speed 3 to 6 m/ s, which is the average wind for the Banda Aceh area. In the best rotation, 3 and 4 blades with a length of 70 and 90 mm, while with 5 blades with a length of 110 mm the best rotation is with a wind speed of 6 m/s and above. This happens because the heavier mass is owned by the longer blades and the number of blades which affect the rotational speed of the rotor.

This relationship is found in Newton's second law equation where force equals mass times linear acceleration, so linear acceleration is equal to force divided by mass. However, in wind turbines what happens is rotational motion, so the equation, in this case, is torque equal to the moment of inertia times angular acceleration, it is found that angular acceleration is equal to torque divided by moment of inertia. The heavier and larger the blade, the greater the moment of inertia so that a greater torque is required to drive the blade. So, it doesn't mean that bigger and heavier is not good, but faster wind speed is needed to rotate the blades, which in the wind energy equation, the greater the cross-sectional area the greater the power that can be generated.

5 Conclusion

The results of the research analysis that have been carried out get the final results which can be concluded in the following conclusion:

1. With a speed of 8 m/s in the test section, the inlet at point 1 of the settling chamber has a speed of 1.02 m/s, in the contraction cone it has a speed of 4.23 m/s, and at the exit at points 1 and 2 the diffuser has a speed of 5.17 m/s. s and 3.81 m/s.
2. The best rotation is seen from the availability of wind where the wind speed of 3–8 m/s is obtained on 5 blades with a length of 110 mm more optimally with a rotation of 279.2 rpm at a wind speed of 6 m/s and a rotation of 526.7 rpm at a wind speed of 8 m/s.
3. The power generated by the rotor is found that the difference in power between the length and number of blades begins to be seen at wind speeds of 6 m/s and above. Taken from the best rotation of 5 blades with a length of 110 mm, the power obtained is 0.39 W at a wind speed of 6 m/s, and 0.64 W is obtained at a wind speed of 8 m/s.

References

1. Damara, D.: Analisa rotor horizontal axis wind turbine (hawt) dengan variasi geometri dan jumlah bilah kapasitas 10 kw turbine (hawt) with variation geometry and number of blade for capacity 10 kw (2016)

2. Haryanti, M., Awaludin, M.: Rancangan Sensor Kecepatan Angin Pada Wind Tunnel. TESLA: Jurnal Teknik Elektro **21**(1), 44 (2019). https://doi.org/10.24912/tesla.v21i1.3246
3. Barat, N.T., Selatan, S.: Pengaruh Sudut Serang Dan Kecepatan Angin Terhadap Kinerja Turbin Angin Heliks Gorlov Dengan Curveplate. Sainteknol : Jurnal Sains Dan Teknologi Penambahan **16**(1), 73–88 (2018). https://doi.org/10.15294/sainteknol.v16i1.14242
4. Mustika, L.: Pengembangan Media Konversi Energi Angin Menjadi Energi Listrik. **3**, 20–23 (2020)
5. Rizianiza, I., Setiorini, D., Djafar, A., Mesin, T.: P-8 Pengaruh Variasi Sudut Serang Dan Kecepatan Angin Terhadap Performansi Turbin Angin Sumbu Horizontal THE Effect of Angle of Attack And Wind Speed To (2018)
6. Sidik, J., Si, M., Wachid, N., Pd, M.: Pengaruh Sudut Serang Terhadap Koefisien Performa Turbin Angin Sumbu Horisontal Skala Mikro Naca 4412 Sholichan Abstrak **1**(1), 1–11 (2020)
7. Soto-Valle, R., Bartholomay, S., Alber, J., Manolesos, M., Nayeri, C.N., 18 Paschereit, C.O.: Determination of the angle of attack on a research wind turbine rotor blade using surface pressure measurements. Wind Energy Sci. (2020). https://doi.org/10.5194/wes-2020-35
8. Wen, B., Tian, X., Dong, X., Peng, Z., Zhang, W., Wei, K.: A numerical study on the angle of attack to the blade of a horizontal-axis offshore floating wind turbine under static and dynamic yawed conditions. Energy **168**, 1138–1156 (2019). https://doi.org/10.1016/j.energy.2018.11.082
9. Li, Y.-G., Yan, J.-H., Li, Y., Xiao, C.-X., Ma, J.-X.: Wind tunnel study of wind effects on 90° helical and square tall buildings: a comparative study. J. Build. Eng. (2021). https://doi.org/10.1016/j.jobe.2021.103068
10. Maurya, N.K., Maurya, M., Tyagi, A., Dwivedi, S.P.: Design & fabrication of low speed wind tunnel and flow analysis. Int. J. Eng. Technol. **7**, 381–387 (2018). www.sciencepubco.com/index.php/IJET
11. Kusumohadi, C.S., et al.: Perancangan awal terowongan angin kecepatan rendah untuk pengujian otomotif. 1(Sens 1), 7– 17 (2015)
12. Fachri, M.R., Hendrayana, H.: Analisa Potensi Energi Angin dengan Distribusi Weibull Untuk Pembangkit Listrik Tenaga Bayu (PLTB) Banda Aceh. *CIRCUIT:* Jurnal Ilmiah Pendidikan Teknik Elektro **1**(1), 1–8 (2017). https://doi.org/10.22373/crc.v1i1.1377
13. Muttaqin: Analisa perbedaan kecepatan turbin angin savonius 2 sudu dengan membandingkan perbedaaan tinggi sudu idzani muttaqin. 2502–4922, 02(02), 92–97 (2016)
14. Susilo, S., Widodo, B., Magdalena Silalahi, E., Priyono, A.: Pengaruh Jumlah Bilah dan Sudut Pasang terhadap Daya Turbin Angin Poros Vertikal Tipe H-Darrieus Termodifikasi sebagai Energi Alternatif Pembangkit Tenaga Listrik Skala Rumah Tangga. Jurnal Energi Dan Manufaktur **12**(2), 92 (2019). https://doi.org/10.24843/jem.2019.v12.i02.p08
15. Hidayat, M.F.: Analisa Aerodinamika Airfoil Naca 0021 Dengan Ansys Fluent. Jurnal Kajian Teknik Mesin **1**(1), 43–59 (2016). https://doi.org/10.52447/jktm.v1i1.332
16. AirfoilTools.com, "Airfoil NACA 4418," 2013. [Daring]. Tersedia pada: http://airfoiltools.com/airfoil/naca4digit. [Diakses: 2-Feb-2021]

Study of Thermal Uniformity and Heat Displacement Characteristics in Multi-stage Drying Equipment

Thaharul Fikri, Ahmad Syuhada(✉), and Razali Thaib

Magister of Mechanical Engineering Program, Department of Mechanical and Industrial Engineering, Faculty of Engineering, Universitas Syiah Kuala, Banda Aceh, Indonesia
thaharul@mhs.unsyiah.ac.id, {ahmadsyuhada, razalithaib}@unsyiah.ac.id

Abstract. Traditional drying still lacks one of the lengths of the curing process that takes 2–4 days depending on the condition of the sun, unlike the modern way of using dryer equipment. Previous research had created a multistoried dryer, but there was still a considerable temperature difference between the shelves. The study aims to reexamine the distribution of temperatures that occur in high-rise dryer equipment. Prior research has designed dryer appliances with a leveling 9 levels of hot gas line and a sharp turning fins, 45-degree bend on the left and right sides of the instruments. Still, they do not achieve the same temperature consistency in the dryer room. Achieving the right temperature in each dryer requires reshaping the fins of a hot gas line. From the long changes to the hot gas line, an efficient fin variation, a size of 35 mm in the gas duct, followed by a 5 mm increase in the last fin at the first deposit of 60 mm, but at the other end of the hot gas line at 7, 8, 9, was shut, with the experiment producing an evenly matched temperature between 1–2 °C.

Keywords: Dryer · Fins · Hot gas line

1 Introduction

Many foods are consumed by communities that do not finish a single meal. Therefore, there is a process whereby people can consume food at a later time. The traditional methods of treatment by society are drying. Drying is a durable method of reducing water levels from objects by using heat energy to make water levels in air conditions normal [1].

People are generally drying food using solar heat. But this method causes problems like the length of the drying time of about 2–3 days in the summer and depending on the weather. On the other hand, there is no guarantee [2, 3].

At present, dry cleaning is used with kiln, heat from fuel sources. On the other hand, the drying equipment is also multistory, which increases the drying capacity and accessibility of humor [4, 5].

Akhyar et al. (Eds.): ICECME 2021, LNME, pp. 237–247, 2023.
https://doi.org/10.1007/978-981-19-3629-6_25

But the problem with these story-drying equipment is that not the uniform temperature is delivered level. Due to the lack of heat distribution of fuel gas or heat displacement in natural convection raises the rack section adjacent to the heat source. Thus, to address the problem, a hot drain at each level is made. It serves to channel the heat from the combustion chamber into each section of the dryer level.

To optimize the heat transfer in the heat distribution channel, it needs to reconsider the characteristic of the heat transfer between heat distribution channels.

Previous studies had devised a multistorey drying system. However, the temperature still stands at 5–7 °C, between rack 1 and rack 9. To tackle that requires an experimental reexamination. The study aims to reexamine the distribution of temperatures that occur in high-rise dryer equipment.

2 Literature Review

2.1 Drying Theory

Drying reduces water levels to a certain level of water, which impedes the rate of damage caused by biological and chemical activity. Drying occurs because the evaporation of water from the material results from the difference in water vapor content between gas and dried material [6].

3 Drying Methods

a. Natural desiccation

   Sun drying, the drying process using sunlight. The drying is applied to large amounts of weathering and is done outdoors, and the gas conditions are dry. This curing method takes two to four days.

   Air drying, this drying process is made possible through the air. The drying process is done by hanging materials where dry gasses blow. This natural drying has some excess and defaults, which does not require expensive equipment since it has an abundance of solar energy. And its lack of control is its temperature. Sometimes its heat is much and sometimes small, it requires vast areas of space to travel, it is not hygienic because it is easily contaminated by outside air, and it takes a long time [7].

b. Artificial drying

The artificial dryer USES artificial heat in addition to solar energy. The heat from arson is used to perform artificial desiccation—the medium of direct contact gas through heating that flows to the surface of the dried material. Temperatures and gas flows can be regulated, more hygenic, and save time.

To generate heat in a multistoried dryer, fuel is used. The energy used for the kiln comes from fossil fuels: coal, petroleum, and natural gas.

Unlike most ovens, the one on the right side of heat energy at the bottom was one with a high temperature among the others, while the dryer was using heat on each side, resulting in every level being evenly produced [4, 8].

3.1 Energy Drying

The curing of heat requires a large amount of heat to vaporize the water content of a material. The thermal energy generated from combustion will produce hot gas. The resulting hot gas transfers heat to the dry object, creating a difference in water levels between the hot gas and the dried thing, hence evaporation [9, 10].

3.2 Heat Transfers on the Drying Process

Heat displacement can be defined as energy transfer from region to region as a result of temperature differences. When a fluid with certain temperatures flows through a perching surface, heat is transferred from one area to another [11].

In the drying process, one of the heat shifts was convection. The heat transfer of convection is a temperature transfer that causes intermolecular movements and bulk motion of the fluid. The faster the fluid moves, the greater the rate of heat displacement will be [5, 12].

The heat displacement was differentiated into two models: free convection and forced convection. Free convection is a natural transfer of heat because of a difference in air pressure. In comparison, force convection was a forced transfer of heat with artificial airflow [5, 13].

The equation to calculate the rate (convection) of heat transfer is:

$$Q = hc \bullet A(T_s - T_\infty) \tag{1}$$

where:

Q = Heat displacement convection (W).
hc = Coefficient convection (W/m^2·K).
A = Vast areas of thermal movement (m^2).
T_s = Fluid temperature (K).
T_∞= Ambient temperature (K).

Convection heat displacement rates can also be expressed:

$$Q = \dot{m}\ C_p(T_{in} - T_{out}) \tag{2}$$

where:

Q = Heat displacement convection (W).
ṁ = The rate of mass flow(kg/s).
Cp = Heat Specific (J/kg · K).
Tin = Water flow inlet temperature (K).
Tout = Water flow outlet temperature K).

The average effluent of heat transfer (h) could be calculated using a combination of Eqs. (1) and (2), thus providing the following equation:

$$h = \frac{\dot{m}\ C_p(T_{in} - T_{out})}{A(T_s - T_\infty)} \tag{3}$$

The heat energy needed for the curing of the banana is obtained in equation [7]:

$$QEvaporation = Mn.Lh \tag{4}$$

description:
Mn = The weight of the object being drained (kg).
Lh = Heat laten (kJ/kg).

4 Method

4.1 Planning a Research System

On research system planning, the retrieval of data at the beginning with the production of multistoried baking equipment (Fig. 1).

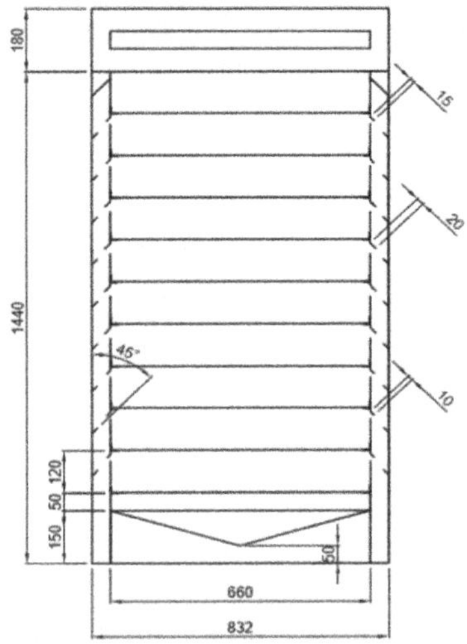

Fig. 1. Front and side by design, dryer space and dimensions

As for the dimensions of the designer space dryer are as follows:
Shelf - type dryer.

Outer length: 832 mm.
Length in: 660 mm.
Width: 820 mm.
Height: 1,440 mm.
Materials: 0.3 mm thick zinc plate.

Tray tray.

Length: 802 mm.

Width: 596 mm.
Thick: 40 mm.
Net size: 14.5 × 14.5 mm.
Rack number: 9 pieces.
Skeletal materials: holo 20 × 30 × 0.85.
Tray's capacity: 10 kg.

After producing dryer equipment, it is modified to measure a fin on a hot gas line. It's done to lower the temperature difference between levels (Fig. 2).

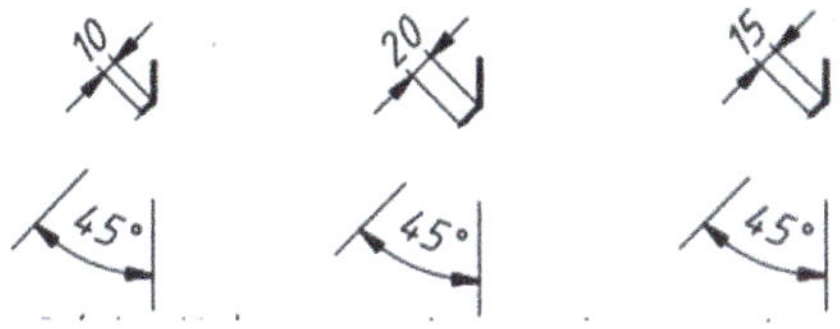

Fig. 2. Variation fins size on the hot gas line

4.2 Data Retrieval

Data retrieval begins by testing temperatures between levels in the drying room with a fin length of an actual hot gas line. Then fins in hot gas lines be variate to get the best results for temperature differences between the levels of dry-cleaning equipment shelves (Fig. 3).

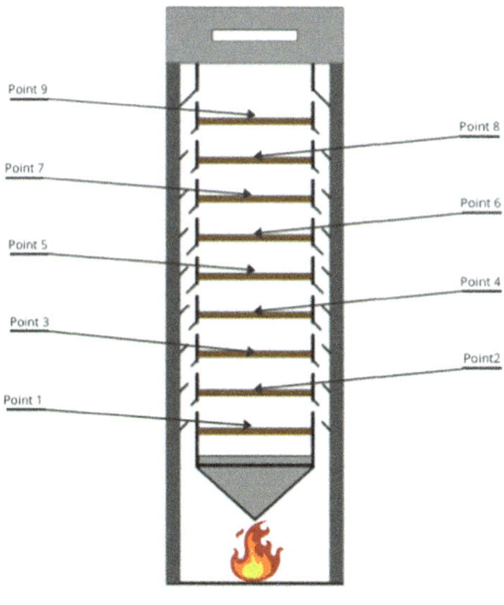

Fig. 3. Data retrieval point on the rack

5 Result and Discussion

10 mm, 15 mm, dan 20 mm fins

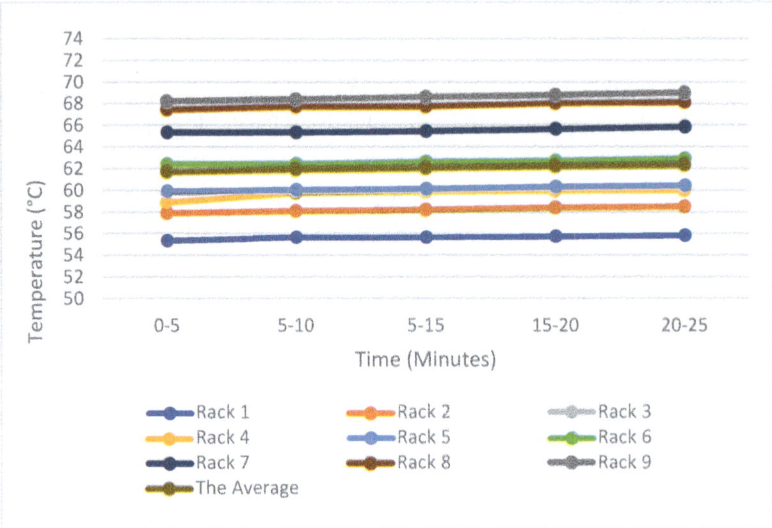

Fig. 4. Temperature charts against time on each shelf with a temperature of 60 °C

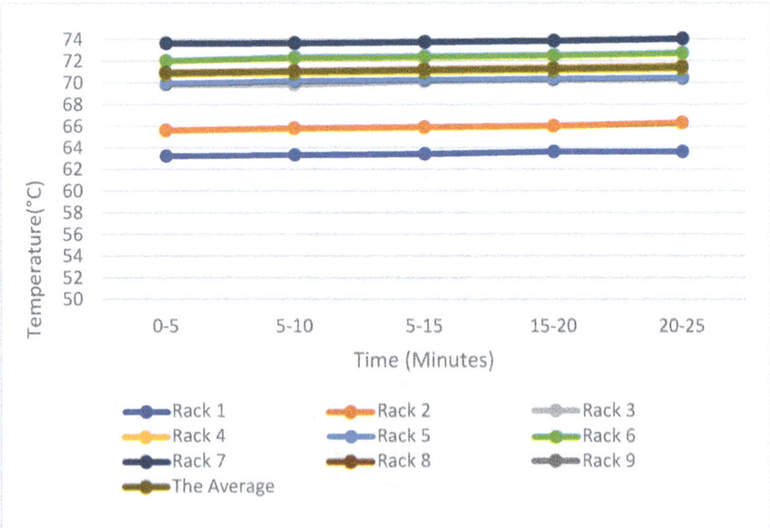

Fig. 5. Temperature charts against time on each shelf with a temperature of 70 °C

At 10 mm, 20 mm, and 15 mm, Fig. 4. And 5. It looks like there's still a massive difference in temperature between the dryer, which is about four to eight degrees celsius.

This is obtained at temperature checks on shelves 1, 2, 3, 7, 8, and 9. Then the fins of the hot gas line need to be remodified.

10 mm fins

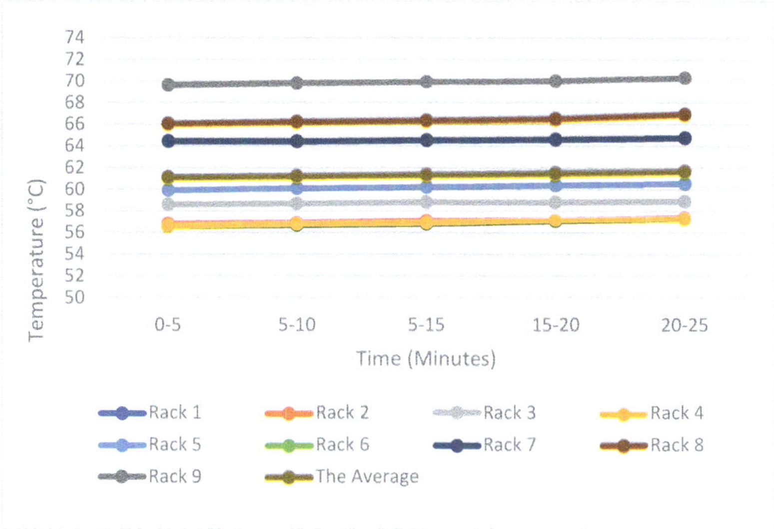

Fig. 6. Temperature charts against time on each shelf with a temperature of 60°C

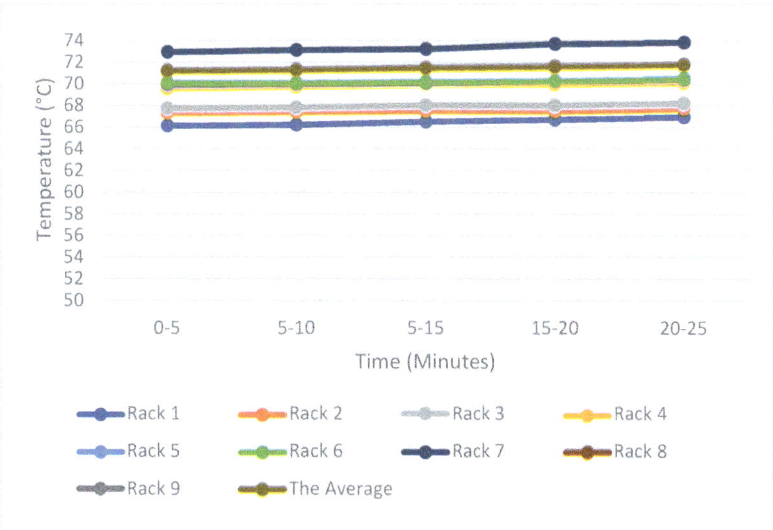

Fig. 7. Temperature charts against time on each shelf with a temperature of 70°C

In Figs. 6 and 7, a 10mm fin at each level, there is still a significant temperature difference in the 9th shelf because it reaches a temperature difference of about 10 Celsius, and rack 1 remains the same as Figs. 4 and 5 as about 4 Celsius

Fins increased by 5 mm per level (beginning with ninth shelf fins in early 20 mm)

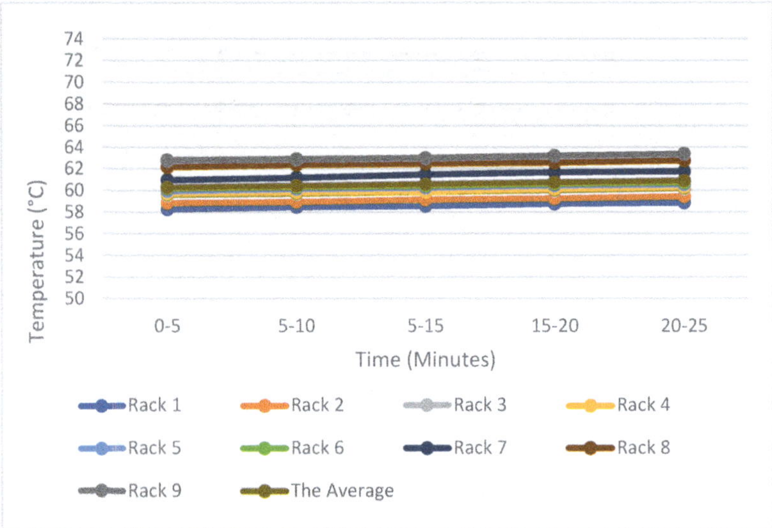

Fig. 8. Temperature charts against time on each shelf with a temperature of 60 °C

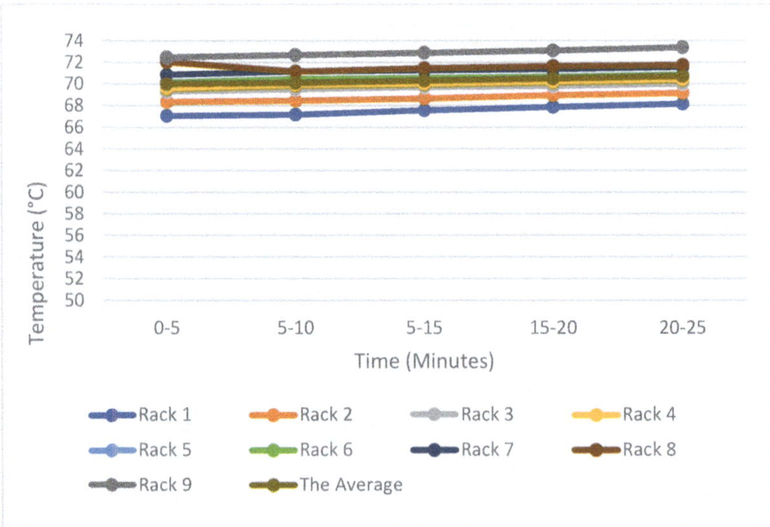

Fig. 9. Temperature charts against time on each shelf with a temperature of 70 °C

Figures 8 and 9 A significant drop in the temperature range from about 2 to 3 Celsius was, because the warm gas fin on rack 9 was reduced in length by the extension of the friendly gas line on rack 1, and thus there was a temperature reduction in each frame.

A flipper of the hot gas line with an increased size of 5mm per level (beginning with a sixth shelf early of 35 mm) but at rack 7, 8, and 9 of the heating line was closed.

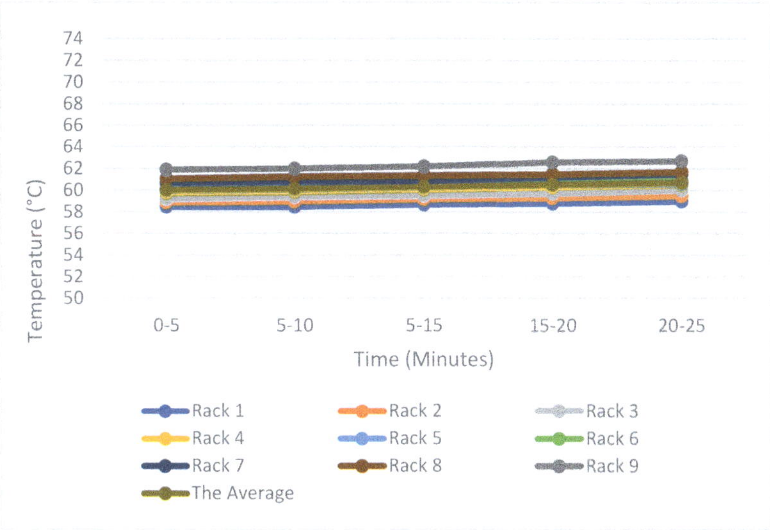

Fig. 10. Temperature charts against time on each shelf with a temperature of 60 °C

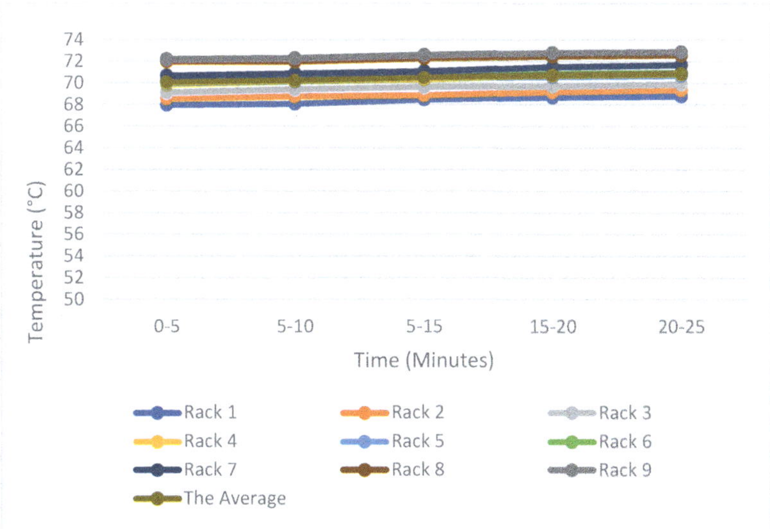

Fig. 11. Temperature charts against time on each shelf with a temperature of 70 °C

In Figs. 10 snd 11, the difference in temperature between shelves ranging from 1 to 2 Celsius is that the gas line on shelves to 7, 8, and 9 is closed, so the 7, 8, and 9 do not receive the heat provided in the hot gas line. The heat received on the shelves came from previous frames, so the temperature on the shelves did not rise. Too, with shelves 1, 2, and 3 fins in the hot gas mains here so long that the heat transmits on these shelves results in an equaling distribution of temperatures, so experiments at variations in heat channel fins produce even continuous temperatures

6 Conclusion

Research has concluded: fins of 10 mm, 20 mm, and 15 mm still have a temperature difference of between 4–8 and Celsius. A 10 mm fins, with a difference in temperature between dry shelves ranging from 4 to 10 °C. A 20 mm long fin at the 9th level followed by a 5 mm prolonged increase in the lower shelves results in a difference in temperature between dry shelves varying from 2–3 °C. Variations fins in the hot gas line at a rate of 5 mm per level (ranging from a sixth shelf at an early age of 35 mm), but on shelves 7, 8, and 9 fins on the hot gas line were closed, resulting in a small sub-heat rack of about 1–2 °C, so the dryer equipment produced a steady temperature differential of about 1–2 °C each shelf.

References

1. Sary, R., Syuhada, A.: Study of fish drying process using multilevel shelves with wood fuel, pp. 9–10 (2019)
2. Sary, R., Dinni, A.: Analisis proses pengolahan pisang sale dengan menggunakan gas elpiji (2014)
3. Nazaruddin: Kajian Sistem Pengolahan Pisang Sale dengan Menggunakan Gas Elpiji, vol. 2, no. 1, pp. 28–33 (2018)
4. Shabri, D.: Studi penyeragaman temperatur pada ruang pengering tipe rak bertingkat banyak (2020)
5. Suherman, S., Trisnaningtyas, R.: Analisa energi dan eksergi pada pengeringan tepung tapioka menggunakan pengering kontinyu unggun fluidisasi getar. Reaktor **16**(1), 24 (2016). https://doi.org/10.14710/reaktor.16.1.24-31
6. Wijiati, L., Widodo, B.U.K.: Studi eksperimen perpindahan panas konveksi paksa pada berkas pin fin berpenampang circular dengan susunan aligned. J. Tek. ITS, vol. 8, no. 1 (2019)
7. Apriadi, S., Amanah, Z.A., Bintoro: Analisis Perpindahan Panas dan Massa Proses Pengeringan Jagung Tongkol Pada Beberapa Metode Pengeringan Sederhana. Pros. Semin. Nasionala Perteta. 21–22 Juli 2011, pp. 319–329 (2011)
8. Syauki, A.: Riandi, "Kajian Tingkat Kemampuan Penyerapan Panas Matahari Oleh Tanaman Hias (Bunga) Disekeliling Bangunan." Tek. Mesin Univ, Syiah Kualah (2015)
9. Tobing, I.F., Agustina, R., Studi, P., Pertanian, T., Pertanian, F., Syiah, U.: Modifikasi Pengering Tipe Tray Dryer Dengan Penambahan Insulator (Modification Of Tray Dryer With Insulator), vol. 4, no. November 2019
10. Zikri, A., Erlinawati, Rusnadi, I.: Uji Kinerja Rotary Driye Berdasarkan Efisiensi Termal Pengeringan Serbuk Kayu Untuk Pembuatan Biopellet. J. Tek. Kim. **21**(2), 50–58 (2015)
11. Dewi, A.K., Satibi, L.: Kajian Pengaruh Temperatur Pengeringan Semprot (Spray Dryer) Terhadap Waktu Pengeringan dan Rendemen Bubuk Santan Kelapa (Coconut Milk Powder). Konversi **4**, 25–31 (2009)

12. Awal, J.S., Awaludin, M.: Pengeringan Bengkuang Dengan Sistem Pengeringan Beku Vakum (Vacuum Freeze Drying System), vol. 1, no. 2, pp. 1–13 (2014)
13. Indriani, I., Novi, N.H., Sarosa, A.H., Nurul'aini, K.: Pembuatan Fluidized Bed Dryer untuk Pengeringan Benih Pertanian Secara Semi Batch. Biomass Chem. Eng. **49**(23–6), 21–28 (2009)

Study of the Ability of Hanging Flower Plants to Absorb Solar Heat

Amalia Harmin, Ahmad Syuhada(✉), and Muhammad Ilham Maulana

Magister of Mechanical Engineering Program, Department of Mechanical and Industrial Engineering, Faculty of Engineering, Universitas Syiah Kuala, Banda Aceh, Indonesia
amalia_hmn@mhs.unsyiah.ac.id, {ahmadsyuhada, lhammaulana}@unsyiah.ac.id

Abstract. The lack of plants in multi-storey housing causes an increase in environmental temperature, causing global heat. Therefore, thermal comfort is not achieved for humans which will reduce the quality of life, to overcome this, air conditioning (AC) is used. However, due to the high ambient temperature causing an increase in the use of air conditioning which will cause electricity consumption to increase as well. One effort that can be done is to multiply hanging flower plants around multi-storey buildings. This solution is expected to reduce the environmental temperature around high-rise buildings. This study aims to determine the heat absorption by fifteen types of plants tested by the hanging method, so that it can be seen which plants are most suitable to be planted to reduce environmental temperatures. The plants tested were Devils Ivy (*Epippremnum Aureum*), Purple Heart (*Tradescantia*), Ant Bag (*Discida Geri*), Velvet (*Gynura Aurantica*), Purple Zabrina (*Trandescantia Zabrina*), *Paperomia Hope*, Spider Lily (*Cholorophytum Comosum*), Purple Spider Lily (*Trandescantia Spathacea*), Nerve Leaf, *Anthurium*, Pigeon Feet (*Altheantera Ficoida*), Creeping Charlie (*P. Nummularifolia*), Candle Flower (*Begonia*), *Zebrakraut*, and *Episcia*. From the results obtained indicate that hanging flower plants affect the decrease in environmental temperature. The most effective among the mentioned hanging flower plants on absorbing heat with the hanging method are Spider Lily (*Chlorophytum Comosum*).

Keywords: Hanging plants · Thermal comfort · Solar thermal energy · Ambient temperature · Multi-storey housing

1 Introduction

The sun is the main source of heat energy needed by all living things on earth [1]. The sun's heat that reaches the earth is not only reflected or transmitted, but also absorbed by living things and objects on earth. Solar energy that is not able to be absorbed by the earth's objects is released back into the earth's atmosphere [2]. This excessive heat energy heats the earth's surface so that the temperature increases and causes global warming [3]. The consequences of global warming will cause the heat energy on earth to become unstable which will affect the quality of human life [4].

Akhyar et al. (Eds.): ICECME 2021, LNME, pp. 248–254, 2023.
https://doi.org/10.1007/978-981-19-3629-6_26

Most of the sun's heat is absorbed by plants as energy for the photosynthesis process [5]. The lack of available plants to absorb solar heat is one of the causes of global heat, especially in urban areas, industrial areas and high-rise housing [1]. Housing in urban areas develops in a vertical direction such as flats due to the lack of land therefore, there are very few green plants to absorb solar heat, and consequently the air condition in the area is no longer thermal comfort for human life [6].

People who live in vertical buildings such as flats cause the building to not meet the ASHRAE 55 standard where the variables that affect thermal comfort are less supportive[6]. To overcome thermal discomfort, air conditioning (AC) is used where the higher the temperature, the greater the power required by the AC and electricity consumption also increases [7].

To reduce global heat that occurs in urban areas, especially in buildings that develop in a vertical direction, is to multiply hanging plants. Due to the lack of public understanding of the absorption of heat by plants, it is necessary to study the ability to absorb heat by each plant. The purpose of this study was to determine the characteristics of each hanging plant.

2 Literature Review

Solar Radiation. The sun is the star at the center of the solar system. Its almost perfectly spherical shape consists of hot plasma mixed with a magnetic field. The core of the sun has a temperature of about 15.7 million kelvins and a surface temperature of about 5,800 kelvins. Radiation is a form of energy emitted by any object that has a temperature above absolute zero, and is the only form of energy that can propagate in the vacuum of outer space.

It is estimated that 35% of the solar radiation received in the upper atmosphere is reflected back into space in the form of short waves by scattering and reflection by clouds, dust particles, and the earth's surface. Thus only 65% of the total solar radiation that comes in and can be used to heat the Earth's atmosphere, 14% is used to heat the atmosphere and 51% is absorbed by the Earth's surface. Solar radiation that reaches the earth is not only reflected and transmitted but also absorbed by earth objects such as green plants for the process of their survival through photosynthesis, lakes, seas, rivers, rocks, buildings, houses and others.

Plants. Plants are living things that have leaves, stems and roots so that they are able to produce their own food by using leaf green substances (chlorophyll) through a process called photosynthesis. Plants will produce oxygen (O_2) and convert carbon dioxide (Co_2) so that it can be useful for other living things. Characteristics that are very easily recognized in plants are greenish in color but can be yellow due to the chlorophyll pigment which plays a vital role in the process of capturing energy through the process of photosynthesis.

Photosynthesis. The process of photosynthesis is the process of survival of green plants, where to produce their own food green plants absorb carbon dioxide (Co_2) and water (H_2O) and under the influence of sunlight converts it into organic compounds. The main organ of green plants where photosynthesis takes place is in the leaves. Sunlight will be captured by chlorophyll cells known as chloroplasts. The chemical reaction for the process of photosynthesis is as follows:

$$6Co_2 + 12H_2O \rightarrow C_6H_{12}O_6 + 6O_2 + 6H_2O \quad (1)$$

The process of photosynthesis cannot take place in every cell in the leaf, but can only take place in cells that contain photosynthetic pigments. On the other hand, cells that do not contain photosynthetic pigments cannot carry out photosynthesis. The solar radiation emitted from the sun can be used but only certain wavelengths that can be used by plants for photosynthesis, namely waves with a length of about 380–700 nm or in the visible light range.

Flower. Flowers are ornamental plants that function as an addition to beauty and aesthetics, especially in the flower part. Flower ornamental plants are the most popular plants in the community, because the name flowers on plants are identical to be used as decorations. Flower plants certainly have various shapes and colors depending on the type. Flowers are usually planted indoors using pots or outdoors using pots, hanging pots or planted directly in the ground. From a biological point of view, the flower is one of the organs of the plant body that functions as a means of generative reproduction which has a different shape and arrangement according to its type. In flowers there are reproductive organs, namely stamens and pistils. Flowers consist of petals, corolla, stamens, and pistil. If a flower has all four parts, it can be said to be a complete flower.

Housing Area. Housing is a collection of houses as part of settlements both in villages and in cities which are equipped with facilities, utilities and infrastructure as a result of efforts to fulfill livable houses. However, due to limited land and high prices in urban areas and the high demand for space for various purposes, high-rise buildings were erected. Multi-storey buildings are buildings that have more than one floor vertically. The more floors that are built, the more capacity that can be used and can increase the efficiency of urban land. One of which is included in the vertically storied building is a flat.

Thermal Comfort. Thermal comfort is a condition where the human body temperature is stable at the comfortable limit, which means the body does not feel disturbances caused by thermal factors which are made possible by the balance of body temperature with the environment and climatic factors that are in the comfort zone. Several factors that affect thermal comfort are as follows: air temperature, humidity, wind speed, clothing insulation, and activity.

3 Methods

This research was conducted by means of experimental observations. This research used 15 types of plants, namely: Devils Ivy (*Epipremnum Aureum*), Purple Heart (*Trandescantia*), Ant Bag (*Dischida Geri*), Velvet (*Gynura Aurantica*), Purple Zabrina (*Trandescantia Zabrina*), *Paperomia Hope*, Spider Lily (*Cholophytum Comosum*), Purple Spider

Lily (*Trandescatia Spathaceae*), Nerve Leaf (*Hemigrafis Altergata*), *Anthurium*, Pigeon Feet (*Altheantera Ficoida*), Creeping Charlie (*P. Nummularifolia*), Candle Flower (*Begonia*), *Zebrakraut*, and *Episcia*. Each test plant was planted in a plastic pot. This research used an isolation room made of transparent plastic with a thickness of 0.25 mm. The frame of the isolation room is made of wood this is done to prevent outside influences. The ambient temperature and the temperature in the isolation room containing the test plants were measured using a calibrated digital thermometer.

Theory of Heat Absorption by Plants. Not all of the solar radiation that reaches the earth is absorbed but is reflected and transmitted. One of the living things on earth that absorbs the sun's heat are plants. Plants absorb sunlight to carry out the process of photosynthesis. If there is a difference in temperature, energy will move from areas of high temperature to areas of low temperature. The energy transferred due to the temperature difference is the heat that can be absorbed by a substance, and the equation is as follows:

$$Q = m \cdot C_p \cdot \Delta T \tag{2}$$

To find the mass of air in the isolation room, use the equation:

$$m = v_{chamber} \cdot \rho_{air} \tag{3}$$

To calculate the amount of heat energy absorbed by plants using the equation:

$$Q_{abs} = Q_{emptychamber} - Q_{plantchamber} \tag{4}$$

4 Result and Discussion

Heat Distribution in Isolation Room in Hanging Plant Method. After doing the research, the test data obtained with the hanging plant method, namely the temperature distribution in each hanging plant isolation room tested so that it can be presented as follows Fig. 1:

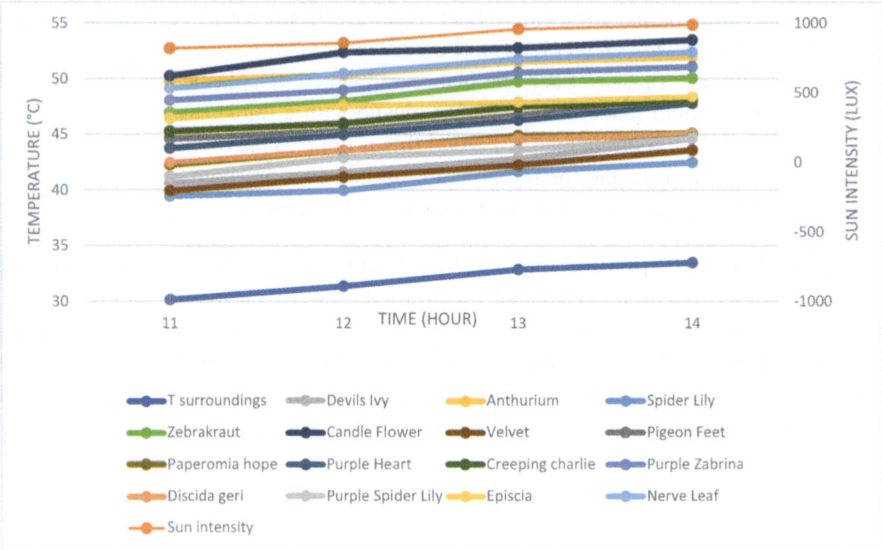

Fig. 1. Heat distribution in the insulation chamber by hanging method

From the Fig. 1 above, it can be seen that the plant test method with the hanging plant method also experience an increase in temperature every hour. The highest temperature was found in the isolation room containing *Episcia* plants and the lowest temperature was found in the isolation room containing Spider Lily.

Distribution of Heat Absorbed by Plants in The Hanging Method. Distribution of heat absorbed by plants in the hanging methode can be in Fig. 2:

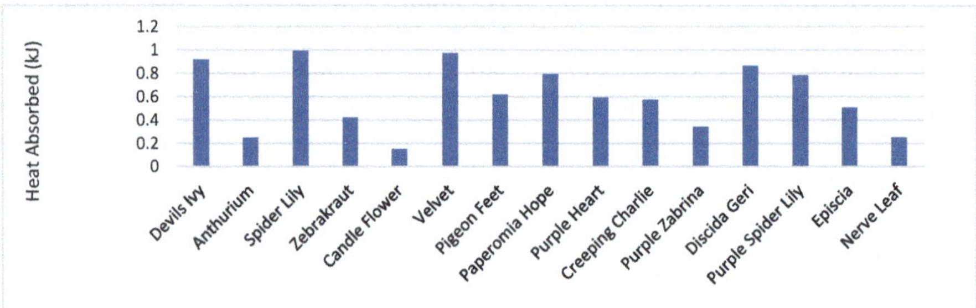

Fig. 2. The distribution of heat absorbed by plants is hanging

From the graph above, it can be seen that the heat absorption absorbed by each plant for the plant test method is suspended. The highest heat absorption was found in Spider Lily, which was 1 kJ and the lowest was found in Candle Flower at 0.158 kJ.

5 Discussion

Distribution of the lowest and highest temperature on the hanging plant research method can be in Fig. 3:

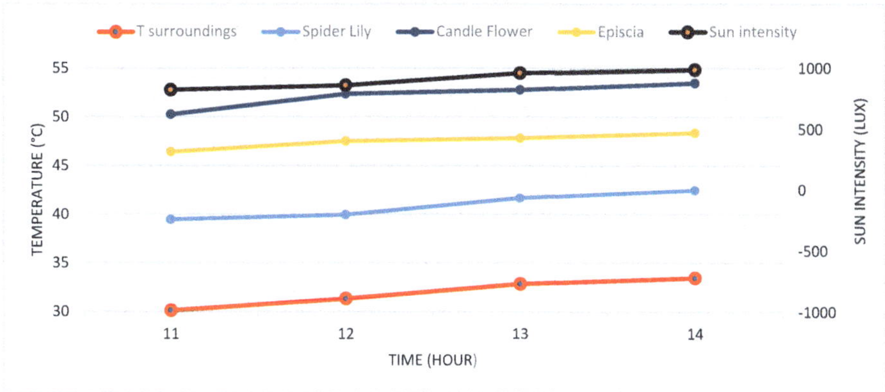

Fig. 3. Distribution of the lowest and highest temperature on the hanging plant research method

Each plant tested has a different ability to absorb heat. In the hanging plant method, the highest heat distribution is found in the isolation room Candle Flower, while for the lowest heat absorption is Spider Lily. In the hanging plant method, the highest heat absorption was found in the isolation room containing Spider Lily with an average temperature of 40.9 °C and the lowest heat absorption was found in the isolation room containing Candle Flower with an average temperature of 52.2 °C. Hanging flower plants are not able to absorb heat optimally because the water content continues to decrease which causes the hanging flower stems to be damaged so that heat absorption does not occur optimally.

6 Conclusion

From the results of the research conducted, it can be concluded as follows:

1. In the hanging plant method, the highest heat absorption was found in the isolation room containing Spider Lily and the lowest in the isolation room containing Candle Flower.
2. Spider Lily absorb heat by 1%.

References

1. Nurul Mukhtar, A.: An Analysis of Occupants Response to Thermal Discomfort in Green and Conventional Buildings in Malaysia (2018)

2. Gou, Z., Gamage, W., Lau, S.S.Y., Lau, S.S.Y.: An investigation of thermal comfort and adaptive behaviors in naturally ventilated residential buildings in tropical climates: a pilot study. Buildings (2018). https://doi.org/10.3390/buildings8010005
3. Sobry Effendi dan Ferdi Affrihatmoko: The Realtionship of Green Open Space with Human Comfort in Urban Areas. J. Agromet **28**(1), 23–32 (2015). http://journal.ipb.ac.id/index.php/agromet
4. Shahzad, U.: Global Warming: Causes. Dureesamin Journal, Riphah International University, Faisalabad, Pakistan, Effect and Solution (2015)
5. . Susilawati, W., Irmasari: The effect of various light intensities on the growth of cempaka seedlings in nurseries. J. Forest Sains Untad (14), pp. 59–66 (2016)
6. Azmoodeh, M., Zamani, Z., Abdollah: Experimental and simulation studies on the vertical greenery system (VGS) AND temperature mitigation in urban space. J. Build. Eng. Azad University, Iran. https://doi.org/10.1016/j.jobe.201807.024
7. Sampurno, Fauzen and Waspodo: Analysis of Efficiency Levels in Air Conditioning Systems at PT. Pelabuhan Indonesia II (PERSERO) Pontianak Branch, Suara Teknik: Scientific Journal (2019)
8. Susilawati, Wardah dan Irmasari: Pengaruh Bberbagai Intensitas Cahaya Terhadap Pertumbuhan Semai Cempaka Dipersemaian. Jurnal Forest Sains Untad (14), hal 59–66 (2016)
9. Baktiar Leu, (2021), *Dampak Pemanasan Global dan Upaya Pengendaliannya Melalui Pendidikan Lingkungan Hidup dan Pendidikan Islami*, Jurnal At tabdir STAI Darul Kamal NW Kembang Kerang NTB, Vol. 5, No.2 2021. https://journal.staidk.ac.id/index.php/tabdir
10. Norman Permana Putra: Asrama Mahasiswa Universitas Panca Bhakti Pontianak. Jurnal Online Mahasiswa Arsitektur Universitas Tanjungpura, vol. 8, No.1, Maret 2020 (2020)
11. Riandi, S.: Study of the Ability to Absorb Solar Heat by Ornamental Plants (Flowers) Around the Building. Syiah Kuala University, Faculty of Engineering (2019)

Automatic Parking Portal Mechanical System Design with DC Motor Development

Rijoi Sinaga, Filbert Sihombing, Bustami Syam(✉), and Maraghi Muttaqin

Faculty of Engineering, University of North Sumatra, Medan 20155, Indonesia
bstsyam@gmail.com

Abstract. Boom gate (parking portal), a tool that functions as an automatic barrier is usually used in parking areas, gates for office buildings, government buildings, police stations, hotels, toll roads, and other public places. The current parking barrier system cannot be separated from the controller system and electric motor as the source of electrical energy to move mechanical parts of the boom gate. However, the parking barrier system which are currently in operated, especially in the Sumatra region, still use an AC motor drive system; in some regions a hydraulic drive system is also utilized. The electricity to drive the system is mostly obtained from the national electricity company (PLN) electrical grid. So, in this study we design a mechanical system of automatic parking portal operated by a DC motor, instead of AC motors. From this study, it was found that the components needed by the boom gate to do the work of opening and closing the bars for 1.5 s, namely: DC motor 24V, 250 W, 2750 rpm, gearbox type 40 ratio 1:60, and sprocket ratio of 1: 4.5.

Keywords: Boom gate · Controller · Hydraulic · Electrical energy

1 Introduction

Electricity needed in the parking system, especially for raising up and down the boom gate beam, can be obtained from the national electricity company (PLN) via AC motors. Recently, our laboratory, the impact and fracture research center (IFRC) has conducted research on energy harvesting using speed bumps [2]. A speed bump is an elevated part of the road in the form of additional asphalt, rubber, or concrete which is installed across the road with the aims to warn drivers to decelerate the speed of the vehicles [3]. Several types of speed bumps made of concrete foam composite has been reported [1, 2, 4].

The electrical energy is obtained from the pressure of the vehicle wheels on the speed bump [2]. The energy is stored in a battery and can be utilized for any needs, including for a boom gate system obtained from this study. At the IFRC the development of mechanical systems for speed bump power generating has been carried out [1, 2, 4]. First, a translational system was produced [1, 4], followed with system using rotational systems [1, 2, 4]. The one operated at the present time has been incorporated in the speed bump power generating stations [1, 2, 4, 5].

Akhyar et al. (Eds.): ICECME 2021, LNME, pp. 255–262, 2023.
https://doi.org/10.1007/978-981-19-3629-6_27

Boom gate (parking portal), a tool that functions as an automatic barrier is usually used in parking areas, gates for office buildings, government buildings, police stations, hotels, toll roads, and other public places. The current parking barrier system cannot be separated from the controller system and electric motor as the source of electrical energy to move mechanical parts of the boom gate. However, the parking barrier systems which are currently in operated, especially in the Sumatra region, still use an AC motor drive system; in some regions a hydraulic drive system is also utilized. The electricity to drive the system is mostly obtained from the national electricity company (PLN) electrical grid. Thus, the boom gate (parking portal) in this study is designed to utilize the electrical energy source obtained from the speed bump power generation, or other resources, e.g., PLN grid using a DC motor. So, in this design, we will substitute one of the components on the parking bar, namely changing the mechanical drive system from AC and hydraulic motors to DC motors.

The current study is carried out to the design and manufacture of a boom gate with a gearbox motor drive system, the aim of using the power that has been generated by the speed bump mechanical system. At the same time to determine the components of the drive system in accordance with what is needed to open and close the bars with a time of 1.5 s. This paper describes the design and manufacture of a boom gate with a gearbox motor drive system.

2 Methods

In designing the boom gate, we start with a conceptual work incorporating results obtained from customer survey. The survey is designed to select ideas and data to be used for a conceptual design, CAD software is used for the conceptual design. The first thing to do is to make a design in the form of an image. Using the software, firstly we figure out an appropriate mechanical system. As we are aware that the boom gate is to

Fig. 1. Boom gate design

be used for comercial. Thus, we need to know the size and the material. The frame size of the boom gate is 1400 mm × 50 mm × 90 mm. The boom gate is made of alternative composites reinforced polymer resin fiberglass material and the weight is 3,4 kg. Next, the assembly process is carried out to unite each drawn part to become a single unit, namely the boom gate. The boom gate design conceptual is shown in Fig. 1.

Design is carried out with the aim of using the power that has been generated by the speed bump mechanical system as well as to determine the components of the drive system that are in accordance with what is needed to open or close the bar with a duration of 1.5 s at an angle of 90°. The reason for determining the duration of time to open or close the bar is based on the commercial boom gate survey that has been carried out. Commercial boom gates, the duration of the opening or closing of the beam is between 1.3 s to 1.8 s. So we choose 1.5 s to open or close the bar. The boom gate mechanical system design in 2D is shown in Fig. 2.

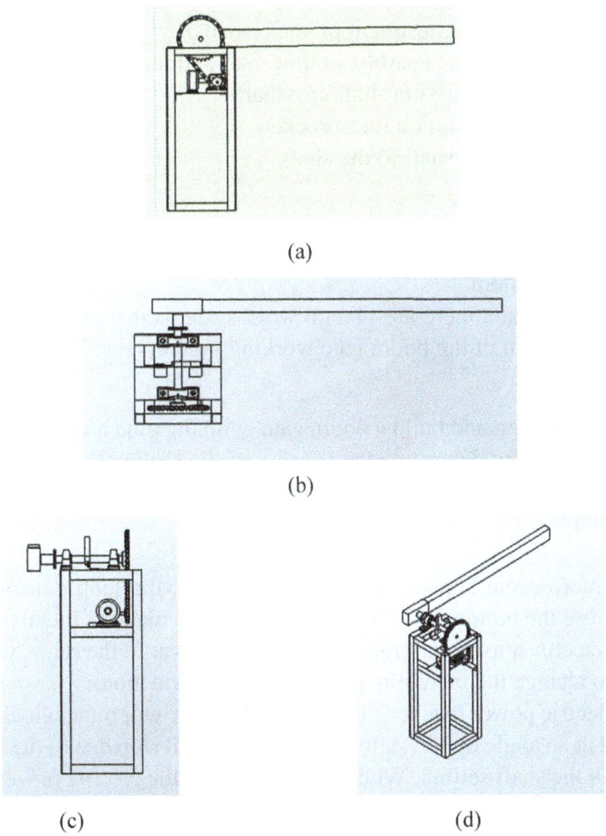

Fig. 2. 2D design of boom gate mechanical system (a) Front view (b) Top view (c) Left view (d) Isometric view

In order to make the boom gate design concept clearer, a 2D design is also carried out which is divided into several views, namely front view, top view, left view, and isometric view as shown in Fig. 2.

After the design process is completed, then the boom gate fabrication process will be carried out according to what has been designed. The following are the stages of boom gate construction:

1. Making the boom gate frame using a hollow iron box measuring 40 mm x 40 mm with a thickness of 2 mm.
2. Cutting the 4 mm plate which is used to place the DC motor, gearbox and also the plate to place the pillow block.
3. Installation of DC motors, gearboxes in the places provided. Where the DC motor shaft is first connected to the input shaft of the gearbox.
4. Mounting the shaft to the pillow block bearing and placing the pillow block bearing on the plate provided.
5. Installation of the axle adjustment on the crossbar.
6. Install the sprocket on the gearbox output shaft and install the sprocket on the shaft connected to the transmission shaft crossbar.
7. Cut the chain and install it on the sprocket.
8. Install the transmission shaft to the shaft.
9. Install the bars.
10. Assemble and install the control system circuit.

Boom gate working system

To operate the boom gate, there are several work systems that need to be done. Below is shown a flow diagram of the boom gate working system (Fig. 3).

Test procedure

This study aims to design and build a boom gate, with the time required to open or close the bar for 1.5 s. Thus, to determine the success of the tool working as expected, a test is carried out to determine the work of the tool and the time it takes to open or close the bar. The procedures carried out include:

1. When the motorcycle approaches the sensor cable, the loop detector sensor will work. Because the motorcycle there are components made of metal.
2. The loop detector sensor will transmit electrical power to the relay, where the relay functions to change the incoming power to the electric motor.
3. After the electric power reaches the motor, the motor will rotate clockwise until the bar is lifted at an angle of 90°. After turning 90° it will stop due to the limit switch in contact with the shaft setting. Where in this process the electric power on the motor will stop.
4. After the motorcycle passes through the loop sensor cable, the loop detector will work again and cut off the incoming power to the relay. So, the relay will transfer the power that is forwarded to the motor. And the motor will rotate counterclockwise.
5. The rotation causes the bar to drop back down to an angle of 90° and will stop due to the work of the limit switch in contact with the shaft setting.

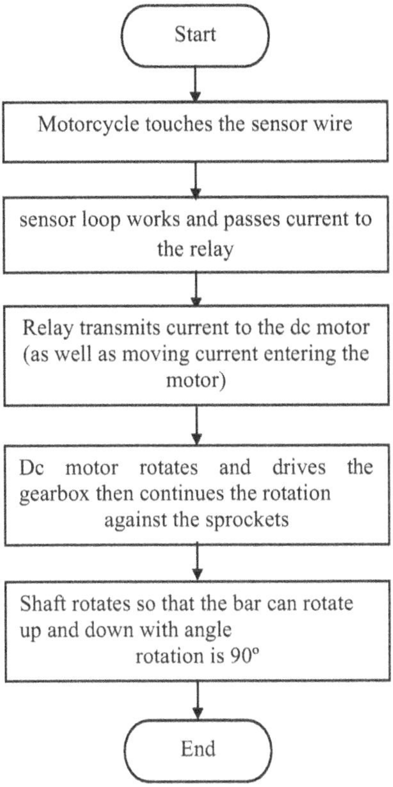

Fig. 3. Flow diagram of the boom gate working system

6. When the bars go up and down, a test will be conducted to determine the time needed to open and close the bars using a stopwatch. And after testing, the time needed is as expected.

3 Results and Discussion

Boom Gate Mechanical System Design

Second sprocket shaft revolution.
In order for the boom gate mechanical system to work as expected, the desired time duration to open or close the bar is 1.5 s. Thus, a calculation must be carried out to determine the required rotation on the shaft. It is also known that in the design of this tool the beam is open and closed with an angle of 90°. From the calculation obtained the required rotation speed is 10 rpm.

Portal Shaft Torque
Before calculating the required power, it must first be known the torque on the cross shaft. In the design of this tool the load that is lifted is a bar. The crossbar used in this tool has a mass of 3.4 kg. After calculating the required torque, it is 46,648 NM.

Planned power
In order for the selected motor to be able to move the bar, it must first be determined the power required by the motor to lift the bar. In this calculation, torque and rotation must be known in advance. From the calculations that have been carried out, the required power is 48.824 W.

Determination of the required components
The next thing to do is to carry out the process of calculating the components needed. With the aim that the components in accordance with the expected rotational speed and torque exceeds required. After the calculations are carried out, the components needed can be determined as expected. These components are:

a. DC motor voltage(V) = 24 V, Power(P) = 250 W, Speed(n) = 2750 rpm
b. Gearbox type 40 with a rotation ratio of 1: 60
c. The selected sprocket is in a ratio of 1: 4.5 (10 teeth and 45 teeth)
d. Pillow block bearing UCP 206 Material: Cast Iron AS: 30 mm Dimension: 42.9 × 165 × 121 mm
e. Shaft (Length: 310 mm, diameter: 30 mm, material: AS White ST41 AScental iron)

The result of making a boom gate mechanical system
In making this boom gate, there are several supporting components. So this tool can work well.

For more details, the components of the boom gate design is shown in Fig. 4.

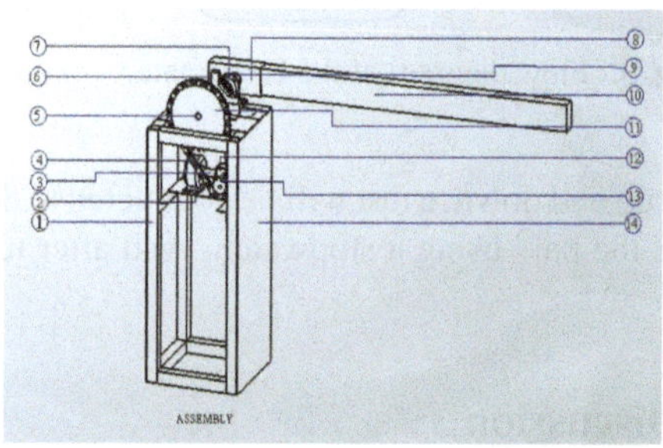

Caption:
(1) Frame
(2) Dc Motor
(3) Gearbox Clutch
(4) Chain
(5) Shaft
(6) Shaft Set
(7) Pillow Block
(8) 36999
(9) Shaft Clutch
(10) Portal
(11) Big Sprocket
(12) Gearbox
(13) Small Sprocket
(14) Cover Plate

Fig. 4. Boom gate components

As mentioned earlier the objective of constructing the boom gate is to utilize the electrical energy produced by the mechanical speed bump system. The electrical power generated will be used to drive the boom gate motor. The result of making the boom gate is shown in Fig. 5.

Fig. 5. Boom gate

Mechanical system testing

The boom gate mechanical system testing was carried out using a beat motorbike and an Nmax motorbike in front of the FT-USU Mechanical Engineering Megister Program. After testing, it was found that the time to raise the bar was 1.5 s. Testing is done using a stopwatch. Testing of the boom gate mechanical system is shown in Fig. 6.

Fig. 6. Mechanical system testing

4 Conclusion

The conclusions obtained from this research are as follows:

1. A mechanical boom gate system has been designed with a DC motor drive and the following components:

    a. Motor DC
        Voltage (V) = 24V
        Power(P) = 250 W
        Revolution(n)=2750 rpm

b. Gearbox
   Type 40
   Ratio 1:60
c. Sprocket
   Ratio of 1: 4.5
   Sprocket 1 = 10 Teeth
   Sprocket 2 = 45 Teeth
d. Pillow block bearing UCP 206
   Material: Cast Iron
   AS: 30 mm
   Dimension: 42.9 × 165 × 121 mm
e. Shaft
   Length: 310 mm
   Diameter: 30 mm
   Material: AS Assental iron ST41 White

2. The boom gate mechanical system fabrication (parking portal) has been carried out and after testing it is found that the duration of time needed to open the bar is 1.5 s and to close 1.5 s.

Acknowledgements. The authors would like to acknowledge the research facility in the IFRC department of mechanical engineering.

References

1. Syam, B., et al.: Analysis of power generating speed bumps made of concrete foam composite. IOP Conf. Ser. Mater. Sci. Eng. **180**, 012033 (2017)
2. Syam, B., et al.: Structural integrity of power generating speed bumps made of concrete foam composite. IOP Conf. Ser.: Mater. Sci. Eng. **308**, 012032 (2018)
3. Budinas, N.: Sigley's Mechanical engineering Design, Eighth Edition
4. Taher, T.: Rancang Bangun Dengan Sistim Mekanik Speed Bump Untuk Menghasilkan Daya Listrik Sebagai Tenaga Pembuka Gerbang Pintu Tol Kota Medan. Universitas Sumatera Utara (2016)
5. Sinambela, J.P.: Study Experimental Sistem Mekanik Pada Speed Bump Generasi Ke-3 Roda Empat Penghasil Energi Listrik. Universitas Sumatera Utara (2019)

Development of Concrete Carving CNC Machine for Wall Decoration

Amir Zaki Mubarak(✉), Syahriza, Asbar Razali, and Arief Nugraha

Department of Mechanical and Industrial Engineering, Universitas Syiah Kuala, Jl. Tgk. Syeh Abdurrauf No. 7 Darussalam, Banda Aceh 23111, Indonesia
{amir_zm,syahriza,asbar}@unsyiah.ac.id

Abstract. CNC machine is a machine that can operate automatically through processing commands performed by a computer. A machine is required for engraving concrete, that the engraving process can be faster and easier. The research is carried out to develop a concrete carving CNC machine. The machine components are grouped into two main parts, horizontal and vertical parts. The design of horizontal component includes analysis and selection of x table, y table, linear drive system, power screw, and motor. The design of the vertical component includes the analysis of the frame, and the spindle arm. Machine production begins with frame welding, spindle arm welding, crafting table, and component purchasing. Assembly begins with the assembly of the base and column frame, the table with its components, and the spindle arm and its spindle. Machine materials are selected from metal, wood and nylon. The workpiece used is concrete with a composition of cement and sand of 1:2. Workpiece dimension is 200 mm × 200 mm × 30 mm. The CNC electronic components are arduino, CNC shield, stepper motor driver and stepper motor. The accuracy of the machine was tested by comparing the results of machining the workpiece with the design on the computer. For testing the age of concrete, the highest accuracy value was obtained when the age of the concrete reached the value of 1 h. For testing patterns, small deviations were shown for testing the sides of each pattern, but slightly larger deviations for the experiment about 0 coordinates.

Keywords: CNC machine · Concrete · Carving · Concrete milling machine

1 Introduction

CNC machine is a machine that can operate automatically through processing commands performed by a computer. CNC machines can perform engraving on a variety of media, especially metal and wood. Recently, CNC machines have been developed to engrave a marble. The main obstacles experienced by marble engraving business are limited human resources (carvers), variations in product design, speed of time, and quality of the engraving results while market demand is getting higher [1]. Considering the material of the workpiece and various types of CNC machines that had been developed, research was carried out on concrete engraving. As the main material for making houses, concrete

Akhyar et al. (Eds.): ICECME 2021, LNME, pp. 263–273, 2023.
https://doi.org/10.1007/978-981-19-3629-6_28

is easy to obtain and has many functions. Then the research was carried out on cutting the wet concrete workpiece.

The research was conducted to make a prototype of a concrete carving machine. Research work is carried out from the design process to product accuracy testing. The decision-making process becomes a major value in the design process that will determine the success of the machine. In this study, the machine accuracy test will only measure basic geometry of the engraving. The geometry is square, cicle, and hexagone.

2 Experiment Method

Machine Production Materials. Machine production material is various of material which needed for fabricating the machine that has been designed. The materials are steel, wood, nylon, lead screw, stepper motor, nut, spindle, milling bits, wire, power suply, arduino, CNC shield, and driver motor stepper.

Workpiece Material. Used workpiece material is a wet concrete with cement and sand ratio 1:2.

Concrete Carving CNC Design. The design of the concrete CNC Milling machine begins with the formulation of the problem and then it is collected by a brainstorming process. From the problems collected, needs and solutions will be formulated. After the best solution is obtained, it is continued by translating the solution at the design stage. The chosen solution may change or be reviewed in line with the design process. The solution is followed by designing the required components. Material selection is carried out in line with the calculation of the load on each component by considering the strength of the material, the magnitude of the load received, and the price of the material. After all calculations are completed, the tool will be produced and assembled. The data collection process for testing will be carried out after the machine movement test can be carried out.

Experiment Data. Based on estimating that the materials needed are wood, concrete and iron. The data focus will be effected on these materials. According to Callister and other sources [2–4], the values of density, compressive strength and coefficient of friction on several materials are listed in Tables 1 and 2.

Table 1. List of design material properties

No	Materials	Yield strengh (Mpa)	Tensile strengh (Mpa)	Density (Kg/m^3)
1	Steel	220	400	7850
2	Stainless Steel	205	515	8000
3	Plywood	–	13.8	400–700
4	Concrete	–	–	2200

Table 2. Material Friction Coefficient List

No	Materials	Coefficient of friction (Static)
1	Wood – Nylon	0, 4
2	Steel – Nylon	0.4
3	Stainless Steel – Brass (*lubricated*)	0.19

Analysis Method. The design is carried out with the assumption of ideal and simple conditions without taking external factors such as temperature, humidity and others. The drive system calculates the compressive/tensile stress, shear stress, frictional force, rotational force, torque, power required to rotate the screw so the motor can handle the load. Calculations on spindle are needed to analyze how much power is needed and the torque that occurs in the spindle during the feeding process.

Quality Control. Quality is needed as a reference whether the tool are successful in carving or not. To determine the quality, quality testing was carried out from several trials. The experiment are execute with the condition of the concrete freshly reacted with water. The data collection process will be carried out by measuring several points with their reference points. After the measurement values on the workpiece, the workpiece will be compared with the values obtained on the computer in the design process. The set of values will provide a comparison of the precision and accuracy of the tool. Precision will see how precise the results are in multiple times of data. Accuracy will see how closely the results are with the design.

3 Results and Discussion

Container Design. The container (Fig. 1) is used to accommodate the workpiece in the form of wet concrete. Because the size of the workpiece is planned to be 200 mm × 200 mm × 30 mm, the inside size of the container is the same as the plan size of the media. A rigid container is needed that can neutralize the fluid properties of wet concrete.

The material of the container was chosen because it did not react with concrete, it was light, rigid, cheap and easy to obtain. The thickness of the wood is adjusted to market availability. If the thickness of the wood is 10 mm, then the outer size of the container is 220 mm × 220 mm x 40 mm (Fig. 2).

Workbench Design. The workbench x receives a compressive load from the container and the workpiece comes from a heavy load of both. The y-axis workbench has more vertical loads, in the form of the weight of the x-axis workbench, wheels, stepper motors, bearings, lead screws, containers and workpieces. Based on Table 1, the density of

Fig. 1. Container design

Fig. 2. Worktable x and y design

plywood can be written around 400–700 kg/m^3. The density of 630 kg/m^3 was chosen because the plywood layers come from different woods. The material for the work table is plywood. The material was chosen because it does not react with concrete, is stiff, straight, cheap, and easy to obtain (Tables 3 and 4).

Table 3. Tensile stress and dimension of worktable

Table	Tensile/compressive stress	Dimension (mm)
x	378,96 Pa	320 × 300 × 8
y	356,601 Pa	550 × 300 × 12

Selection of Linear Motion System. Wheels are used as the main support of each table. Wheels are used because according to the plan they have less damage when exposed to cement compared to linear rail bearings. The wheel will get 2 types of force, namely radial and frictional forces. Radial and frictional forces will be obtained from the weight of the load on the wheels. It takes a wheel that has sufficient strength to withstand the load on it, has 2 degrees of freedom, and is inexpensive.

Table 4. Friction value on each wheel

Wheel	Friction force	Radial style
x	3,797 N	9,909 N
y	6,04 N	14,71 N
z	11,772 N	14,715 N

Power Thread Selection. The power thread is used to convert rotational motion into translation. The power thread is driven by a motor which is connected by coupling to the power thread. To convert the rotational motion of the motor into translational motion on the workbench, the thread requires a nut that is connected directly to the workbench.

Motor Selection. The motor used to drive the power thread must have a force that is more than the static friction on the wheel. The motor must be able to accurately control where the workbench will move. The motor must be able to rotate in 2 rotation directions, clockwise and counterclockwise. Selected feed speed of 240 mm/min. The power thread has a lead l of 8 mm, so 1 rotation will result in a table displacement of 8 mm. Selected stepper motor nenma 17 with 12V and 1.5A and a maximum torque of 0.35 Nm. The motor is equipped with a coupling to connect the motor shaft with the leadscew.

Frame Design. The frame needed in the design of this CNC concrete carving machine must have rigid, strong, tough properties, resistant to vibration, and does not react to concrete. It takes a frame that can support the weight of the load, as a location for mounting the motor, bearing, and base for the wheels. Two main frames are designed, namely the base and the column. The base is used to support vertical loads and provide balance for the machine as a whole. The column frame is designed to facilitate spindle displacement on the z-axis.

The frame material was chosen in the form of carbon steel type L 30 × 30 because the shape and strength are close to the required requirements. The iron is painted so that it does not react with the wet cement to cause corrosion.

Spindle Arm Design.

The spindle arm is expected to be able to withstand the rotating load from the spindle, can reduce vibrations that occur in the spindle, and can be paired with a power screw nut. The arm is designed so that the tool blade can be at a point where each part of the workpiece can pass. The point is located 300 mm from the column. The arm must be able to support the weight of the spindle. The spindle used has a m_{sp} weight of 1.4 kg. The material used is 20 × 3 iron. The cross-sectional area of A_{sp} which works against the weight of the spindle has a value of 2734.049 mm^2. The compressive load on the arm is (Fig. 3)

$$\sigma_{sp} = \frac{1,4 \cdot 9,81}{2734,049}\,\text{MPa}$$

$$\sigma_{sp} = 0,00502\,\text{MPa}$$

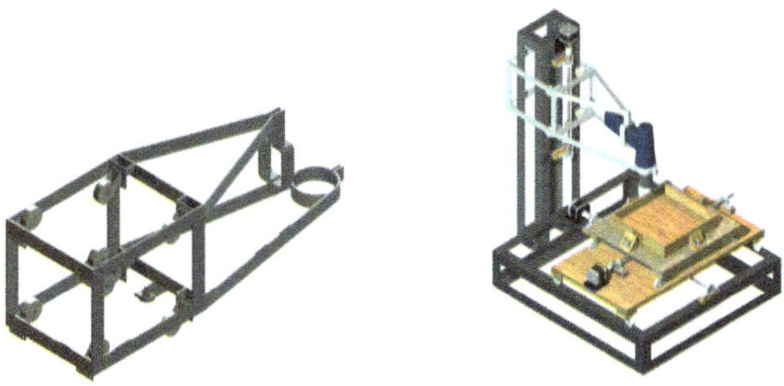

Fig. 3. Machine Spindle Arm Design (right) and Concrete Engraving CNC (left)

Machinery Manufacturing. Machine production begins with the frame and spindle arm. The frame and spindle arms are welded to design and bolted between the column frame and the base. Then the bearings are united with the base frame. Followed by the installation of leadscew and connected to the stepper motor. In line with the installation of the leadscrew, the workbench will be installed and connected with the nuts. The process is repeated on the x table installation. After all the x and y table parts are installed, then the spindle arm is inserted into the column frame. The bearing and leadscew will be joined after the spindle arm is properly installed. New z-axis stepper motor is installed and connected with leadscew. Once everything is installed, the last part is to install the spindle and make sure the spindle is perpendicular to the x table.

The test command is compiled on the computer and sent to the Arduino to be forwarded to the stepper motor to move according to the given coordinates. An electronic circuit is needed to run a CNC machine, which is an arduino, a CNC shield, a stepper motor driver, and a jumper cable. Arduino and its CNC Shield are connected to a computer and a motor. Figure 8 shows the main wiring scheme between PC, power supply, arduino, CNC shield and stepper motor in the study (Figs. 4, 5 and 6).

Description :
1. Base Frame
2. Column Frame
3. Bearing
4. Leadscew
5. Motor Stepper
6. Wheel
7. Workbench Y
8. Workbench X
9. Spindle Arm
10. Spindle

Fig. 4. Machine assembly result

Tool Testing. Testing the accuracy of the machine is done by carving the workpiece in the form of concrete that is undergoing a hydration reaction. Then the dimensions of the engraving on the workpiece will be compared with the design dimensions. In this way, the error value of the machine will be obtained according to formulas 10 and 11. There will be 2 types of accuracy testing. Testing the age of concrete and testing the carving pattern. Testing of the age of concrete is used to take the right time so that the machining can be done neatly. Testing of patterns is carried out to see how much accuracy and precision the tool is.

Machine Testing Against Concrete Age. The first test was carried out to find out at what hour the best value for engraving was on soft concrete. The test was carried out in the first 7 h of cement hardening. The tool used is a 5 mm end mill. The test was carried out with a feeding speed of 240 mm/min, a depth of 5 mm, and a spindle speed

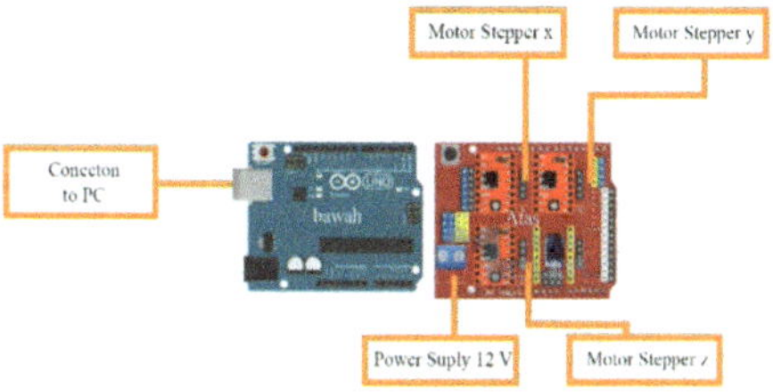

Fig. 5. Arduino and CNC shield wiring schematic

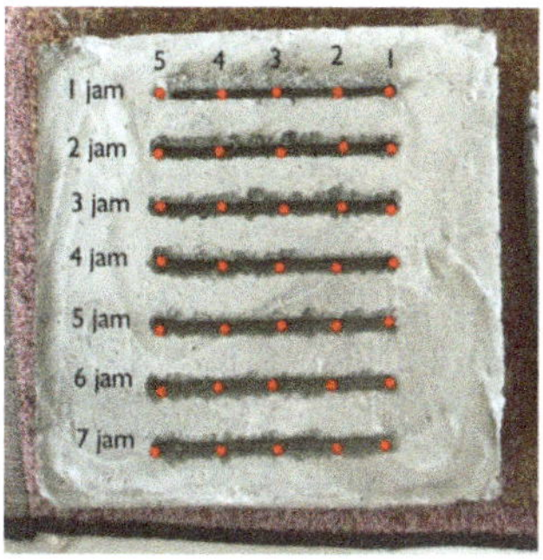

Fig. 6. Machine testing against concrete age.

of 400 rpm. The test is carried out by drawing a line as far as 100 mm. Then from that line, the line width values are taken at several points as in Fig. 7. The point is obtained by dividing the length of the line into 5. So the distance between the points is approximately 25 mm. Data retrieval is done by repeating 1 time.

From Tables 5 and 6 on the measurements made, it can be seen that the smallest error value was obtained in engraving at the age of 1 h of concrete after the hydration reaction occurred.

Table 5. Line width at each point (mm)

	5	4	3	2	1
1 h	4,4	5,15	4,6	5,35	5,45
2 h	6	10,15	5,35	6,5	7,9
3 h	7,55	9,15	10,25	14,2	10,25
4 h	11,85	5,9	10,8	8,3	7,85
5 h	9,7	10	11,45	9,05	11,2
6 h	12,4	10	8,9	8,2	8,35
7 h	7	10,5	11,25	10,9	9,65

Table 6. Error percentage at each point

	5	4	3	2	1
1 h	12%	3%	8%	7%	9%
2 h	20%	103%	7%	30%	58%
3 h	51%	83%	105%	184%	105%
4 h	137%	18%	116%	66%	57%
5 h	94%	100%	129%	81%	124%
6 h	148%	100%	78%	64%	67%
7 h	40%	110%	125%	118%	93%

4 Testing Against Patterns

Testing of the workpiece is carried out on several basic shapes that often exist in the machining process. The shapes used are square, hexagonal and circle. The difig pattern on the computer is then given the CAM program and translated by arduino as movement to the stepper motor. The CNC program for engraving squares, hexagons and circles is shown in Fig. 8 (Table 7).

As a reference to facilitate the research, a workpiece test was taken from Putra's research [5], and a measurement technique approach was given to Jekky's research [6] (Table 8).

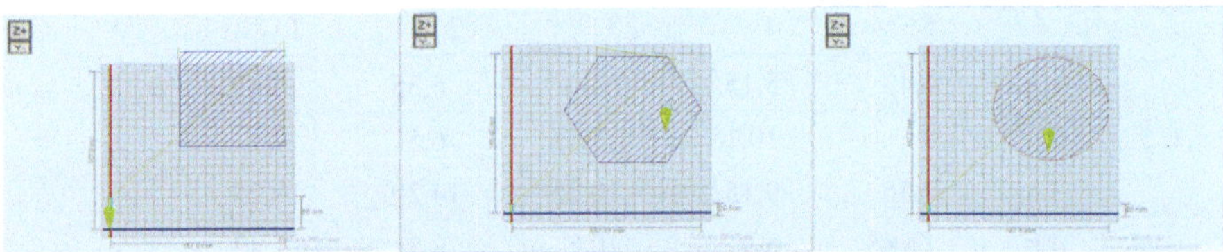

Fig. 7. CNC motion program or pattern testing

The tool used is a 5 mm end mill. The test was carried out with a feeding speed of 240 mm/min, a depth of 5 mm, and a spindle speed of 400 rpm. The test is carried out between 1–2 h after the cement sand and water are combined. The research was conducted with 1 repetition. Measurements were made when the hardening was more than 12 h. The measuring instrument used is a caliper (Table 9).

Based on the pattern test, the average error value of the square to its 0 coordinates is 4.33%. The average error value from the error test on the square side is 0.763%. The average error value of the hexagonal to its 0 coordinates is 1.347%. The average error value of the error test on the hexagonal side is 2.64%. The average error value of the circle to its 0 coordinates is 3.97%. The average error value of the error test on the diameter of the circle is 0.70%. From these values, it can be seen that the machine has better precision performance than accuracy. Because the object side has a much smaller

Table 7. Point error 0 to square point

Number	Plan value (mm)	Actual value (mm)	Error	Error percentage
1	70,71	66,1	0,065	6,52%
2	158,11	151,2	0,044	4,37%
3	212,13	209	0,015	1,48%
4	158,11	150,3	0,049	4,94%

Table 8. Error on square side

Number	Plan value (mm)	Actual value (mm)	Error	Error percentage
1–2	100	100,7	0,007	0,70%
2–3	100	99,45	0,0055	0,55%
3–4	100	99,2	0,008	0,80%
4–1	100	99	0,01	1,00%

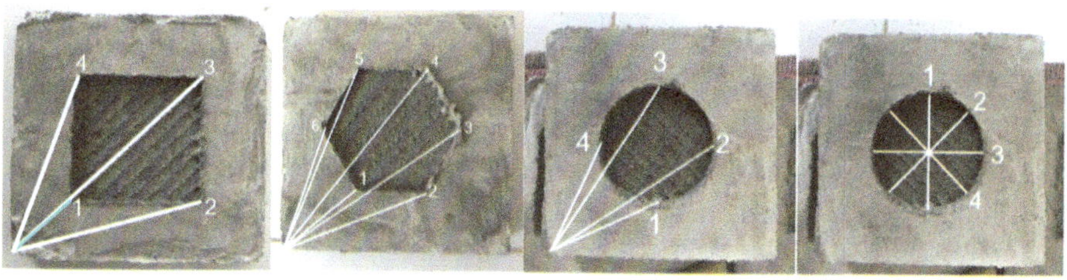

Fig. 8. Square hexagonal and circle carving test

Table 9. Error point 0 to the hexagonal point

Number	Plan value (mm)	Actual value (mm)	Error	Error percentage
1	84,90	82,6	0,027	2,71%
2	138,59	135,34	0,023	2,35%
3	188,68	187	0,009	0,89%
4	199,98	20	0,900	0,01%
5	167,31	169	0,010	1,01%
6	107,70	106,5	0,011	1,11%

Table 10. Error on hexagonal side

Number	Plan value (mm)	Actual value (mm)	Error	Error percentage
1–2	60	58,85	0,0192	1,92%
2–3	60	58,95	0,0175	1,75%
3–4	60	54,9	0,085	8,50%
4–5	60	60,55	0,00916	0,92%
5–6	60	59,05	0,0158	1,58%
6–1	60	59,3	0,0117	1,17%

Table 11. Error point 0 to circle point

Number	Plan value (mm)	Actual value (mm)	Error	Error percentage
1	111,803	105,4	0,057	5,73%
2	180,278	178	0,013	1,26%
3	180,278	178	0,013	1,26%
4	111,803	103,3	0,076	7,61%

Table 12. Error in circle diameter

Number	Plan value (mm)	Actual value (mm)	Error	Error percentage
1	100	99,9	0,001	0,10%
2	100	101,2	0,012	1,20%
3	100	99,3	0,007	0,70%
4	100	100,8	0,008	0,80%

error than the location of the pattern with respect to point 0. Except for hexagonal, there is an undesirable hardening of the concrete on 3–4 sides (Tables 10, 11 and 12).

5 Conclusions

Based on the research, it can be concluded several points, including:

1. The CNC concrete engraving machine has the ability to cut concrete that is still undergoing hydration reactions.
2. Concrete carving CNC machine can move in 3 dimensions.

3. The CNC concrete engraving machine is directly connected to a PC and is capable of engraving according to the input pattern.
4. The best engraving results were found when the concrete was aged 1 h after the hydration reaction started.
5. Engraving produces a pattern that matches the design, but with slightly distorted coordinates.

References

1. Suharto, K.P., Supriyadi, A., Karnowahadi: Grafir batu alam dengan mesin cnc router untuk meningkatkan ragam produk kreatif indonesia. dalam Jurnal Simetrik. No. 2 Vol. 8. hlm 121–125 (2018)
2. Callister Jr, W.D.: Materials Science and Engineering An Introduction, 8th edn. John Wiley & Sons, Inc, Hoboken (2009)
3. Naini, U.: Studi Komparasi Mutu Kayu Jati, Kayu Mahoni, Kayu Johar, Kayu Akasia Dan Kayu Meranti Di Surakarta Antara Hasil Uji Laboratorium Dengan Analisis SNI 7973–2013". 6–11 (2018)
4. Engineering ToolBox, Friction and Friction Coefficients (2004). https://www.engineeringtoolbox.com/friction-coefficients-d_778.html. 2 July 2021
5. Putra, F.W., Febriantoko, B.W., Hariyanto, A.: Kalibrasi Sumbu Y Terhadap (2016)
6. Jekky, B., Syaifurrahman, E.K.: Analisis Nilai Penyimpangan Mesin Bubut CNC Pada Benda Kerja Plastik Nylon dalam Jurnal Teknologi Rekayasa Teknik Mesin Vol. 2, No. 1, 2021: 01–05 (2020)

Numerical Analysis of Small-Scale Horizontal Type Wind Turbine Using the Computational Fluids Dynamics (CFD) Method

Husaini(✉), M. Oki Nugraha Lubis, Darwin, and Akhyar

Department of Mechanical Engineering, Universitas Syiah Kuala, Darussalam, 23111 Banda Aceh, Indonesia
{husainiftm,akhyar}@unsyiah.ac.id

Abstract. The small-scale Horizontal axis wind turbine (HAWT) is presently being used for environmentally friendly power plants, and is found to be affordable at pretty low costs. The system uses aerodynamic blades (airfoils) fitted to a rotor, which is either positioned upwind or downwind. Therefore, the analysis of increasing the performance of the HAWT is important for its overall capability. Based on these conditions, this numerical study aims to determine the effects of the angle of attackcharacteristics on blades ranging from 15–10°, using the NACA 4412 airfoil. This airfoil was applied from the blade base (root) to the tip, against the torque and mechanical power of the turbine. The aerodynamic characteristics produced with the rotor diameter in the HAWT was 4,90625 m^2 along the swept of 2.5 m, as the TSR (tip speed ratio) of 6 on the wind speed ranged between 3–7 m/s. The results showed that 0.32, 54.49%, and 150.9 W of the coefficient, efficiency, and average mechanical powers of the turbine were produced in the simulation process, respectively. This indicated that the mechanical power of the HAWT increased with higher wind speed.

Keywords: Horizontal Axis Wind Turbine (HAWT) · NACA 4412 · Torque · Mechanical power of the turbine

1 Introduction

The limited availability of essential energy sources and increase in demand is found to cause the high costs of fossil efficiency, which is known for influencing air pollution in the world. This is due to the extraction and transportation of the fuel to power plants, for utilization. Based on these conditions, a breakthrough in renewable energy sources, such as wind efficiency, is carried out [1]. The application of this energy source has the advantages of low air pollution and economic use in Indonesia. Indonesia as a country near the equator, has a low annual average wind speed of 2–7.5 m/s, with the utilization of this energy being very wide like a small-scale turbine, through the wind power conversion system [2]. This turbine has the ability to convert wind kinetic energy into mechanical power, which is further transformed to electrical sources through the utilization of generators.

Akhyar et al. (Eds.): ICECME 2021, LNME, pp. 274–286, 2023.
https://doi.org/10.1007/978-981-19-3629-6_29

The study of Nikhade and Thakre (2017) was based on the output power of a 3-blade horizontal axis with a 2.28 m diameter, as the results showed the rotational values of 172, 229, 287, 344, and 401 rpm, at wind speeds of 3, 4, 5, 6 and 7 m/s, respectively [3]. Furthermore, Hugh Piggott researched the analysis of horizontal axis wind turbines (HAWT) with a 2.4 m diameter, which produced a rotation of 150–300 rpm. Based on the research, it produced 100 W at a wind speed of 6 m/s [4]. Jayakrishnan and Dhruv also conducted research based on the analysis of computational methods from NACA 0012, 0014, 4412, and 4418. The results showed that the NACA 4412 produced the most efficient performance of small-scale wind turbines [5].

The study of A.K Wright and D.H Wood reported that most of the initial torque were generated near the hub, based on the performance of the tested method. It also stated that most of the torque extracting power was obtained from the blade tip region [6]. Meanwhile, the study of Abdullateef, concluded that a higher TSR value changed the performance of the wind turbine, as the tip speed ratio of 0.6–11 provided the best efficiency [7].

This study aims to analyze the effect of the angle of attack characteristics ranging from 15–10° on blades, through the use of the NACA 4412 airfoil. Based on the analytical-numerical method, these blades are designed to obtain the optimal mechanical and coefficient power values of the turbine, at wind speeds of 3–7 m/s.

2 Methodology

Preprocessing. The modeling for wind turbine blades was carried out with vertical axis NACA 4412, due to the airfoil having an efficient optimization capability than other series [8]. This process was performed by using computer-aided design software to produce an original model, in order to avoid errors. Based on specifications, the used horizontal wind turbine blade and the dimensions of the rotor rotating region were 1300 mm. This turbine was further modeled with three blades having diameters (d) and lengths (l) of 2500 and 1125 mm, respectively. The design also used the standard profile of The Neighborhood Assistance Corporation of America (NACA) 4412, with Reynold Numbers 1,000,000, as well as maximum thickness and chamber of 12 and 4% at a chord of 30 and 40%, respectively. This airfoil was applied from the blade base to the tip, with variations in the angle of attack (AOA) ranging from 15–10°, as shown in Fig. 1.

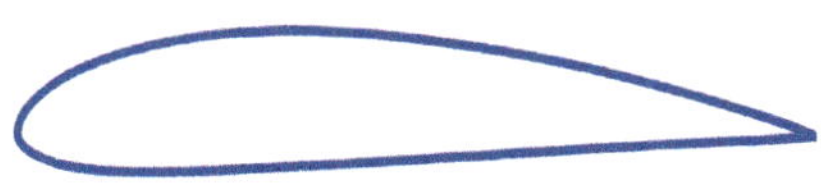

Fig. 1. NACA 4412 profile design [9]

The blade is divided into eleven parts of the airfoil, with variations in chord length and angle of attack as stated above. The division of the AOA on the blade is shown in Table 1.

Table 1. Blade design characteristics

Airfoil section	Radius *r/R* (mm)	Chord length (mm)	Angle of attack (°)
1	250	200	15
2	350	190	14,5
3	450	180	14
4	550	170	13,5
5	650	160	13
6	750	150	12,5
7	850	140	12
8	950	130	11,5
9	1050	120	11
10	1150	110	10,5
11	1250	100	10

Based on the data acquisition in Table 1, the design of a wind turbine blade was modeled, as shown in Fig. 3a, which consisted of a combined airfoil with different dimensions in one blade. Figure 3b also showed three blades, which was a combination of those in Fig. 2a.

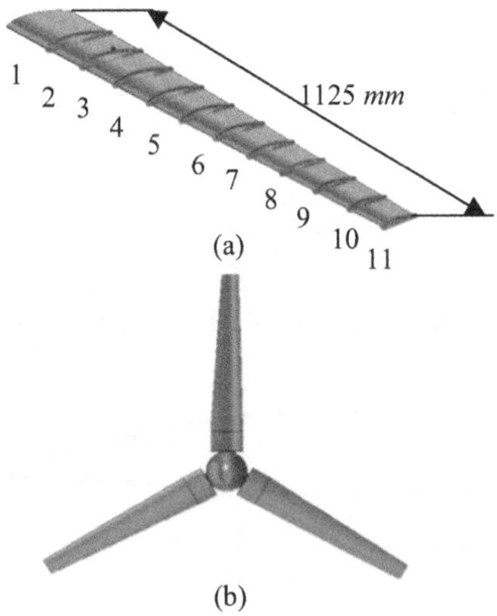

Fig. 2. The design of the blade geometry. (a) The division of the airfoil on the blade. (b) Front view of the blade geometry.

Based on this study, the 3-blade design was selected, due to being superior to other turbine types. This superiority was based on the wind speed efficiency, torque, output power, and relatively low-cost treatments. In the modeling analysis, three geometries were observed in the analytical model, with the first, second, and third being the wind turbine blade, rotating body zone, and boundary condition domain, respectively, as shown in Fig. 3.

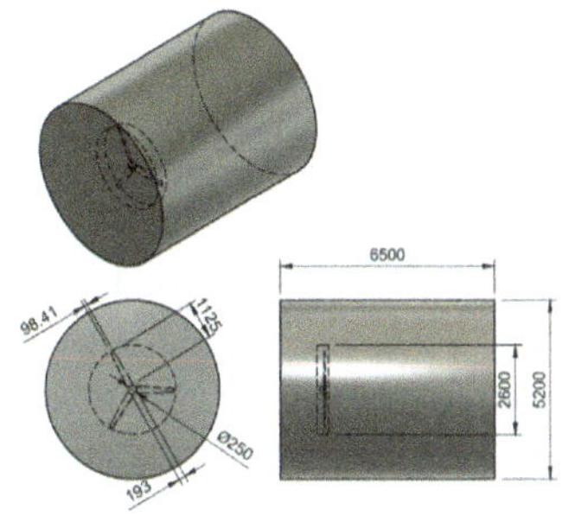

Fig. 3. Wind turbine domain model

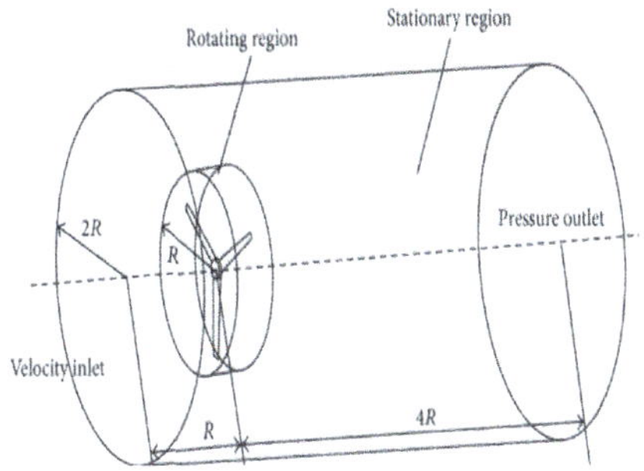

Fig. 4. Dimensions of the analysis model

According to the model dimensions, the Radius *(m)* of the wind turbine blade was obtained, with one parameter also applied to all the dimensional parts of the boundary conditions, as shown in Fig. 4.

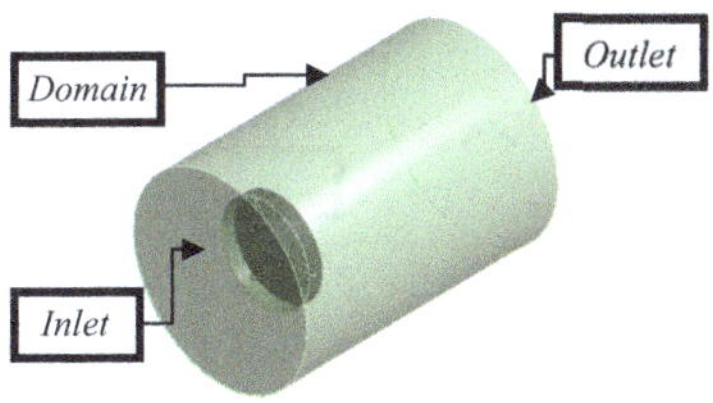

Fig. 5. Determination of boundary conditions.

Postprocessing. The accuracy of the results highly depended on the pre-processing stage, where the inlet, outlet, and domain were specified. The inlet was observed near the turbine angle, while the outlet was located at the rear position, with fluid boundary conditions also found to be 5200 mm in diameter. The determination of these conditions is further shown in Fig. 5.

The analysis was numerically carried out 5 times, with the wind and rotational speed data of different turbines listed in Table 2.

Table 2. Input data for testing

Test	Wind Speed (m/s)	Rotation Speed (rpm)
Numerical Analysis I	3	136
Numerical Analysis II	4	183
Numerical Analysis III	5	229
Numerical Analysis IV	6	275
Numerical Analysis V	7	321
Average speed of rotation of the rotor 229		

Based on the meshing process after the determination of boundary conditions, the model was divided into several small elements having a characteristic mesh tetrahedrons of 129545 nodes and 686674 elements. This contained 122036/7509 nodes and 649511/37163 elements on the rotating body and stationary domain, respectively, as shown in Fig. 6.

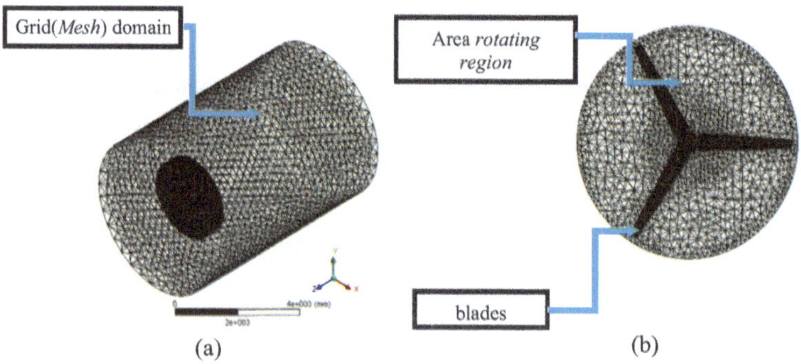

Fig. 6. The meshing stage of the wind turbine model. (a) Domain stage, (b) Blade stage.

Solve Setup. Fluid is defined as a substance that continuously changes shape when followed up by shear stresses of any magnitude [10]. At the processing stage of the flow analysis, the fluid was simulated by using the viscous model equation, realizable k-epsilon, and scable wall functions. The transport equations of the Launder and Spalding models were further described in Eqs. (1), (2), and (3).

$$\frac{\partial}{\partial t}(\rho k) + \frac{\partial y}{\partial x_j}(\rho k u_j) = \frac{\partial y}{\partial x_j}\left[(\mu + \frac{\mu_t}{\sigma_k}\frac{\partial k}{\partial x_j}\right] + G_K + G_b - \rho\varepsilon - Y_M + S_k \quad (1)$$

dan

$$\frac{\partial}{\partial t}(\rho\varepsilon) + \frac{\partial y}{\partial x_j}(\rho\varepsilon u_j) = \frac{\partial y}{\partial x_j}\left[(\mu + \frac{\mu_t}{\sigma_\varepsilon}\frac{\partial \varepsilon}{\partial x_j}\right]\rho C_1 S_\varepsilon - \rho C_2\frac{\varepsilon^2}{k + \sqrt{v\varepsilon}} + C_{1\varepsilon}\frac{\varepsilon}{k}C_{3\varepsilon}G_b + S_\varepsilon \quad (2)$$

$$C_1 = max\left[0, 43, \frac{\eta}{\eta + 5}\right], \eta = \frac{S}{k_\varepsilon}, S = \sqrt{2S_{ij}S_{ij}} \quad (3)$$

where G_K = the turbulent kinetic energy generation due to the average velocity gradient, G_b = the turbulent kinetic energy generation due to buoyancy, Y_M = the fluctuating dilatation contribution in compressible turbulence to the overall dissipation level, C_1 and $C_{1\varepsilon}$ = constants, σ_ε and σ_k = Prandtl numbers turbulent for k and ε, respectively, S_k and S_ε = user-defined source terms [11]. Furthermore, the Cell Zone Conditions were set into the Mesh Motion model, which was in the form of a Multiple Reference Frame (MRF). After this, the Solution Methods and Initialization were arranged and set into a simplec scheme and hybrid mode, respectively.

3 Results and Discussion

Simulation Result. The results on the isosurface pressure contour were shown at wind speeds of 3–7 m/s, as the highest and lowest values occurred at 7 and 3 m/s (773,387 and 138,073 Pa), respectively. This indicated that greater wind speed produced higher pressure through the turbine rotor. The results are further shown in Fig. 7.

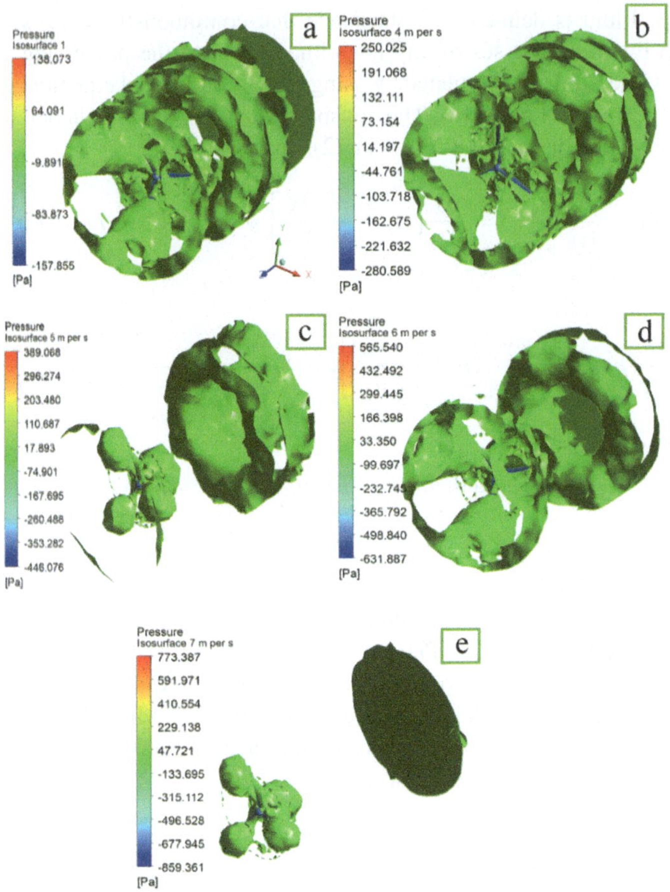

Fig. 7. The contour of pressure on the wind turbine model. (a) At 3 m/s, (b) At 4 m/s, (c) At 5 m/s, (d) At 6 m/s, (e) At 7 m/s.

The results on the blades were further shown as velocity contours at 3–7 m/s, with the highest and lowest values occurring at 7 and 3 m/s (48.050 and 19,814 m/s), respectively. Furthermore, an increase was detected based on the value of the wind speed around the blade. This was due to the collision of the wind with the blade, especially at the rotating tip. In addition, the greater values of wind speed produced higher velocity around the blade. These results are further shown in Fig. 8.

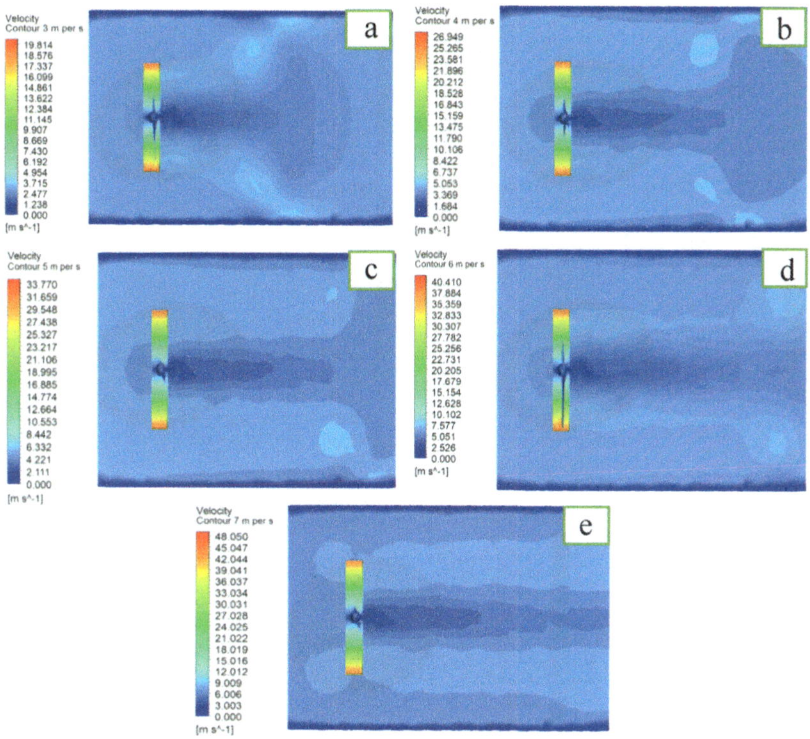

Fig. 8. Velocity contour on the turbine model. (a) At 3 m/s, (b) At 4 m/s, (c) At 5 m/s, (d) At 6 m/s, (e) At 7 m/s.

The results on the blade were also shown based on the contour velocity stn frame of 3–7 m/s, as the highest and lowest values occurred at 7 and 3 m/s (48.659 and 20,433 m/s), respectively. According to the wind speed values around the blade, an increase was also observed. This was due to the collision of the wind with the blade, especially at the rotating tip. Therefore, the greater values of the wind speed produced higher velocity around the blade. These results are further shown in Fig. 9.

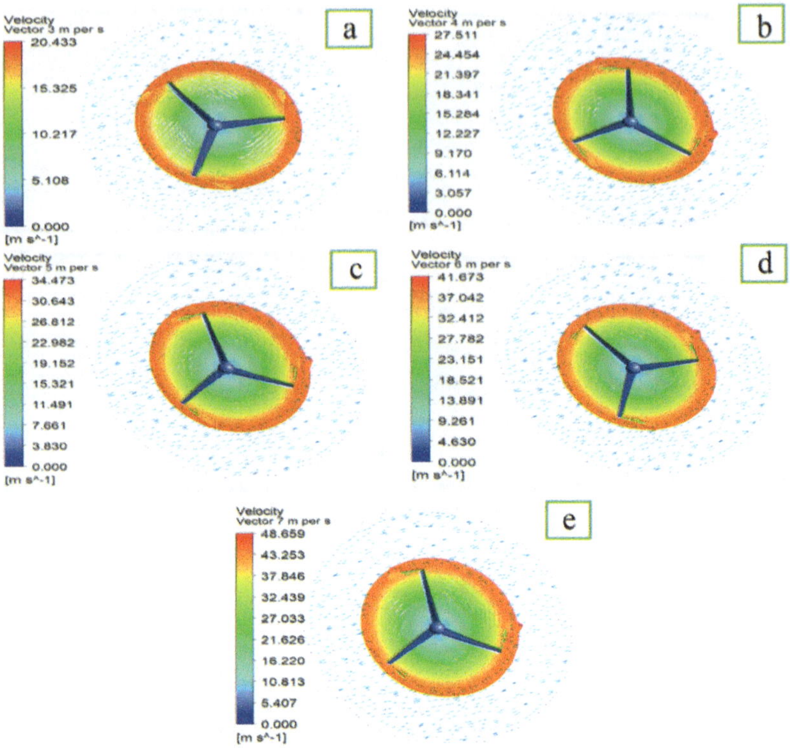

Fig. 9. Velocity vector on the wind turbine model. (a) At 3 m/s, (b) At 4 m/s, (c) At 5 m/s, (d) At 6 m/s, (e) At 7 m/s.

3.1 Wind Turbine Torque

The torque values from the turbine with TSR of 6 were obtained from numerical simulations and wind blade calculator, as shown in Table 3.

Table 3. Torque values of wind turbines

Wind velocity (*m/s*)	Rotating speed		Torque (*N.m*)		
	(*rpm*)	(*rad/s*)	CFD	Wind Blade calculation	Error (%)
3	137	14,24	1,79	2,53	29,2
4	183	19,16	3,28	4,49	26,9
5	229	23,98	5,10	7,02	27,3
6	275	28,79	7,32	10,11	27,5

(*continued*)

Table 3. (*continued*)

Wind velocity (*m/s*)	Rotating speed		Torque (*N.m*)		
	(*rpm*)	(*rad/s*)	CFD	Wind Blade calculation	Error (%)
7	321	33,61	9,92	13,76	27,9
Average value					
5	229	23,95	5,48	7,58	27,76

3.2 CFD and BEM Comparison Results

This indicated the results of both the BEM and numerical simulation method, based on the determination of the mechanical power value of wind turbines. In the analysis of both methods, the turbine mechanical power values were 47.68, 113.28, 221.28, 382.91, and 607.7 W (BEM), as well as 25.48, 62.84, 122.3, 210.7 and 333.4 W (Numerical Simulation Method), at wind speeds of 3, 4, 5, 6, and 7 m/s, respectively. Furthermore, the average CP value in this study was 0.322, through the numerical simulation method. The values of mechanical power, CP, and efficiency of wind turbines at TSR 6 are shown in Table 4.

Table 4. The value of turbine mechanical power, power coefficient, and wind turbine efficiency at TSR 6.

Wind velocity (*m/s*)	Wind power (*W*)	Power turbine BEM (*W*)	Power turbine CFD (*W*)	*CP*	Efficiency η(%)
3	81,13	47,68	25,48	0,314	53,01
4	192,3	113,28	62,84	0,326	55,14
5	375,6	221,60	122,3	0,325	54,94
6	649	382,91	210,7	0,323	54,79
7	1030,7	607,7	333,4	0,323	54,58
Average value					
5	465,7	278,6	150,9	0,322	54,49

Based on this study, numerical analysis was also carried out using a similar method with a different TSR value of 7. The values of mechanical power, CP, and efficiency of wind turbines at TSR 7 are shown in Table 5.

Table 5. The value of turbine mechanical power, power coefficient, and wind turbine efficiency at TSR 7.

Wind velocity (*m/s*)	Wind power (*W*)	Turbine power BEM (*W*)	Turbine power CFD (*W*)	*CP*	Efficiency $\eta(\%)$
3	81,13	47,68	57,28	0,706	119,13
4	192,3	113,28	136,64	0,710	119,81
5	375,6	221,60	267,12	0,711	119,98
6	649	382,91	439,15	0,676	114,07
7	1030,7	607,7	721,28	0,699	117,95
Average value					
5	465,7	278,6	324,29	0,322	118,18

Based on this study, the relationship curve of mechanical power and wind speed is shown in Fig. 10.

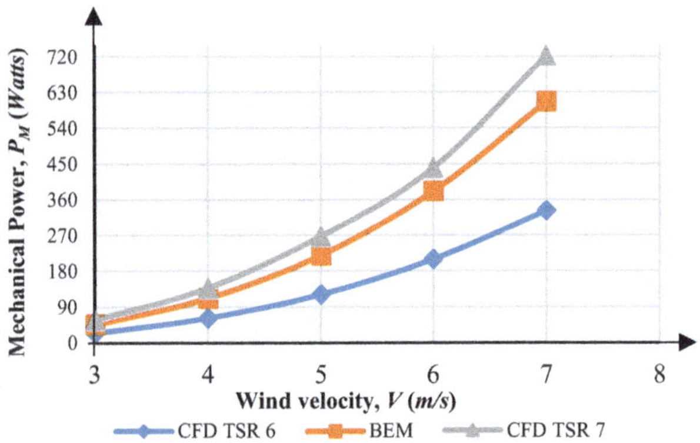

Fig. 10. The curve of the relationship between mechanical power and wind speed.

Furthermore, the plot of the relationship curve between the mechanical power and CP value of the turbine to the effect of wind speed, is shown in Fig. 11.

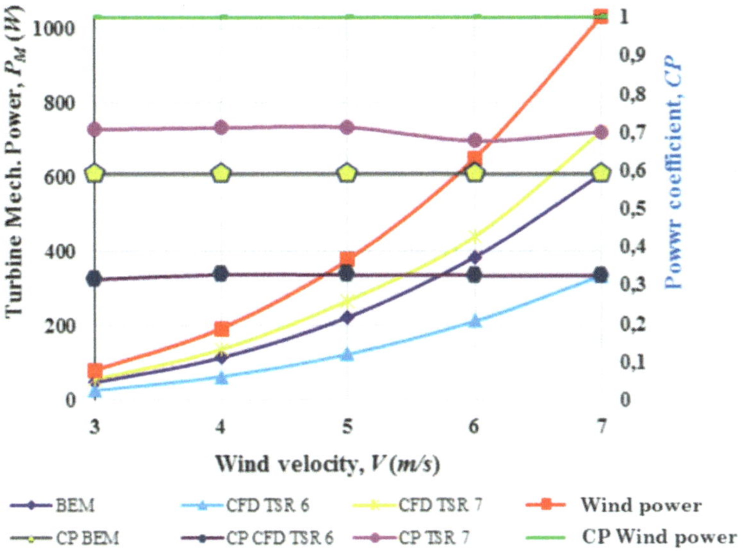

Fig. 11. The curve of the relationship between mechanical power and CP to wind speed.

4 Conclusion

Based on the aerodynamic analysis on HAWT blades through the finite volume method, several conclusions were obtained,

1. The average mechanical power value through numerical simulations was 150.9 W.
2. The average CP value through numerical simulations was 0.322, with the efficiency acquisition of 54.49%. This was acceptable because the CP value did not exceed that of the Betz boundary theory.
3. The turbine blade analysis used NACA 4412 airfoil at AOA (angle of attack) of 15–10°, which produced a low CP value of the wind turbine. This was caused by the angle of attack on the airfoil entering the stall region, which was a critical for its decreased and increased lift and drag forces, respectively.
4. The results of TSR 6 were better than TSR 7. This was because the CP value of TSR 7 was generally greater than that of wind turbines.

Acknowledgment. The authors would like to thank Universitas Syiah Kuala for providing partial financial support to this research through Research Grant No. 14/UN11.2.1/PT.01.03/ PNBP/2021.

References

1. Yu, J., Yu, Y., Liu, P.: Horizontal axis wind turbine numerical simulation of two dimensional angle of attack. Adv. Mat. Res. **619**, 111–114 (2013)

2. Musyafa', A.: Comparative analysis of small-scale wind turbine design for the low rate wind speed. Dept. of Eng. Physic, ITS, Surabaya, Indonesia 1, 5–17 (2013)
3. Nikhade, S., Kongre, S., Thakre, S., Khandare, S.: Comparison of computational fluid dynamics and experimental power output of a micro horizontal-axis wind Turbine. In: Proc. of the Int. Conf. on Sci. & Eng. for Sust. Devel. Bangkok, Thailand, 2017, pp. 11–18 (2018)
4. Piggott's, H.: Scoraig Wind Electric. Windmill building workshop courses, North West Scotland (2003)
5. Jayakrishnan, D.: Design and optimisation of a low reynolds number airfoil for small horizontal axis wind Turbines. IOP Conf. Ser.: Mater. Sci. Eng. **377,** 012053 (2018)
6. Wright, A.K., Wood, D.H.: The starting and low wind speed behaviour of a small horizontal axis wind turbine. J. Wind Eng. Indust. Aerody **92**, 1265–1279 (2004)
7. Abdullateef, A., Dhari, Y.: Zaid, optimal performance of horizontal axis wind Turbine for low wind speed regime. Int. J. Multi. Curr. Res. **2**, 159–164 (2014)
8. Suresh, A., Rajakumar, S.: Design of small horizontal axis wind Turbine for low wind speed rural applications. Proc. Mat. Today. **617**, 1–7 (2019)
9. Vardar, A., Alibas, I.: Research on wind Turbine rotor models using NACA profiles. Renew. Energy **33**, 1721–1732 (2008)
10. Bruce, R., Munson, Y.F., Donald, H.O.: Theodore, Fundamentals of Fluid Mechanics. Printed in the United States of America - USA (2013)
11. Ansys Fluent -2006, Realizable k– Model, from the website https://www.sharcnet.ca/Software/Fluent6/html/ug/node480.html. Accessed 10 Dec 2018

Assess the Level of Temperature Uniformity in Intermediate Level Incubators Type Incandescent Lamp and Heater Plate

Ditya Riswandha, Ahmad Syuhada(✉), and Hamdani

Magister of Mechanical Engineering Study Program, Department of Mechanical and Industrial Engineering, Faculty of Engineering, Universitas Syiah Kuala, Banda Aceh, Indonesia
ditya1999@mhs.unsyiah.ac.id, {ahmadsyuhada, hamdani}@unsyiah.ac.id

Abstract. One of the business activities favored by the villagers is poultry farming. This happened because the harvest period is fast and the market is easy, especially for public consumption in the province of Nanggroe Aceh Darussalam. Due to the large number of people who are trying to raise ducks, chickens, and laying eggs, problems have arisen with the provision of these poultry seeds. So far, poultry seeds in the Aceh Province have been obtained from a small portion of the populationAcehnese themselves and the rest were imported from North Sumatra. This causes the fluctuating supply of seed poultry which will eventually disrupt the community's poultry farming business in Aceh Province. Seeing these problems, this research wants to produce a multilevel incubator heating system that maintains a uniform temperature for better incubator efficiency. The test is carried out by paying attention to the temperature distribution in the egg incubator. Thus, measuring the distribution temperature with a value range of 37 °C–38 °C at certain points through the test channel is the main thing that must be done. The variation in this research is the number of incandescent lamps/heating plates (4 and 6) which differs whether or not the zinc plate is under the incandescent lamp/heating plate with a power of 100 and 112 Watts for each heater. The heater in the incubator room will turn on at 37 °C and will turn off automatically at 38 °C due to the influence of the thermostat equipment. After testing, the temperature was 0.7 °C for 4 (four) heating plates without using a zinc plate under the heating plate and 0.8 °C for 4 (four) incandescent lamps without using a zinc plate under the incandescent lamp. However, there is a small comparison, namely 0.4 °C for 6 (six) heating plates using a zinc plate under the heating plate and 0.5 °C for 6 (six) light bulbs using a zinc plate under an incandescent lamp. Also 96% test results with various methods. It is a good heat transfer and efficient process of hatching egg incubators.

Keywords: Incubator egg · Temperature · Lamp incandescent · Heater plate

Akhyar et al. (Eds.): ICECME 2021, LNME, pp. 287–298, 2023.
https://doi.org/10.1007/978-981-19-3629-6_30

1 Introduction

One of the business activities favored by the village community is a poultry farming business. This happens because the harvest is fast and the market is too easy, especially for public consumption in Province of Nanggroe Aceh Darussalam. As a result of many people who try to raise ducks and chickens, the problem that arises is the provision of seeds for the poultry. So far, poultry seeds in the area of Aceh Province are obtained by a small part of the community of Aceh itself and the rest were imported from North Sumatra. This causes the fluctuating supply of seeds poultry which will eventually disrupt the business community poultry farming in the Aceh area [10–15].

In 2018 (4 688 120.66 tons) 2019 (4 753 382.23 tons) 2020 (5 044 394.99 tons) laying hens egg production at the National and in 2018 (57072.44 tons) 2019 (12312.10 tons) 2020 (13065.87 tons) laying hens egg production in Aceh province. Where as in 2018 (3 409 558.00 tons) 2019 (3 495 090.53 tons) 2020 (3 275 325.72 tons) laying hens egg production at the National was also 2018 (18278.78 tons) 2019 (36748.34 tons) tons) 2020 (34437.67 tons) broiler meat production in Aceh province [1].

Dongo conducted research on an incubator using palm oil as an energy source to produce chicks with a hatching rate reaching 75% [2]. The results of experimental tests conducted by Yadav concluded that the efficiency and hatchability of this automatic egg incubator was 72.22% [3]. Boz's research obtained the value of incubation results ranging from 67.3–85.2% [4]. Sobejana's research resulted in a value of 65–75% hatching eggs [5]. From the results of research by testing tools with setting the incubator room temperature between 38 °C–40 °C, the hatching success rate of 30 eggs was incubated, 24 eggs hatched and 6 eggs failed, with a hatching percentage reaching 80% [6].

However, based on these problems, research was conducted on the effect of branch heat on egg incubators using incandescent lamps as heaters. The aim is to produce an incubator heating system that obtains a uniform temperature to improve the efficiency of the incubator better, by using several variations of lamps and by placing or not having a zinc plate under the lamp which has never been done before, but research on egg incubators has frequently [7–15].

2 Research Method

2.1 Project Development

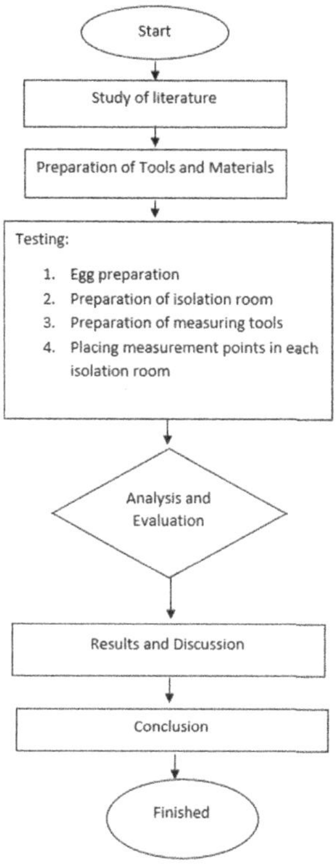

Fig. 1. The flow chart of overall project development

2.2 Egg Hatching

Hatching eggs is a process/effort to incubate poultry eggs. This process includes the development of the embryo inside the egg until the egg breaks and produces chicks. In the process of hatching eggs, it can be done naturally by the broodstock or artificially using an artificial incubator after checking the development of the embryo from day 5 to incubation until 14 days later [16–21].

2.3 Incubator

An Egg incubator/Incubator is a closed room heated by electricity or other artificial heaters used to incubate and incubate eggs. Breeding with incubators is usually done by breeders because relatively many eggs are hatched. Breeders with large capital usually prefer to use incubators because they are more effective and efficient. Usually, the egg incubator is equipped with a heater, an egg player, and a temperature sensor so that the temperature contained in the egg incubator can be stabilized also given the influence of a zinc plate under/above the heating source [22–31].

2.4 Heat Transfer System in Incubator

Heat transfer in the incubator chamber refers to natural poultry brooders in incubating eggs, per-heat transfer from incandescent light bulbs (radiation) to eggs if not using a zinc plate, but when placed zinc plate under the incandescent lamp (conduction) and moveon the egg (convection) [32–37].

2.5 Materials and Materials Used

The following is a list of tools and materials required for the heat transfer process in the incubator to which includes:

1. Digital thermometer,
2. Tube thermometer,
3. Thermostat,
4. Stopwatch,
5. Hygrometer,
6. Heating plate,
7. Incandescent lamps,
8. Partial incandescent lamp plate/cover,
9. Plywood/wood,
10. Glass, and.
11. Egg.

Setting (setting) in the heat transfer process this is the temperature used. The temperature researchers use is 37 °C–38 °C on the thermostat.

2.6 Research System Design

This egg incubator is designed with a size of 120 cm × 80 cm × 125 cm. So that-lam this incubator room can only accommodate as many as 400 hatching eggs, with a capacity of 4 container eggs, and in 1 egg container can accommodate 100 eggs eggs, with zinc plate 70 cm × 50 cm × 0.3 cm and given variations in the number of holes with a diameter of 2 cm. The experimental designs such as Fig. 2, Fig. 3, and Fig. 4.

Caption:

1. Air Vent,
2. Egg Rack,
3. Water Container,
4. Incandescent Lamp / Heating Plate,
5. Air Ventilation, and.
6. Thermal Insulation.

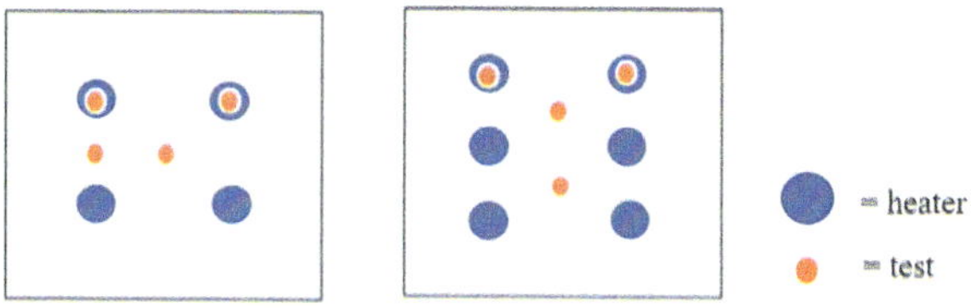

Fig. 2. Top view of the test chamber without using a zinc plate

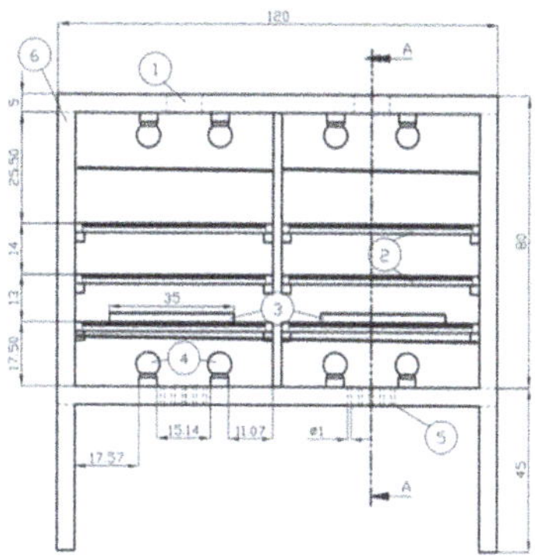

Fig. 3. Front View of Research Equipment

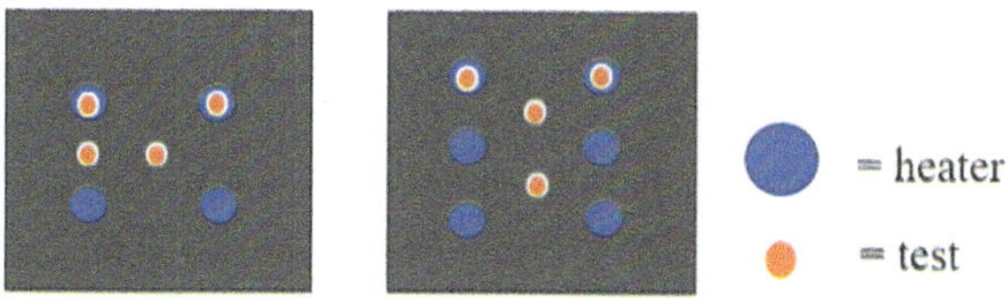

Fig. 4. Top view of the test chamber using a zinc plate

The basic materials for the manufacture of this incubator frame is Plywood/Wood, this is due to the Kon-the thermal conductivity of plywood/wood is smaller compared than aluminum. So, it will be more durable to changes in outside temperature.

3 Research Results and Discussion

Based on the heat transfer test at egg incubator in the Lab. ManipulationThermal, Faculty of Engineering, Syiah Kuala University. There are isseveral temperatures that are almost the same as indicated right in Table 1, Table 2, Table 3, Table 4 and Table 5.

From Table 1, the value of the test results is 0.7C on the number of heating plates as much as 4 without using a zinc plate under the heating plate and 0.6C occurs at conditions by using a zinc plate under the heating plate with a total wattage of 112. Value 0.5 °C and 0.4 °C for the condition of the number of heating plates as much as 6. This is influenced by the convection flow rate from the influence of the zinc plate and the radiation emitted by the heating plate with a total watt of 90.

Table 1. The test results between the number of heating plates and the temperature difference (°C) with or without using a zinc plate under the heating plate

No.	Number of heating plates	temperature difference (C) without using a zinc plate under the heating plate	temperature difference (C) by using a zinc plate under the heating plate
1	4	0.7	0.6
2	6	0.5	0.4

Table 2. The test results between the number of heating plates and the temperature difference (°C) with or without using a zinc plate under the heating plate

No	Number of heating plates	Temperature difference (C) without using a zinc plate on a heating plate	Temperature difference (C) by using a zinc plate on a heating plate
1	4	0.6	0.4
2	6	0.5	0.3

From Table 2, the test result value is 0.6 °C on the number of incandescent light bulbs as much as 4 without using a zinc plate on the heating plate and 0.4 °C occurs at conditions by using a zinc plate on a heating plate with a total wattage of 112. Value 0.5 °C and 0.3 °C for the condition of the number of incandescent bulbs as much as 6. This is influenced by the convection flow rate from the influence of the zinc plate and the radiation emitted by the heating plate with a total watt of 90.

Table 3. The test results between the number of incandescent lamps and the temperature difference (°C) with or without using a zinc plate under an incandescent lamp

No	Number of incandescent bulbs	Temperature difference (C) without using a zinc plate under the lamp	Temperature difference (C) by using a zinc plate under the lamp
1	4	0.8	0.7
2	6	0.7	0.5

From Table 3, the value of the test results is 0.8C on the number of incandescent light bulbs as much as 4 without using a zinc plate above the incandescent lamp and 0.7 °C occurs at condition by using a zinc plate over an incandescent lamp with a total wattage of 100. Value 0.7 °C and 0.5 °C for the condition of the number of incandescent light bulbs as much as 6. This is influenced by the convection flow rate from the influence of the zinc plate and the radiation emitted by the incandescent lamp with a total watt of 90 W.

Table 4. The test results between the number of incandescent lamps and the temperature difference (°C) with or without using a zinc plate under an incandescent lamp

No	Number of incandescent bulbs	Temperature difference (C) without using a zinc plate on the lamp	Temperature difference (C) by using a zinc plate above the lamp
1	4	0.6	0.5
2	6	0.5	0.4

From Table 4, the test result value is 0.6 °C on the number of incandescent light bulbs as much as 4 without using a zinc plate above the incandescent lamp and 0.5 °C occurs at condition by using a zinc plate over an incandescent lamp with a total wattage of 100. Value 0.5 °C and 0.4 °C for the condition of the number of incandescent light bulbs as much as 6. This is influenced by the convection flow rate from the influence of the zinc plate and the radiation emitted by the incandescent lamp with a total watt of 90 W.

Table 5. Hatching percentage (%) of poultry eggs on ball lamps/heating plate with or without using a zinc plate

Test	Number of eggs	Number of hatches	Hatching percentage (%)
No zinc plate under an incandescent lamp	50	48	96
With zinc plate under an incandescent lamp	50	48	96
No zinc plate under the heating plate	50	48	96
With zinc plate under heating plate	50	48	96
Without a zinc plate on an incandescent lamp	50	48	96
With a zinc plate on an incandescent lamp	50	48	96
No zinc plate on the heating plate	50	48	96
With zinc plate on a heating plate	50	48	96

There are 4 (four) measurement points, including T1 (digital thermometer under the incandescent lamp/heating plate), T2 (digital thermometer in the middle of the incandescent lamp/heating plate), T3 (digital thermometer under the incandescent lamp/heating plate), and T4 (thermostat). under incandescent lamp/heating plate) for each treatment (variation of incandescent light bulb/heating plate and the presence or absence of zinc plate under incandescent lamp/heating plate). With a minute the incubator turns on until it reaches the highest temperature (38 °C) is 1 min, and the minute when the incandescent lamp turns off to the lowest temperature (37 °C) is 3 min.

Graph 1. Hatching percentage (%) of poultry eggs on ball lamps/heating plate with or witout using a zinc plate

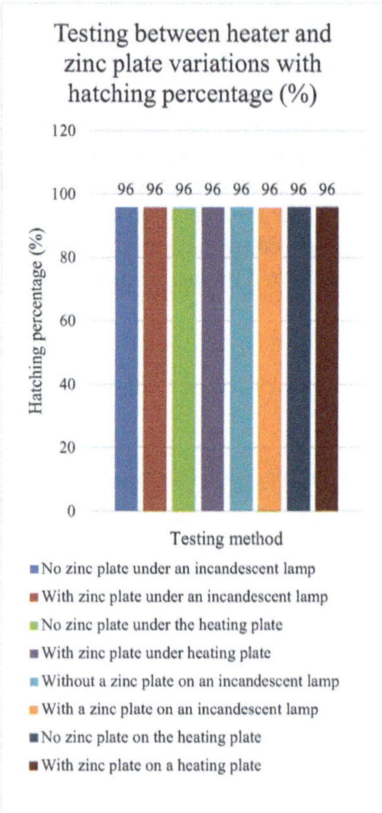

Graph1Hatching percentage (%) of poultry eggs withwithout or using a zinc plate under the heater

Graph1 is a presentation of test results the percentage of hatching in graphic form, the percentage is 96% in the light bulb/heating plate by not using a zinc plate under/above the lamp/heating plate with a total of 50 eggs and 48 eggs successfully hatching perfectly and 96% in the light bulb/plate heating without using a zinc plate under/above the lamp/heating plate, the incubation process of 50 eggs was carried out and 48 eggs hatched perfectly. This is a good setup.

Carry out test samples to complete testing within 14 days of incubation after previously checking embryo development on day 5 with a uniform temperature to be maintained in the egg incubator room with a range of 37 C–38 C.

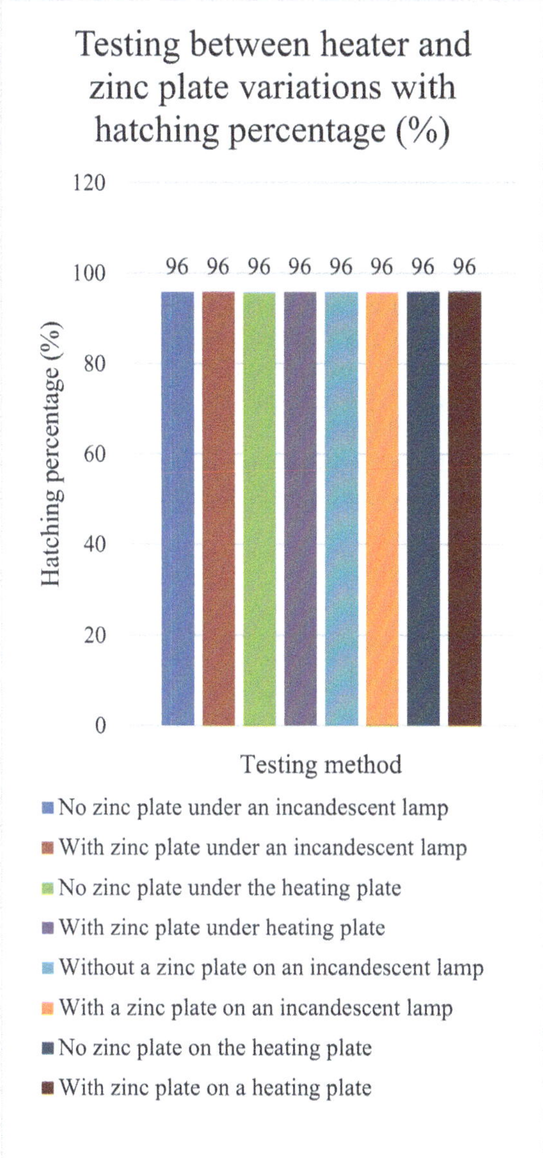

Fig. 5. Front view of egg incubator room

4 Conclusion

From the results of the research entitled Assess the Level of Temperature Uniformity in Intermediate Level Incubators Type Incandescent Lamp And Heater Plate, the following conclusions can be drawn:

1. 1. The results in this study obtained a temperature difference of 0.7 °C on 4 (four) heating plates without the addition of a zinc plate under the heating plate and 0.6 °C on 4 (four) heating plates with a zinc plate added. under the heating plate. However, there is a small ratio of 0.5 °C for 6 (six) heating plates without the addition of a zinc plate under the heating plate and 0.4 °C for 6 (six) heating plates with no addition of a zinc plate under the heating plate. Meanwhile, 4 (four) incandescent lamps with no addition of a zinc plate under the incandescent lamp resulted in values of 0.8 °C and 0.7 °C in 4 (four) incandescent lamps with a zinc plate added under the incandescent lamp. However, there is a small comparison of 0.7 °C for 6 (six) incandescent lamps without the addition of zinc plates under incandescent lamps and 0.5 °C for 6 (six) incandescent lamps without the addition of zinc plates under incandescent lamps. And some other test methods have different values due to the influence of the convection flow rate and the influence of the zinc plate and the radiation emitted by the heat source.
2. Other results in this study obtained a hatching percentage of 96% in several variations of incandescent lights bulbs/heating plates by not using a zinc plate under/above the incandescent lamp/heating plate with 50 eggs and 48 eggs that successfully hatched perfectly.

References

1. Director General of Livestock and BPS: Statistics Animal Husbandry 2020. Director General of Animal Husbandry DepartmentAgriculture Republic of Indonesia (2020)
2. Dongo, P.E., et al.: Design of Egg Incubator Based-Palm Oil in Bokonzi City (Sud Ubangi Province), Democratic Republic of the Congo. Budapest Int. Res. Exact Sci. J. 3(2), hal, 135–142 (2021). https://doi.org/10.33258/birex.v3i2.1883
3. Yadav, B.K., Pokhrel, N., Khatiwada, D., Khanal, M., Bajracharya, T., dan Dhakal, R.: Design, fabrication, and performance analysis of an automatic horizontal egg incubator. J. Inst. Eng. **16**(1), 77–85 (2021). https://doi.org/10.3126/jie.v16i1.36557
4. Boz, M.A.: Effect of egg position in the incubator on broiler hatching results. Rev. Bras. Cienc. Avic. **21**(4), 1–6 (2019). https://doi.org/10.1590/1806-9061-2019-1089
5. Sobejana, N., dan Bacalso, E.J.: Development and Construction of Poultry Egg Incubator Temperature and Humidity Controller (Peitch) With SMS Notification. SSRN Electron. J. (2021). https://doi.org/10.2139/ssrn.3779301
6. Sanjaya, W.S.M., et al.: The development of quail eggs smart incubator for hatching system based on microcontroller and Internet of Things (IoT). In: 2018 International Conference on Information and Communication Technology ICOIACT 2018, vol. 2018-Janua, pp. 407–411 (2018). https://doi.org/10.1109/ICOIACT.2018.8350682
7. Widodo, G.P., Sukandar, E.Y., Adnyana, I.K., dan Sukrasno, S.: Mechanism of action of coumarin against candida albicans by SEM/TEM analysis. ITB J. Sci. **44**(2), 145–151 (2012). https://doi.org/10.5614/itbj.sci.2012.44.2.4
8. Yamak, U.S., Sarica, M., Boz, M.A., dan Önder, H.: Broiler Damızlıklarda Yumurta Kabuk Kalınlığının Bazı Kuluçka Özellikleri Üzerine Etkisi. Kafkas Univ. Vet. Fak. Derg. 21(3), 421–424 (2015). https://doi.org/10.9775/kvfd.2014.12485
9. Keelback, H.B.: Reptiles & amphibians. Hal, pp. 182–184 (2021)

10. Yuhendri, M., Risfendra, M.M., dan Hambali: Development of automatic solar egg incubator to increase the productivity of super native chicken breeds. J. Phys. Conf. Ser. **1594**(1), (2020). https://doi.org/10.1088/1742-6596/1594/1/012033
11. Mariani, et al.: Design modification of a cost-efficient microcontroller-based egg incubator. Indian J. Sci. Technol. **14**(14), 1160–1167 (2021). https://doi.org/10.17485/ijst/v14i14.2289
12. Raja, et al.: The Development of Automatic Forced Air Egg Incubator. e-Academia J. **8**(1), 2289–6589 (2019). [Daring]. Tersedia pada. http://journale-academiauitmt.uitm.edu.my/v2/index.php/home.html
13. Fatimah, F.N.M., Redzwan, N.I., Md. Enzai, Zin, M.F.M.: Development of mobile incubator for quail egg productions in Malaysia, vol. 6, no. 1, pp. 86–98 (2017)
14. Mohanraj, T., dan Periasamy, A.: Designing fully automatic forced air egg incubator Using PIC microcontroller, vol. 2, no. 4, pp. 77–81 (2021)
15. Aher, P.V.A., Dukre, M., Thorat, T., dan Ganesh, A.: Semiautomatic Egg Incubator Machine, hal, pp. 1–3 (2021)
16. Paras, F.O.: Development of a solar heat assisted egg incubation system. IOP Conf. Ser. Earth Environ. Sci. **475**(1) (2020). https://doi.org/10.1088/1755-1315/475/1/012014
17. Tiam Kapen, P., Youssoufa, M., Foutse, M., Manfouo, H., dan Njotchui Mbakop, F.O.: Design and prototyping of a low-cost, energy efficient eggs incubator in developing countries: A case study of Cameroon. Sci. African, vol. 10 (2020). https://doi.org/10.1016/j.sciaf.2020.e00618
18. Ipek, A., dan Sozcu, A.: Comparison of hatching egg characteristics, embryo development, yolk absorption, hatch window, and hatchability of Pekin Duck eggs of different weights. Poult. Sci. **96**(10), 3593–3599 (2017). https://doi.org/10.3382/ps/pex181
19. Dewanti, R., Sudiyono, D., Animal Husbandry, J., Agriculture, F.: Effect of Eggs Weight and Turning Frequency on Fertility, Hatchability and Hatching Weight of Local Ducks. Bull. Breeder. **38**(1), 16–20 (2014)
20. Estela, M., et al.: Sistema de monitoreo y control remoto de temperatura en incubadora de huevo de aves resumen, vol. 42, no. 138, hal, pp. 259–268 (2021)
21. Ganiyat, S., dan Afolake, I.R.: Design of a portable solar powered solar incubator. Int. J. Eng. Adv. Technol. **9**(4), 2366–2369 (2020). https://doi.org/10.35940/ijeat.d7960.049420
22. Ramli, M.B., Lim, H.P., Wahab, M.S., dan Zin, M.F.M.: Egg hatching incubator using conveyor rotation system. Procedia Manuf. **2**, 527–531 (2015). https://doi.org/10.1016/j.promfg.2015.07.091
23. Agbo, D.O., Otengye, O.J., dan Dodo, S.H.: Proposed development of a solar powered automated incubator for chickens. Int. J. Eng. Tech. **4**(1), 517–524 (2018). [Daring]. Tersedia pada: http://www.ijetjournal.org
24. Maaño, R.C., Chavez, E.P., dan Maaño, R.A.: Towards the development of a smart photovoltaic-powered temperature controlled poultry egg incubator. Int. J. Simul. Syst. Sci. Technol. **19**(3), 19.1–19.5 (2018). https://doi.org/10.5013/IJSSST.a.19.03.19
25. Iscan, O., dan Celen, M.F.: The Effects of Egg Weight and Extra Oxygen Addition into the Last Stage of Incubator on Hatching Results in Turkeys at High Altitude, vol. 7, no. 3, hal, pp. 359–365 (2019)
26. Mallam, M., Mangadu, M.M., dan State, G.: The performance of kerosene incubator in hatchability trials of guinea fowl eggs during the rainy season, vol. 12, no. 1, pp. 121–125 (2019)
27. Ya, J., et al.: Development of Thermoelectric Egg Incubator Integrated with Thermal Energy Storage System, vol. 4, no. 1, pp. 13–20 (2021)
28. Miduk Purba, A.M.P.: Utilization of solar panels as an energy source in the egg hatcher temperature control automation system. Mobile-Based Natl. Univ. Online Libr. Appl. Des. **4**(3), 1–7 (2021). [Daring]. Tersedia pada: http://iocscience.org/ejournal/index.php/mantik/article/view/882/595

29. Roihan, I., K. Tjandaputra A., Setiawan, E.A., dan Koestoer, R.A.: Installing and testing the grashof portable incubator powered using the solar box 'be-care' for remote areas without electricity. Evergreen **7**(4), 621–628 (2020). https://doi.org/10.5109/4150516
30. Koestoer, R.A., dan Roihan, I.: Unpatented grashof-incubator as a part of community-enggagement in mechanical engineering university of Indonesia. Pros. SENTRA (Seminar Teknol. dan Rekayasa), no. 2, pp. 285–287 (2016)
31. Koestoer, R.A., Roihan, I., dan Andrianto, A.D.: Product design, prototyping, and testing of twin incubator based on the concept of grashof incubator. AIP Conf. Proc., vol. 2062, no. January (2019). https://doi.org/10.1063/1.5086560
32. Rafiu, O., Olugboji, O., Adedipe, O., dan Okoro, U.: Development of a steam powered incubator with solar supported system. IOP Conf. Ser. Earth Environ. Sci. **730**(1) (2021). https://doi.org/10.1088/1755-1315/730/1/012028
33. Okur, N., Eratalar, S.A., dan Eleroğlu, H.: Effects of egg shell temperature and incubator ventilation programme on incubation results of broiler breeders. Turkish J. Agric. - Food Sci. Technol. **6**(5), 536 (2018). https://doi.org/10.24925/turjaf.v6i5.536-542.1679
34. Tsamaase, K., Zibani, I., Moseki, R., dan Kemoabe, R.: Solar Powered Egg Incubator Design, vol. 14, no. 6, pp. 13–19 (2019). https://doi.org/10.9790/1676-1406011319
35. Kim, A., dan Lee, S.: Study of system using load cell for real time weight sensing of artificial incubator, vol. 11, no. 2, 144–149 (2018)
36. Archer, G.S., Jeffrey, D., dan Tucker, Z.: Effect of the combination of white and red LED lighting during incubation on layer, broiler, and Pekin duck hatchability. Poult. Sci. **96**(8), 2670–2675 (2017). https://doi.org/10.3382/ps/pex040
37. Okpagu, P.E., dan Nwosu, A.W.: Development and Temperature Control of Smart Egg. Eur. J. Eng. Technol. **4**(2), 13–21 (2016)

Analysis of Working Posture on Tofu Factory Operator Activities Using RULA Method and Virtual Engineering Software

Mohd Iqbal[1](✉), Iskandar Hasanuddin[1], Raysa Aleyzia[1], Roja Safitriyawi[1], and Azmi Hassan[2]

[1] Department of Mechanical and Industrial Engineering, Faculty of Engineering, Universitas Syiah Kuala, Banda Aceh, Indonesia
{mohd.iqbal,iskandarhasanuddin}@unsyiah.ac.id
[2] Electrical, Electronic and Automation Section, Universiti Kuala Lumpur Malaysian Spanish Institute (UniKL – MSI), Kulim High Technology Park, 09000 Kulim, Kedah, Malaysia
azmi.hassan@unikl.edu.my

Abstract. This research was conducted to analyze the work posture of tofu factory operators using the Cornell Musculoskeletal Disease Questionnaire (CMDQ) and the Rapid Upper Limb Assessment (RULA) method for operators that work at a tofu factory in Banda Aceh, Indonesia. Data processing begins with direct interviews and closed questions to tofu factory workers toward the activity and posture disorders. From the results of the interview, a score was given based on the CMDQ answers to be integrated with RULA. Based on CMDQ, it can be seen that the operator feels uncomfortable in the upper limbs in sequence starting from the highest, namely the right shoulder (12%), right upper arm (12%), upper back (9%), left shoulder (8%), right and left forearms (8%). RULA measurements were carried out using the Ergofellow software with details of 1 process with a score of 3, 2 processes with a score of 5, 3 processes with a score of 6 and 7 processes with a score of 7. To reduce the critical position of body posture and to reduce the risk of injury to the operator, recommendations for improvement are given in the form of tools for work and determination of rest hours and work time according to standards.

Keywords: Working posture · RULA · Musculoskeletal disorders · CMDQ · Ergonomic · Virtual engineering

1 Introduction

Tofu is one of the popular foods among Indonesia people. The high interest in tofu is not balanced with the number of existing tofu processing factories. The small number of tofu factories makes some workers have to work extra to fulfill the demands of the costumers. Overwork brings up complaints for manufacturing workers. Tofu making in Aceh is also still using the traditional method so that it involves most of the human labor.

Akhyar et al. (Eds.): ICECME 2021, LNME, pp. 299–309, 2023.
https://doi.org/10.1007/978-981-19-3629-6_31

Most of the tofu making processes are still using the traditional method. In its manufacture also requires a lot of water sources and little involvement of the machine. In the manufacturing process, starting from burning wood to obtain steam power to cutting tofu, everything requires a very large human role. At work, tofu making workers are equipped with makeshift tools, high workloads, irregular working hours and poor work postures. The impact that often occurs due to poor work posture can lead to Musculoskeletal Disorders (MSD). According to Sumiati in Joanda and Suhardi [1], the results of research on companies in Indonesia show that Musculoskeletal Disorders (MSD's) are 16%, cardiovascular (8%), nervous disorders (3%) and Ear, Nose and Throat (ENT) disorders are 1.5%. It can be seen that of all occupational diseases that occur in every company in Indonesia, musculoskeletal injuries are the diseases that occupy the most vulnerable position among other occupational diseases with a percentage of 16%.

One of the tofu processing factories in Banda Aceh that still uses the traditional method will be discussed and analyzed ergonomically in this study. The process of making tofu begins with preparing steam which is flowed into the furnaces from wood burning, followed by soaking, milling, cooking I, sifting, cooking II, forming until the last one is cutting. In working, operators often do critical work postures. Such as bending up to 90°, twisting limbs, using two hands to perform two different activities, lifting heavy weights with improvised tools and moving things by hand by dragging while bending over. The unclear division of tasks makes each operator carry out every work process so that they have to go back and forth to see each production process which causes body parts to rotate frequently. Other conditions such as hot temperatures, high noise levels, and narrow spaces also make working conditions uncomfortable. From this problem, it can be said that measurements are needed to see how the category of work posture of the tofu factory operators is.

To analyze the complaints and their relation to the work posture of the tofu factory operators, the Cornell Musculoskeletal Discomfort Questionnaire (CMDQ) and the Rapid Upper Limb Assessment (RULA) were used with the help of virtual engineering software, Ergofellow. Based on Siska and Angrayni in [2] Rapid Upper Limb Assessment (RULA) is a method developed in the field of ergonomics that identifies and assesses the work activities of operators in the upper body. Inputs to this method include the posture of the palms, upper arms, forearms, back and neck, the load being lifted, the power used (static/dynamic), the number of operators. Previous research on work posture in every human activity by Oesman, *et al.* in [3] conducted on 20 workers in the leather craft industry with an analysis of perceived complaints using the RULA and Symtom Questionnaire methods with the aim of knowing the level of risk to workers and the relationship with Upper Extremity Work-Related Musculoskeletal Disorders (UEWMSDs), where the results obtained from these conditions are by designing a cutting table that is evaluated by RULA category 3 with sizes according to the anthropometry of workers. Research on the effectiveness of lifting weights has also been carried out by Zeki *et al.* [4] using a Nordic Body Map questionnaire and this problem interview was solved using a biomechanical approach and a physiological approach with the final result being the design of a tool that can overcome distances in lifting weights. The construction worker posture analysis uses the Ovako Working Posture Analysis System (OWAS) method where the method is used to identify critical levels of workers with the results of research proposals to reduce critical risks to body posture and reduce injuries to workers based on the research of Iqbal *et al.* [5].

2 Research Methods

Place and Time of Research. The research and survey were conducted at the tofu factory by starting direct observations to identify problems that occurred.

Method of Collecting Data. This research is based on the results of a direct survey at the tofu factory in question. Data collection in the study was obtained through documentation of operator activities at each work station and interviews with operators. The data used in this study are primary data and secondary data as well as a closed-question questionnaire from the Cornell Musculoskeletal Discomfort Questionnaire (CMDQ), as shown in Fig. 1.

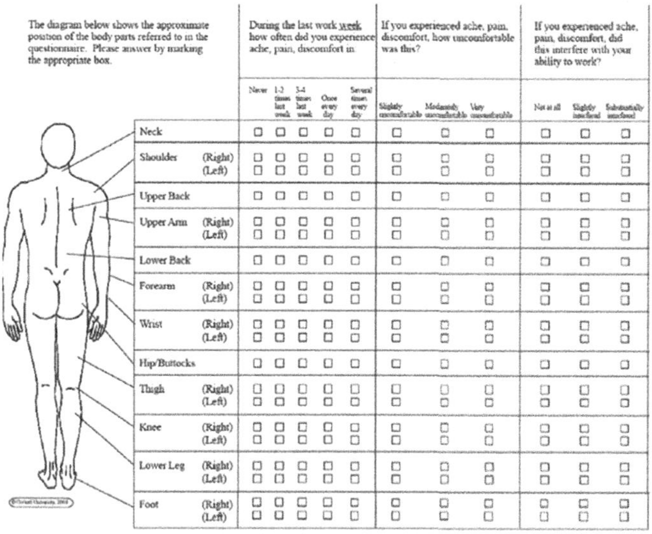

Fig. 1. CMDQ questionnaire [6]

Data Processing Method. After obtaining data through questionnaires and closed-ended interviews, the next activity was observing problems at each work station operator using a camera. These results will be used as supporting data for analysis using the Rapid Upper Limb Assessment (RULA) application on ErgoFellow. In this application, the movement generated by the operator will be analyzed to determine the score that will be generated as a recommendation for prevention or improvement which will be considered for implementation to reduce the risk of injury and prevent work accidents, so that health and safety can be maintained.

3 Results

CMDQ Questionnaire Analysis. In Hedge *et al.* [7] the Cornell Musculoskeletal Discomfort Questionnaire (CMDQ) is a measuring tool in the form of a questionnaire mapping body parts that may experience discomfort when doing a job. This questionnaire was obtained by answering three types of questions with several answer options related to several members of the operator's body. Questionnaires were given to 6 operators working in the tofu factory. The answers obtained through CMDQ are shown in Table 1.

Table 1. Tofu factory operator CMDQ results.

The diagram below shows the approximate position of the body parts referred to in the questionnaire. Please answer by marking the appropriate box.	During the last work week how often did you experience ache, pain, discomfort in :					If you experienced ache, pain, discomfort, how uncomfortable was this?			If you experienced ache, pain, discomfort, did this interfere with your ability to work?		
	Never	1-2 times last week	3-4 times last week	Once every day	Several times every day	Slightly uncomfortable	Moderate uncomfortable	Very uncomfortable	Not at all	Slightly interfered	Substantially interfered
Neck	2	2	1	1		3	3		5	1	
Shoulder (Right)	1	1	2	2		2	4		4	2	
Shoulder (Left)	1	2	2	1		3	3		5	1	
Upper Back		2	3	1		4	2		5	1	
Upper Arm (Right)		3	3			1	5		3	3	
Upper Arm (Left)	1	4	1			4	2		4	2	
Lower Back	1	3	2			4	2		4	2	
Forearm (Right)	1	3	2			3	3		3	3	
Forearm (Left)	1	3	2			3	3		3	3	
Wrist (Right)		5	1			4	2		4	2	
Wrist (Left)	4	2				6			5	1	
Hip/Buttock	6					6			6		
Thigh (Right)	4		2			3	3		6		
Thigh (Left)	4		2			3	3		6		
Knee (Right)	4	1	1			4	2		6		
Knee (Left)	4	2				6			6		
Lower Leg (Right)	3	2	1			4	2		6		
Lower Leg (Left)	3	2	1			4	2		6		
Foot (Right)	4	1	1			4	2		4	2	
Foot (Left)	4	1	1			4	2		4	2	

CMDQ is used for screening purposes or to see in general whether any member of the operator's body feels discomfort during work. The CMDQ score can be calculated by multiplying the frequency score (0, 1.5, 3.5, 5, 10), discomfort score (1, 2, 3) and interference score (1, 2, 3). The results of the calculation of the CMDQ score are shown in Table 2.

From Table 2 it can be seen that the operator feels complaints in the upper limbs sequentially starting from the highest, namely on the right shoulder (12%), right upper arm (12%), upper back (9%), left shoulder (8%), right and left forearm (8%). This is very likely to happen because when the operator is working the upper limb is the most used. Complaints can also occur due to irregular working hours starting from the morning around 9 to the afternoon or it can be until midnight which causes the operator to feel tired until some complain of body aches in the morning. Operators also only have a break of 10–30 min 3–4 times a day. In working, the operator is only equipped with simple tools that require the operator to exert excess energy when working and the last is the operator does not have a standard guide or work assignment so that all of them work on the entire process of making tofu without any division of the work process.

Table 2. CMDQ score calculation results

Part of the body	Frequency	Discomfort	Interference	Total	%
Neck	11.5	9	7	724.5	6%
Shoulder (Right)	18.5	10	8	1480	12%
Shoulder (Left)	15	9	7	945	8%
Upper Back	18.5	8	7	1036	9%
Upper Arm (Right)	15	11	9	1485	12%
Upper Arm (Left)	9.5	8	8	608	5%
Lower Back	11.5	8	8	736	6%
Forearm (Right)	11.5	9	9	931.5	8%
Forearm (Left)	11.5	9	9	931.5	8%
Wrist (Right)	11	8	8	704	6%
Wrist (Left)	3	6	7	126	1%
Hip/Buttock	0	6	6	0	0%
Thigh (Right)	7	9	6	378	3%
Thigh (Left)	7	9	6	378	3%
Knee (Right)	5	8	6	240	2%
Knee (Left)	3	6	6	108	1%
Lower Leg (Right)	6.5	8	6	312	3%
Lower Leg (Left)	6.5	8	6	312	3%
Foot (Right)	5	8	8	320	3%
Foot (Left)	5	8	8	320	3%

Work Posture Analysis Using RULA. Rapid Upper Limb Assessment (RULA) is a method developed in the field of ergonomics that investigates and assesses the working position of the upper body. This equipment does not perform special devices in providing measurements of neck, back, and upper body postures in line with muscle function and external loads supported by the body in Lueder [8]. In this study, the reason using RULA is because the CMDQ results show that the upper limb area gets a lot of complaints so RULA is the choice to measure the posture of the operator based on the research of Purbasari and Haulian [9]. This is done using the ErgoFellow software as shown in Fig. 2. Measurement using ErgoFellow is done by selecting the options that have been provided according to the work posture of the tofu factory operator that has been previously documented. Figure 3, Fig. 4, and Fig. 5 show the RULA scores obtained from several work processes analyzed.

From the measurements that have been made with ErgoFellow, it was found that many work processes fall into the category of high RULA scores. In Fig. 3 with a score of 7 it is shown that the worker uses both hands to put tofu with his left hand and put water in with his right hand. The position of the water reservoir at the back causes the operator's back to rotate several times because he has to take water and put it in.

A score of 7 was also found in the finished tofu extraction process mixed with vinegar as shown in Fig. 5. The operator was required to bend down to about 90° to pick up the tofu in a very hot furnace. Tofu retrieval is carried out repeatedly so that the operator bends his body several times in close proximity. The tofu load weighing about 1 kg also puts pressure on the operator's arm so that his arm feels discomfort. The recapitulation of RULA scores from the work process of making tofu is shown in Table 3.

From Table 3, it can be seen that the RULA scores obtained were mostly in the high category and only a few were in the low category. There is 1 process with a score of 3,

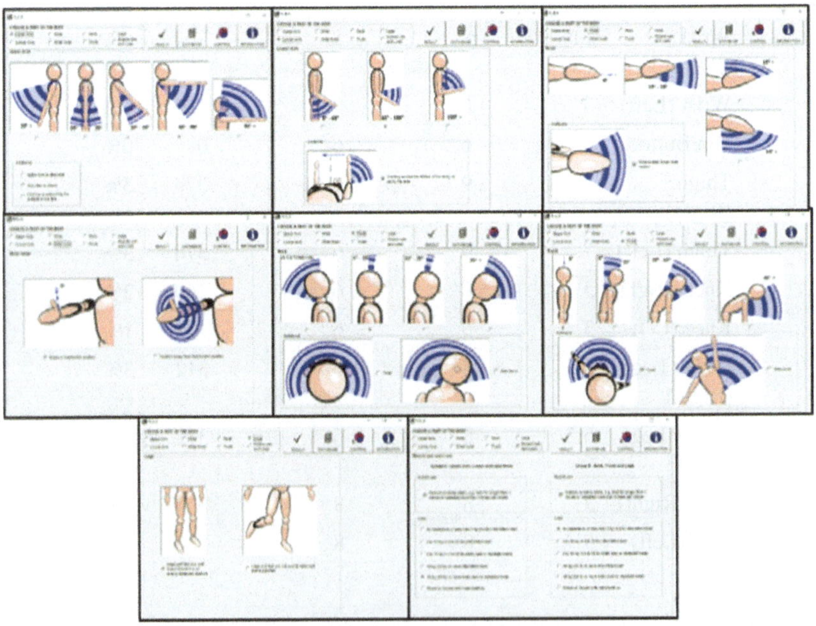

Fig. 2. ErgoFellow display for RULA

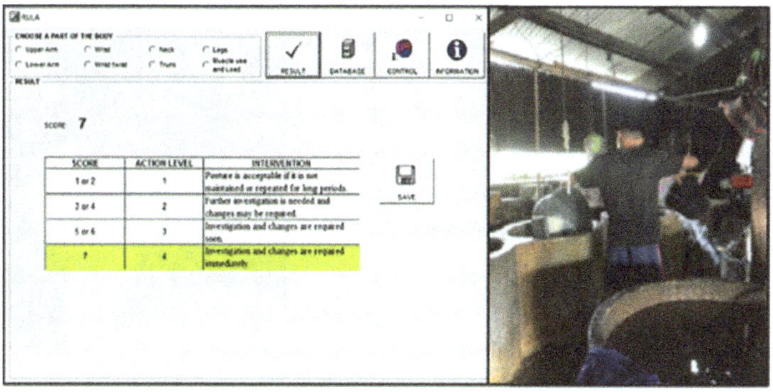

Fig. 3. Putting soybean porridge into the furnace

2 processes with a score of 5, 3 processes with a score of 6 and 7 processes with a score of 7. Based on Iqbal *et al.* [10].

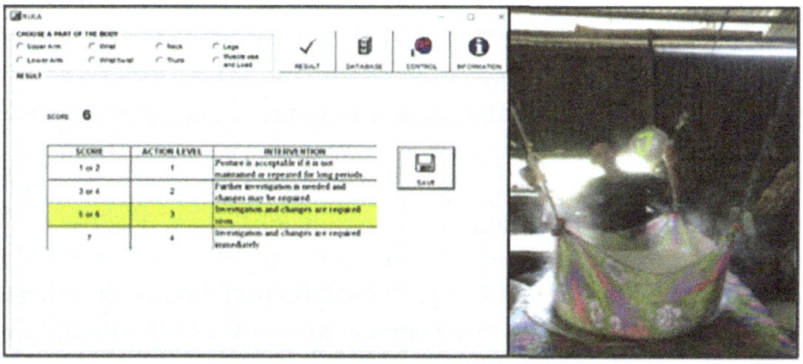

Fig. 4. Sifting

for processes with a score of 3 indicates that the process requires investigation and changes may be needed. For processes with a score of 5–6 indicates that the process requires investigation and changes are needed in the future. And processes with a score of 6 + or 7 require immediate investigation and change.

The work process of the tofu factory requires urgent investigation and changes to address complaints from operators while working. It is feared that jobs with high RULA scores will have an adverse impact on the health and safety of operators such as the occurrence of Musculoskeletal Disorders (MSDs) or even have the potential to lead to burnout as studied by Erwan *et al.* [11, 12].

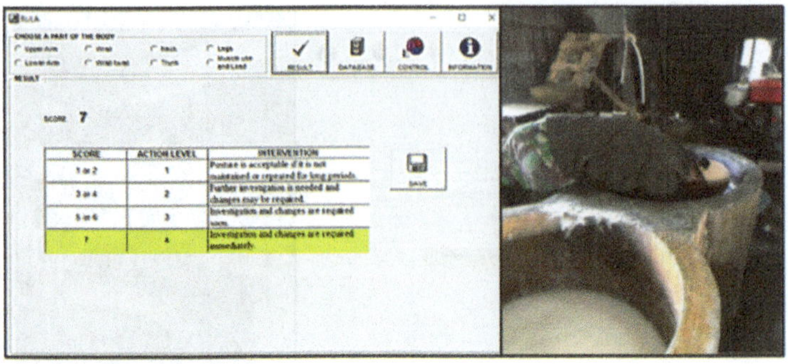

Fig. 5. Taking Tofu

3.1 Improvement Recommendations

There are several improvements that might be made, including providing adequate tools, adequate rest hours and working time according to standards. Previous studies show that the improvement reduce the critical working postures and risk of injury [10], as well as increase the productivity [13, 14], reduce production cost [15] and increase the efficiency of energy consumption [16, 17].

Widodo and Astuti [18] said that to be able to improve a work process, one of them can be done by designing a tool that can make it easier for a worker to do his job compared to the previous way of working. One tool that can be used is a trolley. The trolley can be used to move soybean porridge weighing approximately 20 kg. The use of this trolley will save workers from having to bend over when moving the soybean porridge into the furnace and will reduce the burden on the shoulders, back and arms, especially if the size of the trolley matches the body dimension of the worker [19]. Furthermore, the use of pipes can be an alternative to drain water from one furnace to another to prevent the use of a lot of hands and the waist that often rotates to take water that is in the back position. In addition, the use of a pulley attached to the tube can make it easier for the operator to pick up the tofu from the furnace, so there is no need to bend deeply because it can put pressure on the back and also risk burning limbs due to the hot stove.

Sufficient rest hours and standard working hours are needed for workers. Giving a break of about one hour to one and a half hours is expected to reduce the fatigue faced by the operator and also should not work more than 8 h a day [20]. In Tabita *et al.* [21] it is explained that by regulating working hours and breaks will reduce the condition of the occurrence of Musculoskeletal Disorders (MSD), reduce fatigue and improve the performance of workers. Tofu factories can set work shifts so that the target order is still achieved so that operators do not work more than 8 h a day.

Table 3. RULA score recapitulation operator

Process	RULA Score
Burning	4
Soybean filter	7
Soybean mill	5
Soybean porridge transfer	7
Laying Soybean Porridge into the Furnace	7
Transfer of Tofu to Sieve	7
Tofu Sifting	6
Mixing Tofu with Vinegar	7
Pemisahan Cuka dari Tahu	7
Separating Vinegar from Tofu	7
Tofu forming	6
Tofu pressing	6
Tofu cutting	3

4 Conclusion

Based on calculations using RULA, it was found that 7 of the 13 work processes carried out by the tofu factory operators fall into the + 6 category, namely 7 where the activities are very risky and urgent changes are needed. This is also reinforced by the CMDQ results which show that the most complaints are felt in the upper limbs, namely on the right shoulder (12%), right upper arm (12%), upper back (9%), left shoulder (8%), right and left forearm (8%). The suggestion that can be given is to provide adequate work tools so that the work posture of the operator is not too extreme and reduces the burden on the upper limbs. To reduce complaints, it is also possible to determine adequate rest hours and work hours according to standards.

References

1. Joanda, A.D., Suhardi, B.: Analisis Postur Kerja dengan Metode REBA untuk Mengurangi Resiko Cedera pada Operator Mesin Binding di PT. Solo Murni Boyolali. In: Seminar dan Konferensi Nasional IDEC (2017)
2. Siska, M., Angrayni, S.A.: Analisis Postur Kerja Manual Material Handling pada Aktivtas Pemindahan Pallet Menggunakan Rappid Upper Limb Activity (RULA) di PT. Alam Permata Riau. Jurnal Sains, Teknologi dan Industri **15**(2), 1–10 (2018)
3. Oesman, T.I., Irawan, E., Wisnubroto, P.: Analisis Postur Kerja dengan RULA Guna Penilaian Tingkat Risiko Upper Extremity Work-Related Musculoskeletal Disorders. Studi Kasus PT. Mandiri Jogja Internasional. Jurnal Ergonomi Indonesia **5**(01) (2019)
4. Zeki, M., Iskandar, I., Iqbal, M.: Analisis Efektifitas Kerja Pengangkatan Beban Pada Bagian Pengantongan Di PT. Pupuk Krueng Geukuh. Ind. Eng. J. **8**(2) (2019)

5. Iqbal, M., Angriani, L., Hasanuddin, I., Erwan, F., Soewardi, H., Hassan, A.: Working posture analysis of wall building activities in construction works using the OWAS method. In: IOP Conference Series: Materials Science and Engineering, vol. 1082, No. 1, p. 012008). IOP Publishing (2021)
6. Pratama, T., Hadyanawati, A.A., Indrawati, S.: Analisis Postur Kerja Menggunakan Rapid Office Strain Assessment dan CMDQ pada PT XYZ. Seminar dan Konferensi Nasional IDEC (2019)
7. Hedge, A., Morimoto, S., McCrobie, D.: Effects of keyboard tray geometry on upper body posture and comfort. Ergonomics **42**(10), 1333–1349 (1999)
8. Lueder, R.: A proposed RULA for computer users. In: Proceeding of the Ergonomic Summer Workshop, San Francisco (1996)
9. Purbasari, A., Haulian, B.A.: Analisis Sikap Kerja Terhadap Faktor Risiko Ergonomi Pada Kerja Assembly Manual (Studi Kasus: Laboratorium Teknik Industri Universitas Riau Kepulauan). Profisiensi vol. 6, no.1 (2018)
10. Iqbal, M., Hasanuddin, I., Hassan, A., Soufi, M.S.M., Erwan, F.: The study on ergonomic performances based on workstation design parameters using virtual manufacturing tool. Int. J. Integr. Eng. **12**(5), 124–129 (2020)
11. Erwan, F., Iqbal, M., Hasanuddin, I., Zuhri, S., Maydini, C.M.: Burnout Syndrome Analysis Among Hospital Nurses Using Maslach Burnout Inventory - Human Service Survey (MBIHSS). IOP Conference. Materials Science and Engineering, vol. 931, no. 1, p. 012025. IOP Publishing, (2020)
12. Erwan, F., Iqbal, M., Hasanuddin, I.: The analysis of nursing's work systems in relation to burnout syndrome case study: nurses in RSUDZA, BLUD RSIA, and RSUD Meuraxa, Banda Aceh, Indonesia). In: IOP Conference Series: Materials Science and Engineering, vol. 453, no. 1, p. 012051. IOP Publishing (2018)
13. Iqbal, M., Bamatraf, O.A.A., Tadjuddin, M.: The study of production performance of water heater manufacturing by using simulation method. J. Phys. Conf. Ser. **962**(1), 012019. IOP Publishing (2018)
14. Iqbal, M., Purnomo, M.R.A., Imra, M.A.B.M., Konneh, M., Karim, A.N.: The study on production performances of part transporter in flexible manufacturing system by using simulation method. Adv. Mater. Res. **576**, 714–717. Trans Tech Publications Ltd (2012)
15. Karim, A.N., Hamdan, M.A.H., Alam, M.Z., Iqbal, M.: Economic Scenario of Cellulase Enzyme Production Using Palm Oil Mill Effluent (POME). In: Advanced Materials Research, vol. 576, pp. 723–726. Trans Tech Publications Ltd (2012)
16. Rahmawati, S., Iqbal, M., Sara, I.D.: Economic study in simulating 5 MWp solar farm planning with 2 technologies under Indonesia Feed-in Tariff in Weh Island-Aceh, Indonesia. In: IOP Conference Series: Materials Science and Engineering, vol. 931, no. 1, p. 012014. IOP Publishing (2020)
17. Rahmawati, S., Iqbal, M., Sara, I.D.: Energy analysis of solar farm planning in weh island, aceh, indonesia. In: IOP Conference Series: Materials Science and Engineering, vol. 539, no. 1, p. 012023. IOP Publishing (2019)
18. Widodo, A.W., Astuti, R.D.: Perancangan Alat Bantu Untuk Memperbaiki Postur Kerja Pada Aktivitas Memelitur Dalam Proses Finishing. Seminar Nasional IENACO (2015)
19. Iqbal, M., Salam, R., Hasanuddin, I., Hassan, A.B.: Body segment dimensions of Indonesian male army. In: IOP Conference Series: Materials Science and Engineering, vol. 931, no. 1, p. 012013. IOP Publishing (2020)

20. Purnomo, A., Ridwan, M., Iqbal, M., Sufa, M.F.: Solving multiple routes travelling salesman problem using modified genetic algorithm. In: Advanced Materials Research, vol. 576, pp. 718–722. Trans Tech Publications Ltd (2012)
21. Tabita, A., Adiputra, N., Sutarja, I.N.: Pengaturan Organisasi Kerja Housekeeping dengan Pendekatan Ergonomi Dapat Menurunkan Keluhan Muskuloskeletal, Kelelahan, dan Mempercepat Waktu Kerja. Jurnal Ergonomi Indonesia Indonesian J. Ergonom. **3**(1) (2017)

The Influence of Cutting Speed on the Dimension Quality of Micro Milling Process Results on 6Al-4V Titanium with Pre-cooling of the Workpiece

Muhammad Tadjuddin[1(✉)], Syahriza[2], Said Amir Azan[1], Udink Aulia[1], and Dedy Syahputra[1]

[1] Department of Mechanical and Industrial Engineering, Faculty of Engineering, Universitas Syiah Kuala, Banda Aceh, Indonesia
{m.tadjuddin,saidamirazan,uaulia}@unsyiah.ac.id
[2] Department of Civil Engineering, Faculty of Engineering, Universitas Syiah Kuala, Banda Aceh, Indonesia
syahriza@unsyiah.ac.id

Abstract. The machining process is one of the production technologies that is almost required for every product. In this modern era, the industry continues to develop. Along with these developments, the miniaturization of machining processes is micro-milling. This research was conducted to determine the significance of cutting speed on the workpiece dimensions on the micro-milling process with pre-cooling of titanium 6Al-4V using the Anova test. The selected cutting speeds are 15, 20, and 25 m/min, and the initial cooling temperatures of the workpiece are 26 °C, 0 °C, and -3 °C. The tools used are solid carbide HRC63 with a diameter of 0.4mm. The results obtained the cutting speed with 15, 20, and 25 m/min variation. The research found that all of those three cutting speeds affect the quality of the dimension of the workpiece with preconditioning temperature. The results obtained that the lowest deviation is at a temperature of −3 °C occurs at the three cutting speeds. It means that the cutting speed matters, and pre-cooling the workpiece at a temperature of −3 °C is the best.

Keywords: Micro milling · Dimensional quality · Titanium alloy · Pre-cooling

1 Introduction

One of the most popular materials today is titanium. Titanium is a material that is widely used in industry. Titanium is also a metal that is very familiar with human life. Many people know that titanium can be used in jewelry, electronics, medical devices, and other products that require material strength [1]. Titanium is also an attractive metal because it is a light metal but has strong properties. In industry, titanium is used as a component for aircraft construction, pipeline systems in the petrochemical industry, offshore oil refineries, and sports equipment. Titanium is also used as artificial bone or implants in the medical field because of its bio-compatible properties [2].

Akhyar et al. (Eds.): ICECME 2021, LNME, pp. 310–318, 2023.
https://doi.org/10.1007/978-981-19-3629-6_32

However, the characteristic of titanium is that it is easy to react with the cutting tool material so that it tends to form a build-up edge and adhere to the cutting tool surface. The low thermal conductivity of titanium causes surface damage to the cut material, resulting in poor surface quality, abrasion, and microstructural damage. Therefore, titanium is categorized as a material that is difficult to machine [3]. The low thermal conductivity of titanium alloys will increase the temperature at the cutting tool tip resulting in rapid tool failure. Flaking at the cutting tool tip results in a damaged surface condition of the machined material, characterized by a rough and non-uniform surface [3]. Not only causes high surface roughness values, but build-up-edge also causes microstructural damage, which also contributes to surface damage. The surface roughness of the workpiece is one of the parameters in determining the quality of the product. The surface of the workpiece that is too rough will cause the component's function to be less than perfect, so it is considered a damaged product (reject). The workpiece is too rough, causing the friction on the workpiece to be too high or a problem in terms of aesthetics [4]. Titanium is considered a complex material to process using conventional milling machines because of the poor thermal conductivity of the material, so that in the machining process, the high temperatures generated in the cutting zone will cause plastic deformation of the cutting tool and increased chemical reactivity in titanium [1]. On the other hand, the machining process is an indispensable production technology for many types of products [5]. In the modern era, the industrial world continues to develop along with technological advances [6].

The machining process is to change the shape of raw materials into workpieces by producing chips (chips). One of the most widely used machining processes is the milling process [7]. In the milling process, the tool has more than one cutting edge. The tool moves around while the workpiece is straight [8]. The development of micro-machining has also increased with these developments. There are many studies in micro-drilling [9], micro-milling [11], and others. Micro milling uses mini-cutting tools that have been used from macro, and a lot of work has been done to design and manufacture micro milling equipment. Complex 2D and 3D formation can be done with micro-milling components and micro components made from engineering materials such as plastics [10], aluminum [5], brass [6], stainless steel [9], and titanium [1]. High-speed micromachining can be used for hard metals or alloys that are difficult to achieve at low speeds [10]. This process is flexible and is used to fabricate more complex microstructures where the surface topography and cutting forces depend on the cutting parameters [11]. Micro milling can produce attractive surface geometries with good surface finishing and higher accuracy rates, where the material removal rate is one of the most important aspects [12]. The rate of material transfer varies significantly with changes in cutting parameters in micro-milling [13]. The rate of material transfer shows the processing time of the results and is an essential factor that affects the level and cost of production [14]. Reliable prediction of forced reduction capability in micro-milling is necessary for cutting tool design in micro-milling operations, planning, and machining efforts to achieve maximum productivity and quality. Micro milling is a flexible method for creating 3D three-dimensional features. The increasing popularity of micro-milling has prompted these researchers to study micro-milling to improve quality and reduce time to achieve higher productivity, including finished products. The ratio of feed per tooth to cutting radius is much better than micro milling [10]. Conventional milling often leads to a prediction of errors in forced cutting. The loss of the tooltip, even in small microns, also significantly affects the accuracy of the milling process [14] at micro-size compared to conventional milling processes. Miniature systems can provide consumables, low materials, low power consumption, low sample requirements, high heat transfer, and better process integration capabilities than automation [13] because the demand for micro-size products used in the biomedical and electronics industries has increased in recent years. A micro milling machine and a microlaser machine are used to make micro-sized parts made of engineered materials. Micro milling is a popular way to make micro parts with three-dimensional surfaces. The main application of this micro-milling in the manufacturing process is to create micro molds used in the microplastic injection process for the mass production of micro parts [15]. Now high-speed micro-milling is gaining popularity because of its precision and good surface finish. So research on the micromachining process with titanium material becomes very interesting.

2 Methodology

The machining process of this research was conducted in Laboratorium Desain dan Manufaktur Universitas Syiah Kuala. The workpiece material titanium 6Al-4V is placed on a vise which a thermoelectric is attached below for temperature preconditioning. Milling tools of 0.4mm were fixed at a highspeed spindle, as the figure below (Fig. 1).

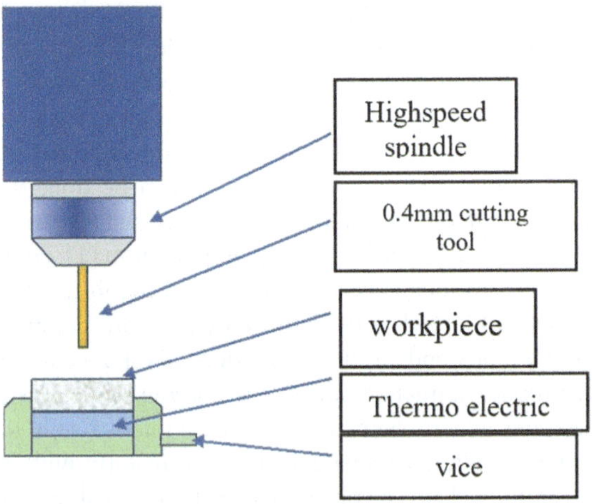

Fig. 1. Experimental setup

The cutting tool used in this study is a Japan Standard Technology-CUT End mill with a diameter of 0.4 mm, flute length of 4 mm, and a total length of 50 mm. The parameters used in the study areas can be seen in Table 1. below;

The setup of the workpiece on the milling machine is as shown in Fig. 3. After the micro-milling process is carried out, the following process measures the quality of the

Table 1. Machining parameters

Material	Titanium 6Al-4V								
Tool diameter (mm)	0.4								
Cutting Speed (m/min)	15			20			25		
Spindle Speed (rpm)	11900			15900			199000		
Feed (mm/ref)	0.005								
Workpiece temperature (oC)	28	0	−3	28	0	−3	28	0	−3

dimensions on each test object. This measurement was repeated three times. Size of the dimensions of the test object using a measuring microscope (Fig. 2).

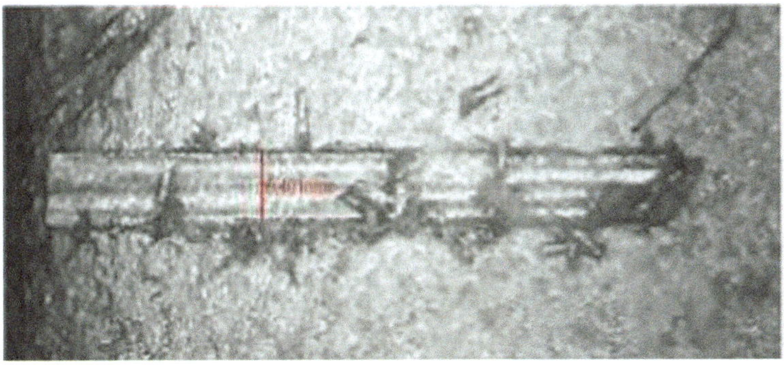

Fig. 2. Measurement of dimensions with a measuring microscope

3 Results and Discussion

The machining process uses the Agma A8 CNC milling machine with the addition of a high speeds spindle. The results were measured using a measuring microscope. ANOVA test carried out the measurement data to determine the level of significance of the variables used. After measuring the quality value of the micro-milling dimension on the test object, the value of the dimension quality of the micro-milling on the test object is obtained. The data of experiments are shown in Table 2 below.

Table 2. Experimental data

No	Workpieces temp. oC	Cutting Speeds Vc, m/min	Milling path, mm			Average
			1	2	3	
1	26	15	0.413	0.42	0.413	0.4153
2	0		0.407	0.413	0.413	0.4110
3	−3		0.407	0.415	0.407	0.4097
4	26	20	0.407	0.411	0.416	0.4113
5	0		0.407	0.413	0.407	0.4090
6	−3		0.407	0.407	0.407	0.4070
7	26	25	0.403	0.405	0.401	0.4030
8	0		0.4	0.407	0.4	0.4023
9	−3		0.401	0.401	0.4	0.4007

The data are taken from measurements then processed by ANOVA test to find out the significance of cutting speed on product dimensions in the milling process with temperature precondition.

1. Cutting Speed 15 (m/min.)

The consistency test of a temperature effect (26, 0, and −3°C) on a cutting speed of 15 (m/min), a one-way ANOVA test was performed with a significance level of 0.05%. The results obtained are as follows

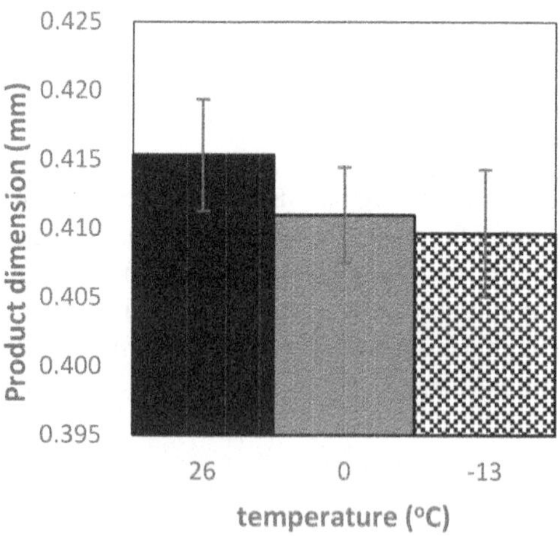

Fig. 3. The effect of variations in cutting temperature on the dimensions of the cutting width (mm) at a Cutting speed of 15 (m/min.).

Table 3. Results of One-Way ANOVA analysis for cutting speed 15 (m/min.).

Sources	*SS*	*df*	*MS*	*F*	*P value*	*F crit*
Between Groups	0.00	2	0.0000	1.5906	0.279096419	5.14325285
Within Groups	0.00	6	0.0000			
Total	0.00	8	0.0000			

Based on Table 3, it can be seen that F < Fcrit, which means Ho is accepted, meaning that the temperature effect model (26, 0, and −3 °C) for cutting speed 15 affects the dimensions of the cutting width (mm) (Fig. 4).

2. Cutting Speed 20 (m/min.)

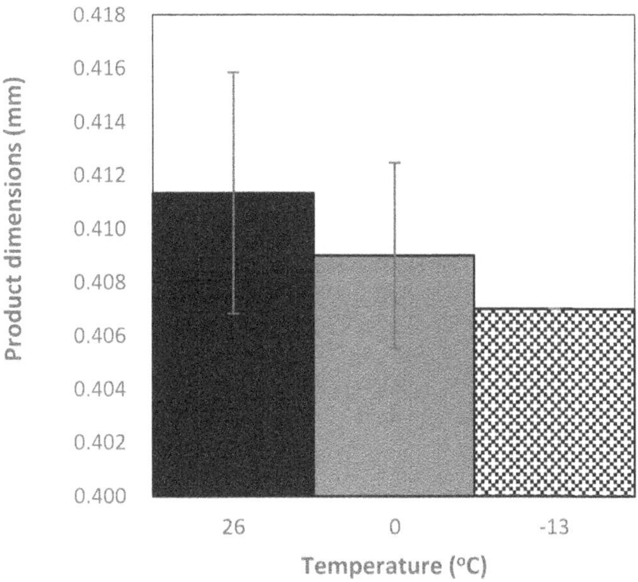

Fig. 4. The effect of variations in cutting temperature on the dimensions of cutting width (mm) at a Cutting speed of 20 (Vc) (m/min.).

To test the consistency of the effect of temperature (26, 0, and −3 oC) for a cutting speed of 20 (m/min.), one-way ANOVA test was performed with a significance level of 0.05%. The results obtained are as follows:

Based on Table 4, it can be seen that F < Fcrit, which means Ho is accepted, meaning that the temperature effect model (26, 0, and −3 oC) for cutting speed 20 affects the dimensions of the cutting width 0.4 (mm) (Fig. 5).

3. Cutting Speed 25 (m/min.)

Table 4. Results of One-Way ANOVA analysis for cutting speed 20 (m/min.).

Sources	*SS*	*df*	*MS*	*F*	*P value*	*F crit*
Between groups	0.00	2	0.0000	1.3093	0.337403918	5.14325285
Within groups	0.00	6	0.0000			
Total	0.00	8	0.0000			

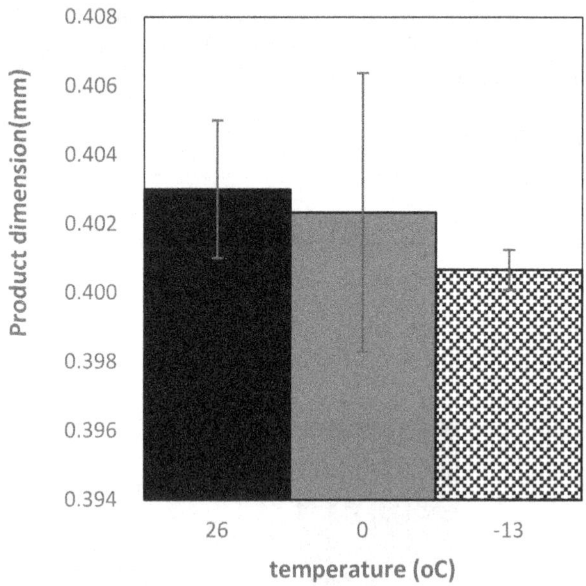

Fig. 5. The effect of variations in cutting temperature on the dimensions of the cutting width (mm) at a Cutting speed of 25 (Vc) (m/min.).

The consistency test of a temperature effect (26, 0, and −3oC) on a cutting speed of 25 (m/min.), one-way ANOVA test was performed with a significance level of 0.05%. The results obtained are as follows:

Table 5. Results of One-Way ANOVA analysis for cutting speed 25 (m/min.).

Sources	*SS*	*df*	*MS*	*F*	*P value*	*F crit*
Between groups	0.00	2	0.0000	0.6290	0.56492563	5.14325285
Within groups	0.00	6	0.0000			
Total	0.00	8	0.0000			

Based on Table 5, it can be seen that F < Fcrit, which means Ho is accepted, meaning that the temperature effect model (26, 0, and −3 oC) for cutting speed affects the dimensions of the cutting width 0.4 (mm).

4 Conclusion

The results of the machining and measurement processes were presented. The Anova test has been carried out on the effect of cutting speed on the quality of the dimensions of the milling workpiece with workpiece temperature precondition. The results obtained that all three variations of cutting speeds (15, 20, and 25 m/min) affected the dimensions of the workpiece with a temperature precondition (26 °C, 0 °C, and −3 °C). The results obtained that the lowest deviation is at temperature −3 °C occurs at all three cutting speeds. The cutting speed has an effect, and the pre-cooling of the workpiece at temperature −3 °C has the best accuracy.

Acknowledgments. We gratefully thank LPPM Syiah Kuala University, which funds this research with the Penelitian Lektor scheme, and the Design and Manufacturing Laboratory that provided research facilities.

References

1. Ganguli, S., Kapoor, S.G.: Improving the performance of milling of titanium alloys using the atomization-based cutting fluid application system. J. Manuf. Process. **23**, 29–36 (2016)
2. Windajanti, J.M., Santjojo, D.J.D.H., Abdurrouf, A.: Formation of Titanium Nitride (TiN) by nitriding process on pure titanium using high density plasma. Mech. Eng. **8**(2), 83–90 (2017)
3. Ibrahim, G.A.: Effect of Dry Machining on Surface Roughness and Hardness of Titanium Alloys. Mechanical **5**(2) (2014)
4. Iqbal, M., Konneh, M., Md Said, A.Y.B., Zaini, A.F.B.M.: Surface quality of high speed milling of silicon carbide by using diamond coated tool. In: Applied Mechanics and Materials, vol. 446, pp. 275–278. Trans Tech Publications Ltd. (2014)
5. Dirhamsyah, M., Tadjuddin, M., Udink, A., Yusuf, Z., Saiful, H.Y.: The effect of cutting speed on dimension accuracy and burr development of high-speed micro drill process on aluminum. In: IOP Conference Series: Materials Science and Engineering, vol. 523, No. 1, p. 012074. IOP Publishing, May 2019
6. Firsa, T., Tadjuddin, M., Udink, A., Hasanuddin, I.: Effect of machining parameters on hole accuracy and burr formation in micro-drilling of brass. In: Defect and Diffusion Forum, vol. 402, pp. 73–80. Trans Tech Publications Ltd. (2020)
7. Iqbal, M., Konneh, M., Ani, M.H., Abdallah, A., Binting, M.F.: Cutting temperature in high speed milling of silicon carbide using diamond coated tool. Int. J. Mech. Prod. Eng. **3**(4), 62–66 (2015)
8. Iqbal, M., Ariff, T.F., bin Mat Roseh, M.S.F., Al-Hazza, M.H.F., Hilmy, I.: Effect of microwave sintering treatment to the flank wear of titanium carbide tools in milling operations. In: Advanced Materials Research, vol. 1115, pp. 59–63. Trans Tech Publications Ltd. (2015)
9. Tadjuddin, M., Dirhamsyah, M., Udink, A., Rahmatsyah, F.: The effect of spindle speed and feed rate on hole diameter of high-speed micro-drilling for micro-screen manufacture. In: Defect and Diffusion Forum, vol. 402, pp. 125–130. Trans Tech Publications Ltd (2020)

10. Ibrahim, G.A., Saputra, E., Harun, S., Supriyanto, E.A., Patihawa, A.: Surface roughness in Ti 6Al-4V ELI (extra low interstitial) micro milling machining. Turbo: J. Mech. Eng. Study Progr. **8**(2) (2019)
11. Fang, F.Z., Liu, Y.C.: On minimum exit-burr in micro cutting. J. Micromech. Microeng. **14**(7), 984 (2004)
12. Yun, H.T., Heo, S., Lee, M.K., Min, B.K., Lee, S.J.: Plowing detection in micromilling processes using the cutting force signal. Int. J. Mach. Tools Manuf. **51**(5), 377–382 (2011)
13. Kathavate, V.S., Cheke, P.R., Adkine, U.S.A.: An Experimental Investigation of Micromilling. Int. J. Tech. Enhance. Emerg. Eng. Res. **3** (2015)
14. Konneh, M., Iqbal, M., Kasim, M.A.B., Isa, B.M.: High speed milling of silicon carbide with diamond coated end mills. In: Advanced Materials Research, vol. 576, pp. 535–538. Trans Tech Publications Ltd. (2012)
15. Oliaei, S.N.B., Karpat, Y.: Experimental investigations on micro milling of Stavax stainless steel. In: Proceedings of the 6th CIRP International Conference on High Performance Cutting, HPC 2014, vol. 14, pp. 377–382. Elsevier (2014)

Experimental Study of Waste Plastic Oil - Pertamax (RON 92) Blend on Spark Ignition Engine Performance at High-Speed Rotation

Muhammad Rikhwan Siregar, Khairil(✉), and Sarwo Edhy Sofyan

Department of Mechanical Engineering, Universitas Syiah Kuala, Jl. Tgk. Syech Abdurrauf No. 7 Darussalam, Banda Aceh 23111, Indonesia
khairil@unsyiah.ac.id

Abstract. Pyrolysis of plastic waste is an attempt to convert waste into alternative fuels. Besides being able to solve the problem of the energy crisis currently faced, it is also a solution in overcoming plastic pollution, especially HDPE, LDPE and PET. This study aims to investigate the effect of pyrolysis waste plastic oil (WPO) of HDPE; LDPE; PET as a mixture of pertamax (RON 92) fuel on the performance of a spark ignition (SI) engine at high speed rotation. The mixture of WPO-pertamax is varied in percentage namely 10:90, 20:80 and 30:70, respectively. Based on the experimental results, the engine performance including the engine power, specific fuel consumption, and thermal efficiency, is then analysed. It was found that the test results on each plastic waste oil mixture obtained that PET-10 at 2500 rpm rotation where the power and specific fuel consumption it produces are 2.98 HP and 0.32 kg/Kwh. Thermal efficiency of 18.95%.

Keywords: Pyrolysis · Waste plastic oil · SI engine performance

1 Introduction

Plastic as a material has an important role in everyday life and has been used in food, electrical, and automotive industries [1]. This is due to its superiority in toughness, lightweight, water and chemical resistance, ease of manufacture, durability, etc. Apart from its benefits environmental problems also arise due to the accumulation of plastic waste. According to the World Bank [2], plastic waste accounts for 8–12% of the total urban solid waste produced in various countries around the world. Plastics are estimated to take 100 to 500 years to decompose completely [3]. Management of plastic waste by burning can cause negative impacts on the environment in the form of air pollution, especially dioxin emissions, which are carcinogens. Therefore, another alternative is needed to handle this volume of plastic waste [4].

One of the alternatives and a promising method for recycling plastic waste is pyrolysis, which involves the thermochemical decomposition of organic and synthetic materials to produce fuel temperature increases in the absence of oxygen [5]. Currently, Indonesia

Akhyar et al. (Eds.): ICECME 2021, LNME, pp. 319–327, 2023.
https://doi.org/10.1007/978-981-19-3629-6_33

largest fuel imports are gasoline at 68.6%, followed by diesel oil (25.3%), avtur (4.4%), fuel oil (1.1%), and the remaining kerosene [6]. According to data from BPS in 2018 [7], total number of vehicles in Indonesia is 146,858,759 units. Fuel demand, including domestic biodiesel, in 2018 reached 465.7 million barrels/year, which was met from domestic refinery production and imports. Fuel production from domestic refineries was an average of 278.1 million barrels and imports an average of 165.4 million barrels [8].

The objective of this study is to examine the suitable of waste plastic oil-pertamax blend fuel on SI engine. To determine the performance of the SI engine using a mixture of plastic waste oil (HDPE, LDPE, PET)-pertamax with several variations of plastic waste oil: pertamax, 10:90 called WPO-10, WPO-20, WPO-30 compared to pure pertamax at certain rotations.

2 Research Method

2.1 Production of Waste Plastic Pyrolysis Oil

The waste plastic used is obtained from the landfill (TPA) in Kampung Jawa, Banda Aceh. The plastic waste is then washed and cleaned and then dried with the help of the sun. After drying, the plastic is cut into small pieces and weighed 1 kg per test. For the preparation of testing equipment, namely providing a condenser by putting water and ice cubes into the container to facilitate the condensation process, the condenser temperature ranges from 14 °C–22 °C, then installing a thermocouple and activating a digital thermocouple which is connected to the pyrolysis reactor room to monitor temperature stability in the pyrolysis process that is taking place. The final equipment preparation for testing is installing an iron pipe and a heat-resistant hose to the copper condenser pipe as shown in Fig. 1.

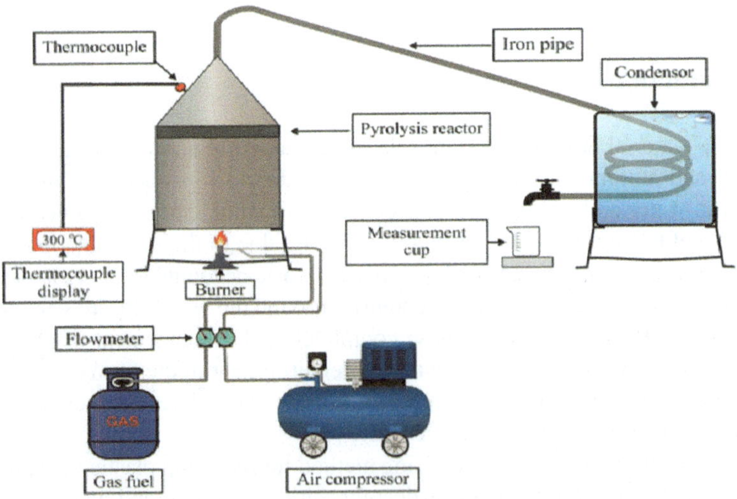

Fig. 1. Schematic diagram for converting plastic waste to oil via pyrolysis [9]

The testing process was carried out three times. The testing stages consist of several stages. The first stage is to enter 1 kg of plastic waste that has been cut into the pyrolysis reactor and then lock the pyrolysis reactor tightly by placing a fire packing so that no leakage occurs. The second stage of testing begins by turning on a combustion device connected to LPG to burn the plastic inside the pyrolysis reactor. The third stage of the test is to adjust the temperature to 250 °C, 300 °C.

3 Fuel Preparation

Three mixtures of WPO (HDPE, LDPE, PET) and pertamax (RON 92), with different percentages of WPO where 10%, 20%, and 30% used in the experiment. WPO – pertamax mixtures were placed in a 1 L beaker, then blending was carried out using a magnetic stirrer for 10 min at a speed of 300 rpm at room temperature. The resultant blends were stored in different bottles, separately. The physicochemical properties of the oil mixture of WPO – pertamax blends, such as viscosity, density, and calorific value, are obtained by using the Eq. 1.

$$NC_{\text{blend}} = X_1 \,.\, NC_{\text{WPO}} + X_2 \,.\, NC_{\text{Pertamax}} \tag{1}$$

where NC_{blend} is the calorific value of fuel blends (kJ/kg), X_1 is the percentages volume of WPO, X_2 is the percentages volume of pure pertamax, NC_{WPO} is the calorific value of WPO, NC_{Pertamax} is the calorific value of pure pertamax.

4 Engine Performance Test Procedure

Engine testing is carried out at various speeds from 1500 to 3000 rpm with 500 rpm intervals for each type of fuel. The tachometer is used to determine engine speed, the voltmeter is used to determine the voltage and the clamp meter is used to measure the

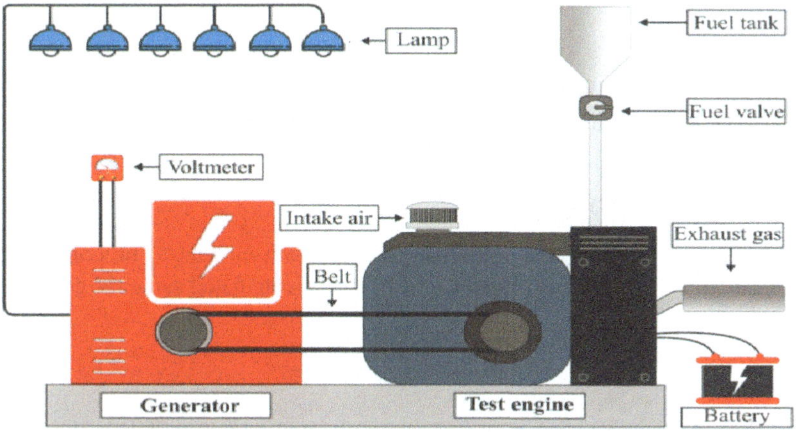

Fig. 2. Schematic diagram of engine experimental setup [9]

electric current of the engine load indicating the experimental setting of the machine. The test was carried out using a Honda GP 160 H-SD engine repeated three times to determine engine performance such as power, specific fuel consumption and thermal efficiency using a mixture of waste plastic oil as fuel with several mixed variations (WPO-10, WPO-20, WPO-30) compared to pure pertamax fuel (Fig. 2).

4.1 Engine Performance

The SI engine performance includes the engine power (*P*), specific fuel consumption (*SFC*), and thermal efficiency (ηth).

4.2 Engine Power

The power generated by an engine connected to a single-phase alternating current (AC) generator can be calculated by the equation [9],

$$Nb = \frac{E.I.pf}{746eg} \tag{2}$$

where N_b is the engine power (HP), *E* is the voltage (Volt), *I* is the current (Ampere), *pf* is the power factor for one single phase = 1, *eg* is the efficiency of electric generators for engines under 50 kVA = 87–89%, for generators using V belts, the power generated is divided by $\eta b = 0.96$.

4.3 Specific Fuel Consumption

The *SFC* is defined as the amount of fuel consumed by the engine to produce power denoted in kW/h. This *SFC* of an engine is usually measured by the mass of fuel consumed for the power output. The *SFC* can be calculated by the following equation [9]:

$$sfc = \frac{mf}{N_b} \tag{3}$$

Where *sfc* is the specific fuel consumption (kg/kWh), *mf* is the total fuel consumption (kg/hour), N_b is the engine power (HP).

4.4 Thermal Efficiency

Thermal efficiency is defined as the heat-use efficiency of fuel to be converted into mechanical work. The thermal efficiency (ηth) can be calculated using the equation [9]:

$$\eta th = \frac{N_b \ x \ 632,5}{mf \ x \ LHV} \tag{4}$$

where ηth is the thermal efficiency (%), N_b is the engine power (HP), *mf* is the fuel consumption (kg/hour), LHV is the fuel calorific value (kcal/kg).

5 Results and Discussion

5.1 The Characteristics of Physicochemical Properties of WPO

Table 1 present the characteristic values of various waste plastic oil (e.g. HDPE, LDPE, PET) and pertamax. As illustrated in Table 1 that the kinematic viscosity of WPO-HDPE tested was 1,574 mm^2/s, lower than WPO-LDPE kinematic viscosity of 6.9 mm^2/s and WPO-PET of 3,5 mm^2/s. The HDPE viscosity value tested was also lower than that produced by Kumar [10] who analyzed the performance and emission of a mixture of used plastic waste HDPE oil in a diesel engine of 2,1 mm^2/s. The density of WPO-HDPE test results is 710–739 kg/m^3, which is slightly lower than the density of WPO-LDPE and PET. The WPO-HDPE density value produced by Kumar [10] was 790 kg/m^3. The calorific value obtained from testing for WPO-HDPE was 44621,98 kJ/kg higher than the WPO-LDPE heating value, that is 42354 kJ/kg and lower than the WPO-PET heating value of 45019,6 kJ/kg. As for Kumar [10] yielded a heating value of 40170 kJ/kg for WPO-HDPE. The test results for the RON value of WPO-HDPE were 81,2, greater than the WPO-PET value of 78.8 and slightly lower than the WPO-LDPE value of 81,6.

Table 1. The physicochemical properties of WPO.

NO.	Properties	HDPE	HDPE	LPDE	PET	RON 92	Unit	Method
1	Kinematic Viscosity	1,574	2,1	6,9	3,5	9,7	mm^2/s	ASTMD4052
2	Density at 15 °C	710-739	790	789	750	715-770	Kg/m^3	ASTMD445
3	Calorific Value	44621,98	40170	42354	45019,6	45240	kJ/kg	ASTMD240
4	RON	81,2		81,6	78,8	92	-	ASTMD2699

Value measured, Kumar, S.et al., Khairil et al.

5.2 Engine Performance WPO – Pertamax Blends

The parameters measured in this study are the electric current and voltage generated by a generator coupled to the SI engine, and the time it takes to spend 20 ml of fuel. In this study, engine testing was carried out at various rotational speeds from 1500 to 3000 rpm with 500 rpm intervals for each type of fuel.

5.3 Engine Power

Figure 3 present the relationship between engine speed and power fueled by WPO – pertamax blend. It can be seen that the power generated by SI engine increases with increasing of engine speed. Where power is the amount of energy consumed per unit time. It can be seen that the effective power produced increases with increasing engine

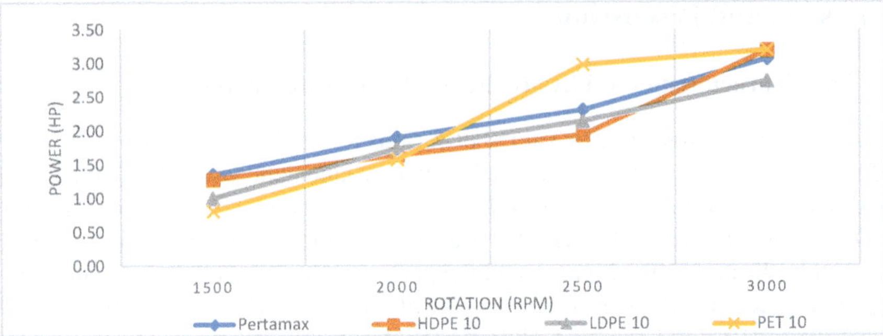

Fig. 3. The relationship between rotation and engine power produced by Pertamax fuel and WPO-10

speed. This may be because an increase in engine speed will increase the number of combustions per minute.

From the figure it can be shown that the highest power produced by Pertamax is 3.06 HP at 3000 rpm. The highest power in the plastic waste oil mixture was also produced at 3000 rpm where each mixture produced 3.19 HP (HDPE-10 and PET-10) and 2.72 HP (LDPE-10). The addition of 10% plastic waste oil mixture has no significant effect on the power produced compared to the power produced by pure pertamax. However, the increase shown by PET-10 and the decrease in HDPE-10 at 2500 rpm is an unusual trend, this may be caused by many variables that require further research.

This situation is different from that reported by Pratama [11] who tested the characteristics of plastic waste pyrolysis oil and its application as an alternative fuel in four-cylinder diesel engines, where the engine power produced by pure biodiesel fuel is still slightly higher than all plastic oil blends. this is because the calorific value and viscosity of waste plastic oil is lower than pertamax. Arismunandar [12] said that the performance of the combustion engine first of all depends on the power it can generate. Where the higher the rotating frequency of the engine, the higher the power given, this is because the greater the frequency the more work steps are experienced at the same time. However, at a certain rotation the power generated will reach the peak point and experience a decrease in power as the engine rotation increases.

5.4 Specific Fuel Consumption

In Fig. 4 where the relationship between engine speed and specific fuel consumption is based on the variation of the waste plastic oil mixture. Where the specific fuel consumption of the waste plastic oil mixture has decreased with increasing engine speed, this is different from pure pertamax which tends to be stable with increasing engine speed.

At 1500 rpm rotation PET-10 produces the highest specific fuel consumption of 0.59 kg/Kwh, a mixture of HDPE-10 plastic waste oil produces a specific fuel consumption of 0.37 kg/Kwh, LDPE-10 0.55 kg/Kwh while pure pertamax of 0.31 kg/Kwh which is the lowest value resulting from the test. In the picture, it is shown that the plastic oil mixture fuel requires more fuel than Pertamax at the same rotation.

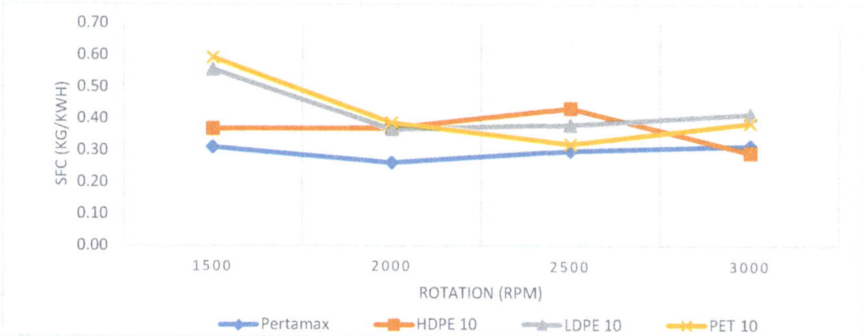

Fig. 4. The relationship between the rotation of the specific fuel consumption produced by the Pertamax and WPO-10 fuels

This is in line with research conducted by Yohannan [13]. Test the performance and emission characteristics of an automotive diesel engine fueled by fuel oil from plastic pyrolysis oil and its diesel mixture, where the specific fuel consumption decreases as the engine speed increases. This is due to the calorific value of waste plastic oil which is lower than the calorific value of pure pertamax. This is also consistent with Pulkrabek [14] which states that the specific fuel consumption decreases as the engine rotating speed increases to a minimum, then increases at high speed as shown in the picture above. Fuel consumption increases at high speeds because of the greater friction losses. At low engine speeds, the time per cycle is longer, leading to excess heat loss and increased fuel consumption.

5.5 Thermal Efficiency

In Fig. 5 where the relationship between engine speed and thermal efficiency is based on variations in the mixture of plastic waste oil.

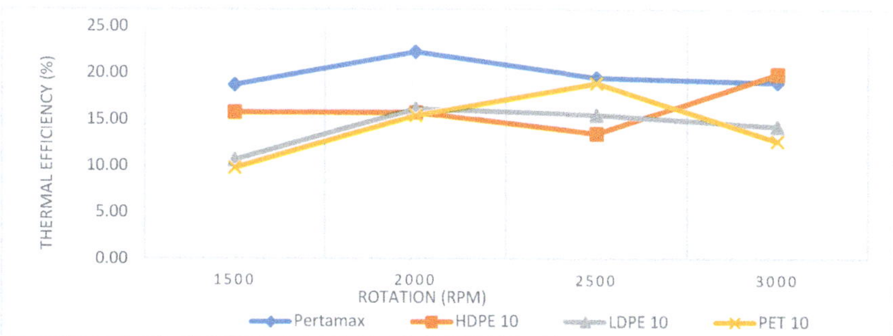

Fig. 5. The relationship between rotation and the thermal efficiency produced by the Pertamax fuel and WPO-10

It can be seen that the resulting efficiency has decreased with increasing engine speed. The highest thermal efficiency is produced by pure pertamax oil of 23,31% at

2000 rpm, 18,77% at 1500 rpm, then 19,55% at 2500 rpm and 19,05% at 3000 rpm. In the waste plastic oil mixture, the highest efficiency was obtained by the HDPE-10 oil mixture at 19,95% at 3000 rpm, slightly higher than the Pertamax efficiency value at the same rotation. The lowest thermal efficiency resulted in a mixture of PET-10 waste plastic oil of 9,85% at 1500 rpm where there was a difference of 48% when compared to the value obtained by pertamax at the same rotation. From the graph, it can be seen that the thermal efficiency of pure pertamax and LDPE-10 has increased from 1000 rpm to 2000 rpm then decreased from 2000 rpm to 3000 rpm rotation. There is a difference in the mixture of PET-10 and HDPE-10 oils, where PET-10 has increased from 1500 rpm to 2500 rpm and has decreased from 2500 rpm to 3000 rpm. The opposite happened to HDPE-10 which decreased from 1500 rpm to 2500 rpm and experienced an increase at 2500 rpm to 3000 rpm.

In general, the figure shows that efficiency increases with increasing engine speed. However, at higher revolutions the rate of increase in efficiency decreases. This indicates that there will be maximum efficiency. This is also shown in the research of Norsujianto [15] where the thermal efficiency of the fuel is very dependent on the calorific value of the fuel. The higher the calorific value of the fuel, the higher the thermal efficiency. The increase in shaft rotation at constant load tends to reduce thermal efficiency, for constant loads, the power generated tends to be constant and the increase in shaft rotation will shorten the mixing process time of fuel and air, so that combustion takes place less well, this will result in smaller combustion energy and tends to reduce efficiency. The higher viscosity of mixed fuels reduces thermal efficiency [16, 17].

6 Conclusion

Waste plastic pyrolysis oil is very feasible to be developed as an alternative fuel. From the WPO characteristic test results, the WPO HDPE density value is similar to the pertamax density value, with the HPDE oil density of 710–739 kg/m^3 and the density of pertamax 715–770 kg/m^3. Also, the WPO HDPE calorific value is 44621,98 kJ/Kg which is close to the heating value of pertamax oil of 45240 kJ/Kg. WPO HDPE has RON 81,2 which is lower than pertamax which has RON 92. The resulting viscosity is 1,574 mm^2/s lower than pertamax 9,7 mm^2/s. The ideal waste plastic oil mixture for mixing is PET-10 at 2500 rpm because the power it produces is 2,98 HP, 29% higher than that produced by pure Pertamax of 2,31 HP at the same rotation. The specific fuel consumption produced by PET-10 at 2500 rpm rotation is 0,32 kg/Kwh less than that produced by pure Pertamax which is 0,30 kg/Kwh. PET-10 thermal efficiency at 2500 rpm rotation is 18,95% compared to pure pertamax which produces an efficiency of 19,04% at the same rotation.

References

1. Joshi, R.A., Punia, R.: Chapter 41 Conversion of Plastic Wastes Into Liquid Fuels – A Review. Recent Adv. Bioenerg. Res. **III** 2014 (2014)
2. Wong, S.L., Ngadi, N., Abdullah, T.A.T., Inuwa, I.M.: Current state and future prospects of plastic waste as source of fuel: a review. Renew. Sustain. Energy Rev. **50**(2015), 1167–1180 (2015)

3. Karuniastuti, N.: Bahaya Plastik terhadap Kesehatan dan Lingkungan. Forum Teknologi **3**(1), 6–14 (2016)
4. Wahyudi, J., Prayitno, H.T., Astuti, A.D.: Pemanfaatan Limbah Plastik sebagai Bahan Baku Pembuatan Bahan Bahan Bakar Alternatif. Jurnal Litbang **XIV**. No.1 Juni 2018, 58–67 (2018)
5. Yuliansyah, A.T., Prasetya, A., Ramadhan, M.A.A., Laksono, R.: Pyrolysis of plastic waste to produce pyrolytic oil as an alternative fuel. Int. J. Technol. **7**, 1076–1083 (2015). ISSN 2086–9614
6. Badan Pengkajian dan Penerapan Teknologi: Outlook Energi Indonesia 2018, Energi Berkelanjutan untuk Transportasi Darat (2018). ISBN 978–602–1328–05–7
7. Perkembangan Jumlah Kendaraan BerengineMenurut Jenis (Unit) 2016–2018, 11 Januari 2021. https://www.bps.go.id/indicator/17/57/1/jumlah-kendaraan-bermotor.html
8. Sekretaris Jenderal Dewan Energi Nasional: Outlook Energi Indonesia (OEI) 2019. ISSN **2527**, 3000 (2019)
9. Khairil, Riayatsyah, T.M.I., et al.: Experimental Study on the Performance of an SI Engine Fueled by Waste Plastic Pyrolysis Oil–Spark ignition Blends. Energies 2020 **13**, 4196 (2020). https://doi.org/10.3390/en13164196
10. Kumar, S., Prakash, R., Murugan, S., Singh, R.K.: Performance and emission analysis of blends of waste plastic oil obtained by catalytic pyrolysis of waste HDPE with diesel in a CI engine. Energy Convers. Manage. **74**(2013), 323–331 (2013)
11. Pratama, N.N., Saptoadi, H.: Characteristics of waste plastics pyrolytic oil and its applications as alternative fuel on four cylinder diesel engines. Int. J. Renew. Energy Dev. **3**(1) 2014, 13–20 (2014)
12. Arismunandar, W.: Penggerak Mula EngineBakar Torak. edisi kelima. Penerbit ITB (1988)
13. Yohannan, A., Roy, R.G.: Performance and emissions characteristics of an automotive diesel engine fueled with petroleum-based fuel from plastic pyrolysis oil and its diesel blends. Int. J. Trend Sci. Res. Dev. (IJTSRD) **4** Issue 6, e-ISSN: 2456–6470 (2020)
14. Pulkrabek, W.W.: Engineering Fundamentals Of The Internal Combustion Engine, p. 07458. University of Wisconsin. Prentice Hall. Upper Saddle River, New Jersey (1997)
15. Norsujianto, T., Saptoadi, H.: Performa EngineDiesel Menggunakan Bahan Bakar Campuran Minyak Hasil Pirolisis Limbah Plastik Dan Biosolar Sebagai Bahan Bakar Alternatif. Naskah Publikasi Seminar Nasional Thermofluid V Jurusan Teknik Mesin dan Industri Fakuftas Teknik Universitas Gadjah Mada Tangal 28 Agustus 2013
16. Padmanabhan, S., Ganesan, S., Arputhabalan, J.J., Varun, C., Ganesh, P.B.: Performance test on compression ignition engine by blending ethanol and waste plastic pyrolysis oil with cetane additive. In: IOP Conf. Series: Materials Science and Engineering, vol. 197, p. 012023 (2017). https://doi.org/10.1088/1757-899X/197/1/012023
17. Anup, T.J., Watwe, V.: Waste plastic pyrolysis oil as alternative for SI and CI Engines. Int. J. Innov. Res. Sci. Eng. Technol. (An ISO 3297: 2007 Certified Organization) **3**. Issue 7(2014). ISSN: 2319–8753

Cu/CuSO$_4$ Solid-State Reference Electrode for Potential Corrosion Measurement on the Reinforcing Steel

M. Ihsan[1,4], S. Fonna[2(✉)], S. Huzni[2], N. Islami[3], and A. K. Ariffin[4]

[1] Faculty of Engineering, Universitas Gajah Putih, 24552 Takengon, Indonesia
[2] Department of Mechanical and Industrial Engineering, Universitas Syiah Kuala, Banda Aceh, Indonesia
syarizal.fonna@unsyiah.ac.id
[3] Department of Material Engineering, Universitas Malikussaleh, 24351 Aceh Utara, Indonesia
[4] Department Mechanical and Manufacturing Engineering, Universiti Kebangsaan Malaysia, 43600 Bangi, Selangor, Malaysia

Abstract. Potential measurement on rebar reinforcing steel plays a vital role in the reinforced concrete corrosion investigation. One of the essential factors in the measurement is the reference electrode. Therefore, improving reference electrodes (RE) for potential collecting data is recommended for corrosion analysis. Liquid-based RE was used frequently by many researchers for this matter. However, the liquid-based RE performance reduction will occur due to the mechanical phenomenon on the RE, i.e. leakage of the electrode solution and range distance between rebar and the concrete cover. Thus, the research focuses on the solid-state RE developed for the potential distribution measurement on the reinforced concrete analysis. By using the method, the limitation of liquid-based RE was solved. Copper-copper Sulfate (CCS) was prepared for the reference solution. The Open Circuit Measurement (OCP) was applied for the stability performance analysis of the solid-state RE. Furthermore, Potential corrosion analysis on the rebar concrete was carried out and validated with the commercial Ag/AgCl RE. The result shows the stability performance of the CCS solid-state reference electrode and similar OCP behavior with Ag/AgCl commercial RE. The influence of distancing between rebar to reference electrode location was performed, and actual potential data from the rebar was measured on the reinforcing rebar. The real-time potential data was collected in order to reinforce concrete corrosion investigation. The result shows the high-risk corroding that occurred on the rebar and was validated by Ag/AgCl potential distribution value. Directly visualization from the rebar concrete specimen proved the high-risk corrosion phenomenon. Therefore, the results obtained allow us to conclude that solid-state RE is recommended for the rebar concrete corrosion analysis.

Keywords: Copper-Copper Sulfate (CCS) · Solid-state · Reference electrode · Reinforced concrete

Akhyar et al. (Eds.): ICECME 2021, LNME, pp. 328–339, 2023.
https://doi.org/10.1007/978-981-19-3629-6_34

1 Introduction

At present, the protection of the reinforcing steel strength structures from various aspects of failure plays a vital role in building the stability, safety and security of infrastructure [1]. However, the penetration of ions and a decrease in the pH value in the concrete will initiate the emergence of corrosion structures. The phenomenon will cause a degradation of the strength of the building structure [2]. Therefore, an evaluation is needed in order to control the corrosion phenomenon that occurs in the concrete structure. The corrosion phenomenon is an electro process, monitoring and evaluating the corrosion phenomenon intensively in reinforced concrete structures using electrochemical methods [3].

Many researchers have recently introduced electrochemical techniques such as half-cell potential mapping, dynamic polarization, open circuit potential measurement, electrochemical impedance spectroscopy and linear polarization. The method provided adequate information on reinforced concrete corrosion status [4]. A reliable RE is required to obtain an accurate measurement result. Saturated calomel electrodes (SCE) like Ag/AgCl, Hg/Hg_2Cl_2 and $Cu/CuSO_4$ is a liquid-based reference electrode (RE) which researchers widely use as one of the leading reference electrodes equipment for potential corrosion measurement of reinforced concrete [5]. Nevertheless, there are some limitations to the established reinforced concrete corrosion assessment method. Most potential measurements were carried out based on ASTM C876 by surface mounting techniques in the current method. This technique may produce non-reliable potential value due to the considerable distance of steel in concrete structures and concrete cover. The leakage of RE electrolyte frequently contaminates concrete structures and induces localized corrosion.

To overcome the limitation of the liquid-based RE, the solid-state RE has been introduced by several researchers. The solid-state RE is considered appropriate for the potential measurement of reinforcing steel. Due to using the embedded solid-state RE, the influence of mechanical factor can be reduced and the potential of collecting data directly on the rebar survace. The solid-state RE can be usable for long-time conditions. The ideal solid-state RE must have challenging characteristics and low maintenance periodically. The solid-state RE must also be able to measure the minimum potential value of rebar in concrete under various conditions of concrete treatment in the field. The solid-state RE must also have long-term durability and stability. Furthermore, the results of potential measurements using solid-state embedded RE are also not affected by contamination of the concrete mix composition and can conduct low currents with minimum polarization.

Many researchers have fabricated the solid-state RE, such as Duffo et al. [6] and Manjakkal et al. [7] derived solid-state RE based on metal-metal oxide electrodes. The result shows the long-term stability RE for potential measurement. Mn_3O_4 based pellet electrode was also fabricated by Maruthapandian et al. [8]. The embedded Mn_3O_4 based RE were used for corrosion monitoring, and the result shows the Mn3O4 based characteristics suitable for the high alkaline concrete environment. Solid embedded RE using graphene-cement composites have been investigated by Jin et al. [2]. The synthesis and properties of graphene oxide were derived by Zhu et al. [9]. The corrosion investigation reveals the advantages of graphene content in solid-state RE composites for better stability and RE reliability.

In this research, the prepared RE embedded into the cement environment above the reinforcing steel was investigated. The potential measured by the solid-state RE on the reinforcing steel was analyzed. By the result, it can be inferred that solid-state RE is considered able to solve the contamination and solution leaking problem of the liquid-based RE. The solid-state Cu/CuSo4 RE for reinforcing steel potential measurement was developed for the purpose. The proposed RE is relatively unobtrusive for most types of engineering problems [10, 11].

Furthermore, the shape of RE adjustment for the embedded process is also an important study. The RE embedded in the concrete has a different cover thickness for each use in reinforced concrete. Therefore, the size of the Cu/$CuSO_4$ reference electrode is important to consider in the rebar potential measurement in the embedded reinforcement. Therefore, the process of re-shape the liquid-based reference electrode to the solid-state reference electrode is one method of miniaturization or reducing the size of the reference electrode. The process is carried out with several rules based on commercial reference electrodes with a mixture of Ag/AgCl which is widely used [12, 13]. In addition, this solid-state reference electrode method has several advantages, such as a simple assembly process, then this solid-state RE can be easily integrated with other measurement devices, sensors and electronic measuring instruments can be mass-produced and low cost. The size of the electrode can be adjusted as needed. Therefore, with the development of Cu/$CuSO_4$solid-state reference electrodes, the potential collecting data process of rebar reinforcement in real-time as input for corrosion detection analysis in reinforced concrete can be carried out.

2 Liquid Reference Electrode

Electrochemical measurements in aqueous solutions are usually measured against a reference electrode (RE), such as saturated calomel electrode (SCE), silver/silver chloride (Ag/AgCl) and copper-copper sulfate (Cu/CuS4). The Ag/AgCl Reference Electrode is widely used for electrochemical measurement due to its simplicity, stability, capability. Combining coated silver wire with a thin layer of silver chloride either by electroplating or dipping the wire in molten silver chloride was fabricated commercially for the measurement as shown in Fig. 1. The porosity concept developed in the liquid reference electrode for the ions movement detecting process makes it susceptible to leakage in the liquid in the reference electrode. Furthermore, the copper sulfate (Cu/$CuSO_4$) electrode is a standard reference that is widely used by researchers for corrosion control in the field. This is due to the level of stability and robustness for field use. The use of (Cu/$CuSO_4$) is commonly used to measure the potential for material requirements in soil and water.

The composition of the reference electrode is easy to obtain. However, the tests carried out by corrosion control researchers are currently widely used to measure the material potential interacting with the familiar environment such as soil, air or water. The surface of material measured directly carries out the widely potential measurement process. However, the liquid reference electrode based on Cu/$CuSO_4$ is prone to leakage at the tip of the porous area of the reference electrode. Thus, the development of the solid-state reference electrode can be used to overcome the weaknesses of the current reference electrode based on Cu/$CuSO_4$ for potential data value detecting in certain reinforced concrete corrosion cases.

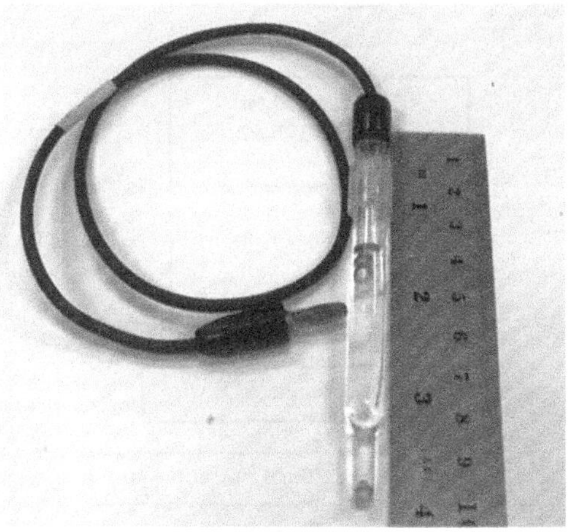

Fig. 1. Commercial Ag/AgCl Liquid Reference Electrode

3 Solid Reference Electrode

Although the stability and performance of the classical construction were accurate to approximately a millivolt, it cannot be used in applications where robustness and ruggedness are required. Potential measurement on the reinforcing steel influenced by the concrete cover thickness and rigid concrete composition. The distance of a reference electrode with the rebar also influences the potential distribution on the rebar. Therefore, an embedded reference electrode in the concrete was obtained for potential measurement. However, the leakage of RE electrolyte frequently contaminates concrete structure and induces localized corrosion. Therefore, the solid-state reference electrode was developed to overcome the limitation. The materials cost for the reference electrode assembly is also shown in Table 1.

Table 1. Solid-state Reference electrode assembly cost

No	Material		Price
1	Cu crystal powder	50 mg	$ 2
2	PTFE tube cylinder	1 m	$ 5
3	Copper Plate	$10 \times 10 \times 0.2$ cm	$ 10
4	PVC Powder + DoP	50 mg	$ 14
5	Arabic gum	1 pack	$ 5
6	Epoxy resin	1 pack	$ 2
Total assembly cost			**$ 38**

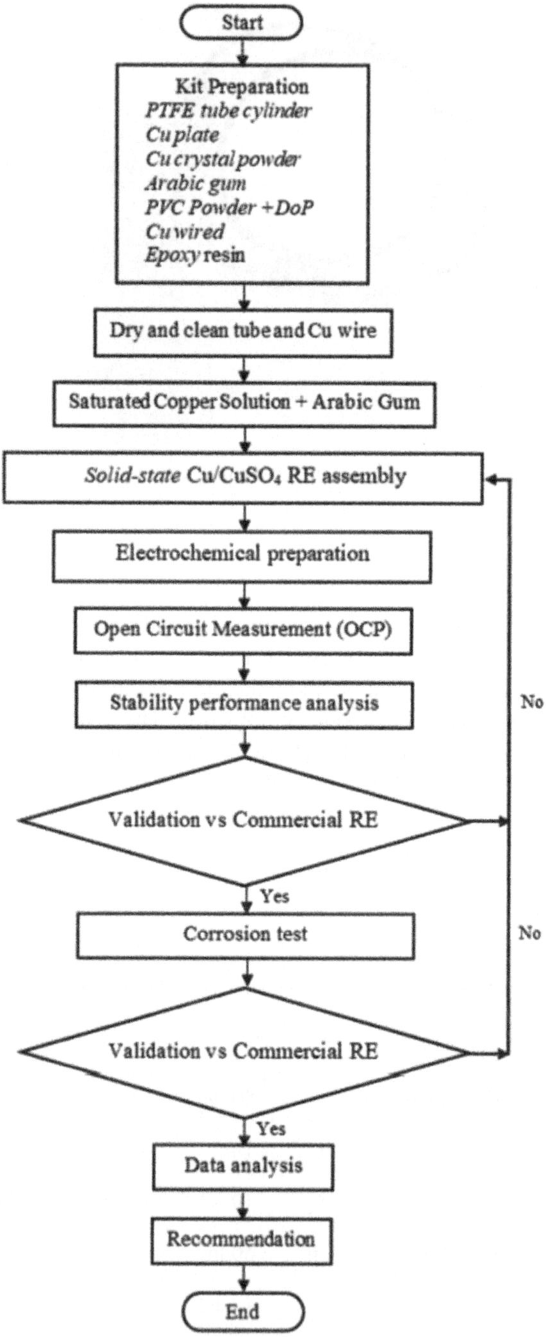

Fig. 2. Flow chart Solid-state Reference Electrode assembly

The solid-state RE was obtained from a cleaned Copperplate 10 × 10 mm and 2 mm thickness. The copper plate was cleaned with Deionised water and dried. The polytetrafluoroethylene (PTFE) tube ID 6 × OD 8 mm was paired on the copper plate with epoxy resin gum and dried at room temperature. Once the dried PTFE tube and copper plate form, 5 mg crystal copper sulfate was infused to distribute the movement of the ions. The crystal copper sulfate is protected by the 35 mg + 65 mg Polyvinyl chloride (PVC) and Doc Dissolved in Tetrahydrofuran (THF) solvent for the cube cover. The selected RE bare and left one night at room temperature to dry. After drying, the selected RE was connected with Cu wire to distribute the potential value to the performance machine test. Later on, above the reference surface was mounted by epoxy resin, excluding the PVC + Dioc membrane. The solid $Cu/CuSO_4$ reference electrode's flow chart is shown in Fig. 2, and the solid-state Cu RE assembly and analysis design is shown in Fig. 3. After the solid reference was performed, the OCP and corrosion investigation was measured for the reference electrode performance analysis.

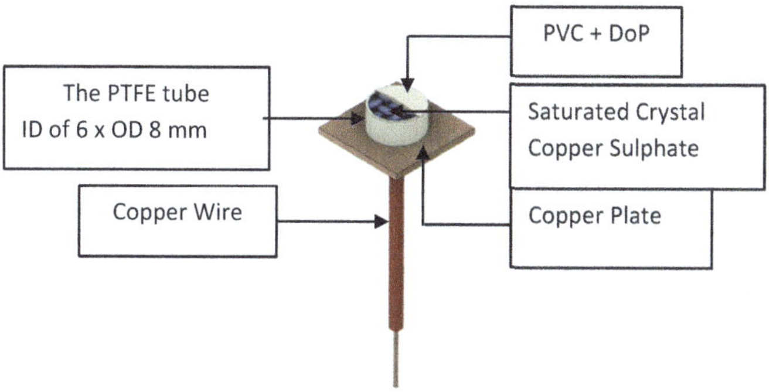

Fig. 3. Solid-State Reference Electrode design

4 Result and Discussion

4.1 $Cu/CuSO_4$ Solid-State Reference Electrode

The $Cu/CuSO_4$ solid electrode has been constructed, as shown in Fig. 4. The developed reference electrode can reduce the research cost significantly against the commercial reference electrode commonly used based on the assemble cost, as shown in Table1. Several potentiodynamic tests were analyzed to perform the solid electrode quality. The solid-state $Cu/CuSO_4$ electrode testing against the commercial Ag/AgCl electrode. A portable model of ZIVE PPI Potentiostat type connected to an electrochemical cell was used for data investigation. A solid-state Cu/CuSo4 electrode and commercial Ag/AgCl was utilized for performance analysis. Corroded reinforced concrete was the working electrode (WE), and a graphite bar was prepared as the counter electrode (CE).

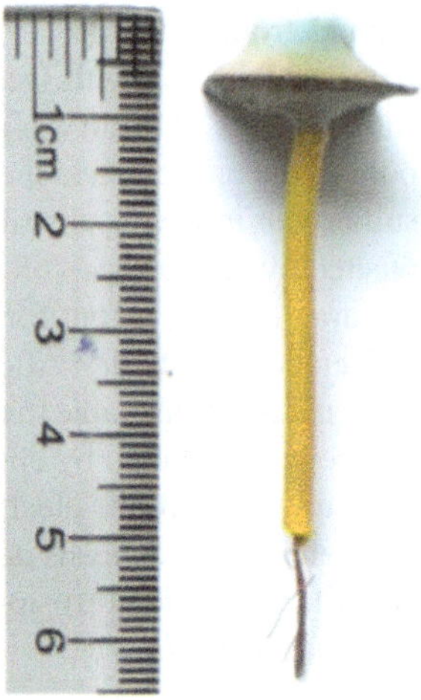

Fig. 4. Cu/CuSO4 Solid State Reference Electrode

5 Experimental Work

The experimental work analyzed the stability performance, the reference electrode's distance with the rebar and the reinforcing steel's potential corrosion. For the stability performance purpose, the potential distribution on the reinforcing steel was analyzed by using Open Circuit Measurement (OCP). Meanwhile, the influence of reference electrode location distance was analyzed by embedded reference electrode specimen as shown in Fig. 5

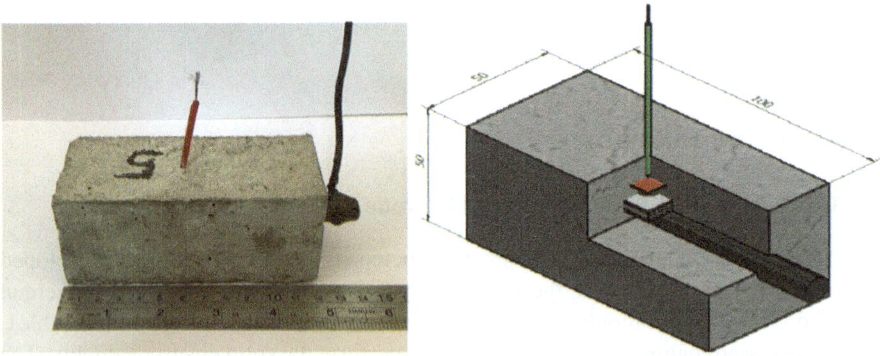

Fig. 5. Embedded reference electrode reinforcing steel for OCP analysis

Furthermore, the corroded concrete rebar specimen was prepared for reinforced concrete corrosion analysis. Tafel Measurement on the concrete rebar will be analyzed using ZIVE PPI Potentiostat. For reinforced concrete corrosion analysis, the specimen was immersed for 34 weeks and exposed to the environment for one year. The specimen is shown in Fig. 6.

Fig. 6. Corroded reinforcing steel for corrosion potentail analysis

6 Experimental Results

Analysis potential distribution was performed with no-current injection by using both reference electrodes. 600 time cycle potential measurement was done during the analyzing process, and a similar experiment setup was treated between solid-state RE and commercial Ag/AgCl RE. The result shows a similar potential value behavior by using both reference and the potential distribution difference value between reference electrodes less than range 0,15 V, indicating that the potential distribution value by using both reference electrodes was recommended. The stability of the Open Circuit Potential (OCP) cycle potential value is shown in Fig. 7.

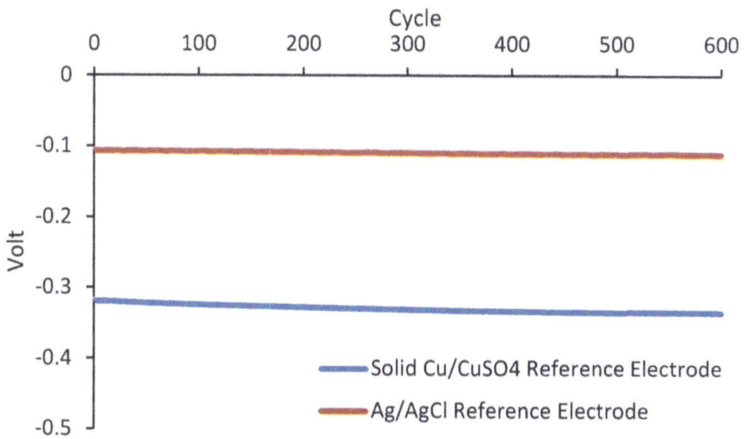

Fig. 7. Normal potential distribution by using solid state and commercial reference electrode

Regarding Table 2, potential value collected by both reference electrodes has been reported. The potential distribution result has been converted to Standard Hydrogen Electrode (SHE) based on the conversion table as shown in Fig. 8. SHE is the standard

Table 2. Potential Range value of the OCP measurement (SHE Conversion)

Potential	Solid RE	AgCl RE
Value	Volt	Volt
Min	−0,3179	−0,1054
Max	−0,3339	−0,1106
Range	0,0160	0,0052

reference for potential electrode analysis. It is also a highly recommended electrode for use in an aqueous solution. The result shows the range of potential data by using the solid-state Cu RE less than 0,0016 V in which the minimum potential value was −0,3179 V, and the maximum potential value was −0,3339 V. Meanwhile, the minimum potential value by using Ag/AgCl RE was −0,1054, and the maximum potential value was −0,1106. It was inferred that the range of potential value by Ag/AgCl RE is only 0,0052 V. The solid $Cu/CuSO_4$ showed stability performance during the OCP measurement.

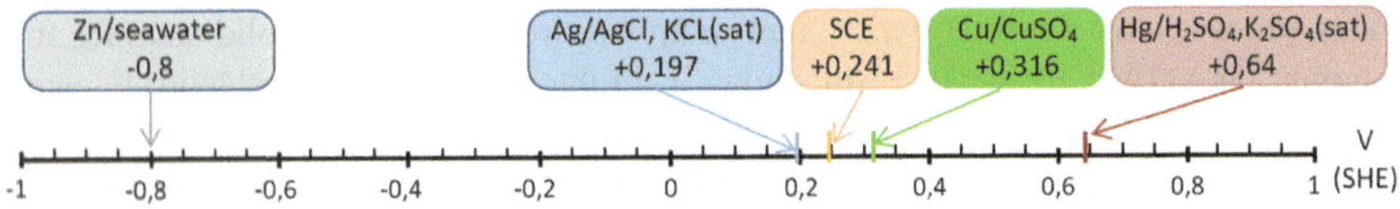

Fig. 8. Conversion of Electrode Potential

Regarding the ASTM table as shown in Table 3, the corrosion risk of the reinforcing steel specimen indicates an uncertain and low-risk corrosion phenomenon based on potential collected value from both reference electrodes. Nevertheless, the rebar concrete specimen for the solid-state RE stability performance analysis was not immersed in the corrosion environment.

Table 3. Reinforcing steel potential range for corrosion risk level (ASTM C876–91).

Rebar potential (mV)		*Qualitative risk of corrosion*
$Cu/CuSO_4$ electrode	Ag/AgCl electrode	Likely corrosion condition
> – 200	> –106	Low (10% risk of corrosion)
– 200 to – 350	– 106 to – 256	Intermediate corrosion risk (uncertain)
< – 350	< – 256	High (> 90% risk of corrosion)
< – 500	< – 406	Severe corrosion

The influence of reference electrode location during collecting potential data has been reported. As shown in Fig. 9, the result shows the differences in potential distribution value between embedded solid-state RE and potential measurement on the concrete cover.

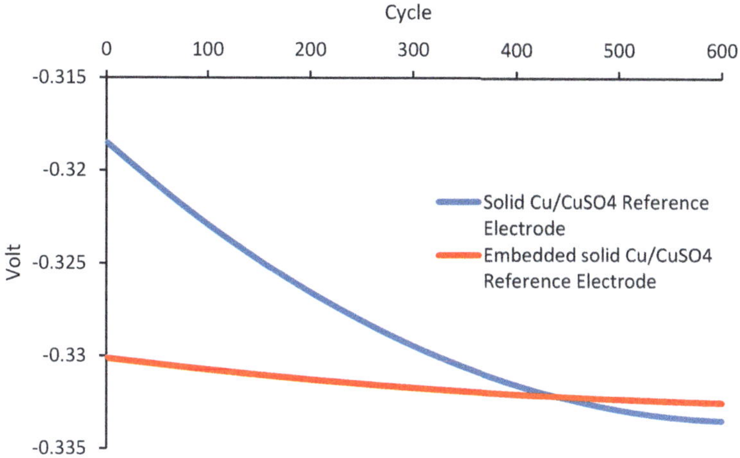

Fig. 9. The potential distribution by embedded reference electrode vc concrete surface reference electrode

The potential value result shows the stability performance of the embedded solid-state reference electrode rather than the concrete surface potential measurement. The result inferred the influence of potential measurement location distance. Cement composition and ion chloride penetration influence the differences of potential distribution trendline by both methods. Therefore, collecting the potential by embedded reference electrodes is preferable to reducing the cement solution influence and environmental factors.

Moreover, the polarization behavior from both RE (Solid-state RE and Ag/AgCl RE) can be seen from the Tafel analysis using the corroded specimen as shown in Fig. 6. Similar potential behavior was expressed by solid-state Cu/CuSO4 and commercial Ag/AgCl reference electrodes, as shown in Fig. 10. Both references reported although the potential value that measured by Commercial Ag/AgCl is less negative than the solid-state Cu reference electrode, there similar corrosion risk level was shown by the result regarding ASTM C 876–91 for corrosion risk criteria by Ag/AgCl reference electrode and corrosion risk by Cu/CuSO4 reference electrode. The solid-state Cu/CuSO4 reference electrode shows the corrosion potential (Ecorr) at −469,159 mV and the (Icorr) at the current density in the condition were 2,120 μA. Meanwhile, the differences in potential value were shown by the commercial Ag/AgCl reference electrode, the corrosion potential (Ecorr) at -687,341, and the (Icorr) at current density were 8,759 μA. Both of the reference electrodes indicate that the rebar was intensively corroded. The visualization of the corroded rebar is also shown in Fig. 6. It can be concluded that the new model of solid-state Cu/CuSO4 reference electrode can be used for the embedded reference electrode for potential collecting data in the reinforcing steel corrosion in the various conditions.

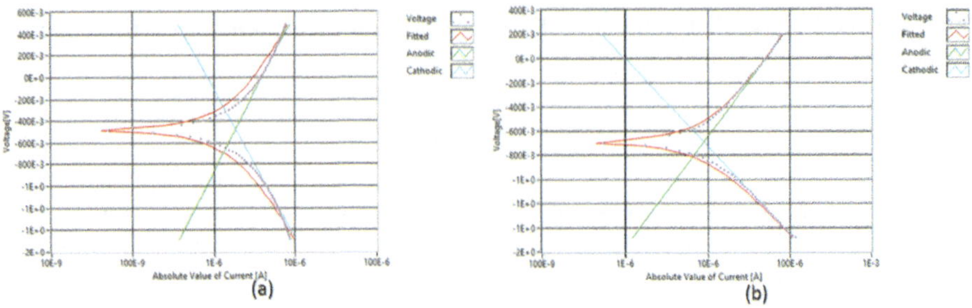

Fig. 10. (a) Tafel slope solid State CuSo4 RE and (b) Tafel slope Commercial Ag-AgCl RE

7 Conclusion

The solid-state reference electrode is a solution to solve the limitation of the liquid-based reference electrode to collect potential data on the reinforced concrete rebar. As a result, the solid-state Cu/$CuSO_4$ that developed intensively decreases the research cost. The Solid-state Cu/$CuSO_4$ embedded reference electrode can reduce the influence of cement solution influence, ion chloride penetration and mechanism process of the reinforced concrete corrosion. The result shows the similar behavior of the solid-state Cu/$CuSO_4$ reference electrode with the commercial Ag/AgCl reference electrode for the corrosion potential data. Therefore, the solid-state Cu/$CuSO_4$ reference electrode can be used for collecting data in reinforcing steel and solve the limitation of the liquid-based and the commercial reference electrode.

Acknowledgement. This work was supported by the Ministry of Education, Culture, Research, and Technology, Republic of Indonesia, Penelitian Dasar (PD) No. 3/UN11.2.1/PT.01.03/DRPM/2021.

References

1. Ihsan, M., Susanto, H.: Efisiensi Model Mesh Kasus Deteksi Simulasi Korosi Beton Bertulang Menggunakan Metode Axisymmetric Mesh Problem. **6**(2) (2020)
2. Jin, M., Jiang, Y., Jiang, L., Chu, H., Zhi, F., Gao, S.: Fabrication and characterization of pseudo reference electrode based on graphene-cement composites for corrosion monitoring in reinforced concrete structure. Constr. Build. Mater. **204**, 144–157 (2019)
3. Luo, D., Li, Y., Li, J., Lim, K.S., Nazal, N.A.M., Ahmad, H.: A recent progress of steel bar corrosion diagnostic techniques in RC structures. Sensors (Switzerland) **19**(1). MDPI AG (2019)
4. Mangaiyarkarasi, G., Muralidharan, S.: Electrochemical protection of steel in concrete to enhance the service life of concrete structures. Procedia Eng. **86**, 615–622 (2014)
5. Fonna, S., Gunawarman, Huzni, S., Ariffin, A.K.: The influence of some solution candidate on the performance of boundary element inverse analysis in detecting rebar corrosion. In: IOP Conference Series: Materials Science and Engineering, September 2019
6. Duffó, G.S., Farina, S.B.: Development of an embeddable sensor to monitor the corrosion process of new and existing reinforced concrete structures. Constr. Build. Mater. **23**(8), 2746–2751 (2009)

7. Manjakkal, L., Szwagierczak, D., Dahiya, R.: Metal oxides based electrochemical pH sensors: Current progress and future perspectives. Prog. Mater. Sci. **109**, 100635 (2020)
8. Maruthapandian, V., Saraswathy, V., Muralidharan, S.: Development of solid state embeddable reference electrode for corrosion monitoring of steel in reinforced concrete structures. Cem. Concr. Compos. **74**, 100–108 (2016)
9. Zhu, Y., et al.: Graphene and graphene oxide: synthesis, properties, and applications. Adv. Mater. **22**(35), 3906–3924 (2010)
10. Stern, H.A.G., Sadoway, D.R., Tester, J.W.: Copper sulfate reference electrode. J. Electroanal. Chem. **659**(2), 143–150 (2011)
11. Szabó, S., Bakos, I.: Reference electrodes in metal corrosion. Int. J. Corros. vol. **2010** (2010)
12. Alva, S., Aziz, A.S.B.A., Syono, M.I.B., Jamil, W.A.B.W.: Ag/AgCl reference electrode based on thin film of Arabic gum membrane. Indones. J. Chem. **18**(3), 479–485 (2018
13. Sophocleous, M., Atkinson, J.K.: A review of screen-printed silver/silver chloride (Ag/AgCl) reference electrodes potentially suitable for environmental potentiometric sensors. Sens. Actuat. A Phys. **267**(December), 106–120 (2017). https://doi.org/10.1016/j.sna.2017

The Analysis of Crankshaft Failure in a 2,500cc Diesel Engine Vehicle Using Experimental Method

Husaini(✉), Osama, and Teuku Edisah Putra

Department of Mechanical Engineering, Universitas Syiah Kuala, Darussalam, Banda Aceh 23111, Indonesia
{husainiftm,edi}@unsyiah.ac.id, osamaa@mhs.unsyiah.ac.id

Abstract. Crankshaft is a major engine component susceptible to mechanical failure. The purpose of this study is to determine the cause of crankshaft failure in a 2,500cc diesel engine pick up vehicle after 3 months of operation. Distance covered by the truck to convey commercial loads on a daily basis lies between 110–150 km. The study adopted an experimental approach encompassing visual observation, stress analysis, SEM fracture surface inspection as well as the evaluations of chemical composition, microstructure and hardness. Based on the chemical composition test with material standards ASTM A536 grade 80-55-06, nodular cast iron was discovered as the applied sample metal type. The SEM results showed an initial crack on the crank arm with subsequent extension leading to the final fracture, due to dynamic loading. Meanwhile, the stress analysis obtained a higher shear stress of 113.75 MPa, compared to the allowable estimate of 48.125 MPa. As a consequence, the crack propagation triggered the crankshaft failure.

Keywords: Failure analysis · Cast iron · Crankshaft · Diesel engines

1 Introduction

Crankshaft is a primary component in internal combustion engines used to convert the up-and-down piston motion to a cyclic nature. The design involves the use of a solid material capable of withstanding heavy load at relatively high engine speeds. Weights peculiar to crankshaft include torsional, bending and centrifugal loads. Therefore, the role of the tool appears very significant in optimizing transmission power, because serious accident is possible in the event of failure.

Crankshaft is a working engine component that alters the reciprocating nature of piston into a rotary motion, using a piston rod. During this process, a crank pin is required with an additional cushion positioned at the end drive rods on each cylinder [1]. Under real conditions, crankshaft is expected to demonstrate very substantial resilient to withstand the bottom safe power stroke, without excessive bending. Therefore, the life of combustion engine is internally dependent on the strength [2]. The failure of a material or structure represents an error across multidisciplinary characteristics. There

Akhyar et al. (Eds.): ICECME 2021, LNME, pp. 340–348, 2023.
https://doi.org/10.1007/978-981-19-3629-6_35

is no particular universal definition, but in a broader perspective, failure depicts the behavior where a component is no longer capable of fulfilling its specific functions [3].

In a simpler opinion, failure refers to a condition believed to instigate direct loss, commonly due to design, material and maintenance errors [4]. Generally, cracks begin from the initial repetition mostly on weak surfaces or stress concentrated areas, such as scratches and pits, or due to repeated loading. This original fault develops into micro-cracks or rapid propagation to form macro deformations that trigger the breakdown [5]. The study by Kakade & Pasarkar serves as a reference where the crankshaft failure occur in the form of cracks, defects and vibrations from load variations [6]. Another similar study on broken screw spring front car suspension indicated a fatigue failure, due to the presence of crack initiation and benchmark (Husaini *et al.*) [7]. Furthermore, steel crankshaft is typically produced by forging, die forging or casting if the material is cast iron [8].

The occurrence of crankshaft failure is commonly dependent on its functions, as massive power or load are always involved, in addition to very high speeds. This circumstance is also due to voltage mechanic appearing in small fillets, sharp angles, grooves, splines and scratches. Subsequently, the shafts often break at the edges where a high level of stress concentration exists. There are supporting factors instigating failure, including the material design and selection, heat treatment, fabrication, machining as well as assembling [9].

The 2500cc diesel engine pick up vehicle operated for 3 months before a complete collapse at the crank arm. This period is currently categorized as a short service life and there is need for further studies, based on the underlying observations. Therefore, the purpose of this study is to determine the cause of the crankshaft breakdown.

2 Material and Methods

The 2500cc pick up diesel engine crankshaft served as the study object, while the failure occurred on the crank arm region. Pick up vehicles cover a distance between 110–150 km on a daily basis, with varying loads (Banda Aceh-Jantho). This study employed an experimental method, starting with visual observations to examine the fracture surface under macroscopic and microscopic conditions. Subsequently, a hardness test was applied to determine the strength distribution on the material overlay, followed by chemical composition analysis of the standard reference material. Finally, conclusions are presented on the cause of the crankshaft failure.

Material. Figure 1 represents the crankshaft where hardness, chemical composition and microstructure assessments were conducted.

Visual Observation. These observations were based on two successive forms, termed macro (visible) and micro (invisible) and also served as initial steps in failure identification. The objective is to investigate the possible occurrence of crack locations and other indications, including the fracture surface pattern. Furthermore, the failure mechanism typically occurs in three cracks stages, termed initiation, propagation and final [10].

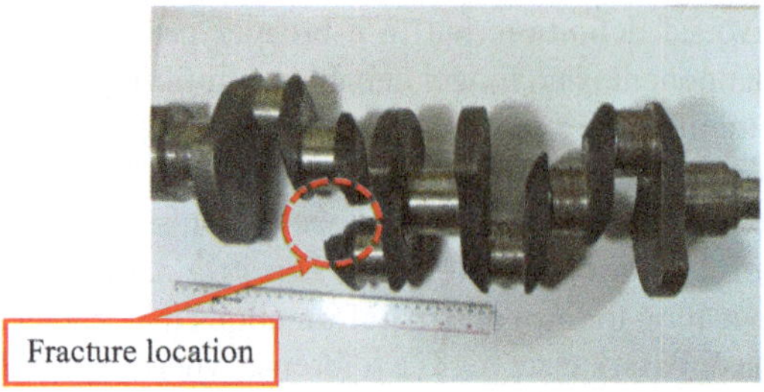

Fig. 1. Failure Crankshaft

Hardness Testing. This phase involved the determination of hardness value on the broken crankshaft surface, viewed from its distribution and compared with the standard applied material. The testing tool is the Rockwell Type Zwick/Roell ZHR, based on ISO 6568 and ASTM 18 standards with a load of 100 Kgf, while the applied eye press (ball) is 1/16 indenter. Figure 2 shows the hardness test performance at 80 points, where 40 were each distributed along the x and y axes, at a uniform distance of 1 mm.

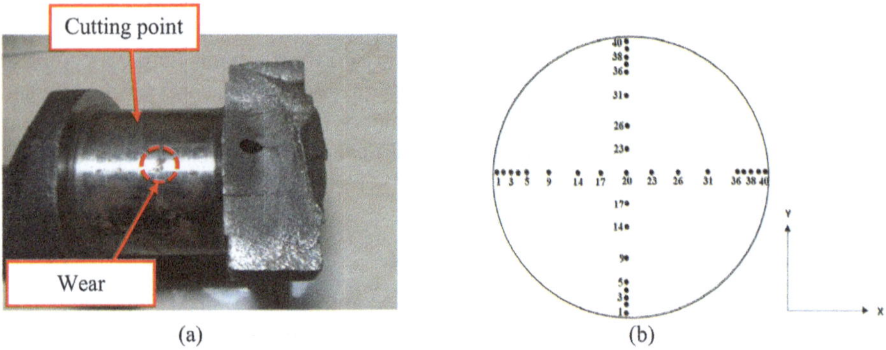

Fig. 2. (a) Hardness Testing Location, (b) Sketch Hardness Testing Point

SEM Observation. The scanning electron microscope (SEM) assessment was used to clarify the fracture surface, with magnification extending to 500x. This method effectively describes the location of crack initiation and propagation.

Chemical Composition Testing. This process helps to identify the chemical components in the crankshaft. The resulting data were compared with the standard material, and also served as a reference for review purposes, depending on the mechanical properties.

Microstructure Testing. Microstructure describes the collection of phrases by metallographic techniques, including the use of microscope in the case of cast iron.

Stress Analysis. This analytical method was aimed to determining the stress value under normal crankshaft conditions, including its operation without defects or damage. The analysis was used to obtain both von-misses and shear stresses for this particular failure event.

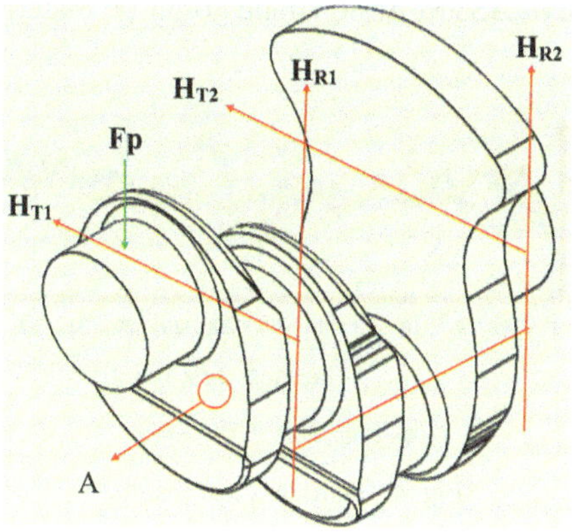

Fig. 3. Distribution of Load and Reaction Force on Crankpin.

Figure 3 shows the force reaction Fp (piston compressive force), H_T (tangential force) and H_R (radial force) in the load distribution. The value "A" represents the shear stress calculated using Eq. 1 [11]:

$$\tau = \frac{T_e \times 16}{\pi \times \left(d_{cp}\right)^3} \tag{1}$$

Where:
τ = Crankpin shear stress (N/mm^2).
T_e = Equivalent torsional moment (N.mm).
d_{cp} = Crankpin diameter (mm).

The equivalent value of the torsional moment T_e was evaluated analytically to obtain approximately 2,034.343 N.m, with a 45 mm crank pin diameter, d_{cp}. Meanwhile, the von-misses stress was used to predict the yield stress of the loading material, by applying Eq. 2:

$$M_{eV} = \frac{\pi \times \left(d_{cp}\right)^3}{32} x\sigma_v \tag{2}$$

Where:
M_{eV} = Equation of bending moment (N.mm).
σ_v = Von-Misses Stress (N/mm^2).

Based on Eq. 2, the bending moment value, M_{eV}, was specified at 2,111.211 N.m. Furthermore, the results of the von-Misses stress analysis is compared with the standard material yield strength σ_y [12].

3 Results and Discussion

Hardness Test. Figure 2 shows varying results of the crankshaft specimen hardness test at each point. The average values in the horizontal and vertical directions were approximately 83.9 and 84.2 HRB, respectively. These estimates closely matched the standard hardness value of ASTM A536 Grade 80-55-06 material at 89.5 HRB.

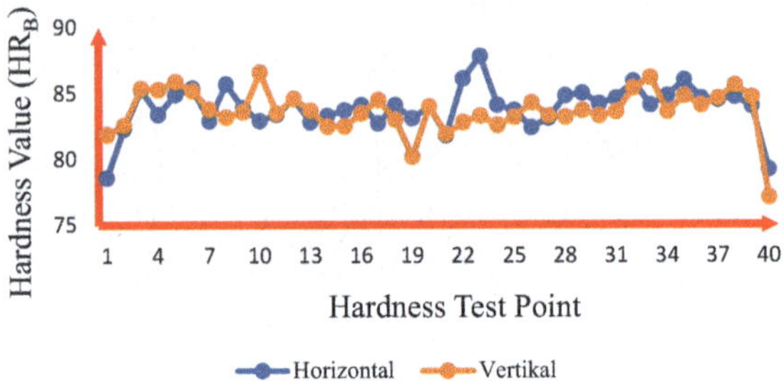

Fig. 4. Hardness test results

Figure 4 indicates a lower hardness value at the surface edge of 4 mm depth. These results are possibly due to a softer crank pin surface commonly prone to frictional effects, as illustrated in Fig. 2.

Chemical Composition. Tables 1 and 2 represents the chemical composition and mechanical properties of the crankshaft, respectively.

Table 1. Chemical Composition of Crankshaft [13]

Element	Test results	ASTM A536 Grade 80–55-06
C	> 2.00	3,5 – 3,9
Si	0,945	2.25 – 3.0
Mn	0,141	0,15 – 0,35
S	0,322	0,25 maks
P	0,150	0,05 maks

Table 2. Mechanical Properties ASTM A536 Grade 80–55-06

Mechanical properties	Value
Tensile strength	551 MPa
Modulus of elasticity	172,4 GPa
Poisson's ratio	0,2925–0,2975
Yield strenght	379,2 MPa
Hardness, Rockwell B	89,5 HRB
Allowable shear stress	48,125 MPa

The chemical analysis was conducted with a spectrometer and the percentage of carbon >2.00 confirmed the use of nodular cast iron as the standard material, with rating as ASTM A536 grade 80-55-06.

These findings also refer to the study by S Chidambaram & Ashok Kamaraj [14], where several automotive industries were known to produce crankshafts using nodular cast iron.

Failure Surface Analysis. Figure 5 represents the results of fracture surface analysis. Based on the illustration, crankshaft experiences three stages before breakdown. The first involved crack initiation and subsequently followed by the propagation phase due to increasing dynamic loading, until the final fracture.

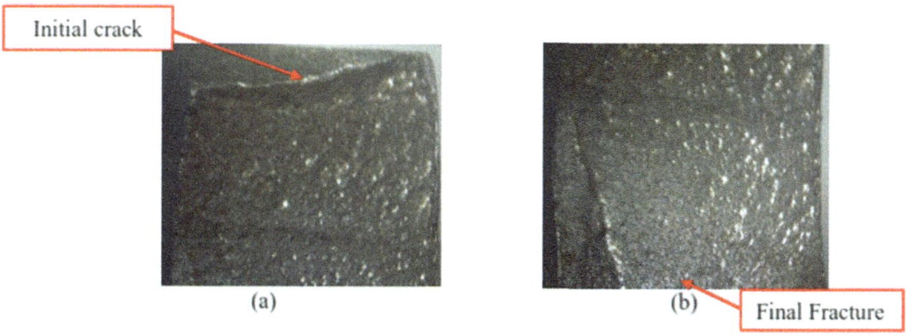

Fig. 5. (a) Enlargement of Crack Initiation, (b) Enlargement of the Final Fracture Area

Figure 6 shows the fractographic test results of the cracked surface using scanning electron microscope (SEM), with a magnification of 500x. The initial crack and the direction of the propagation appear visible.

Figure 7 shows the area of a progressive flat fatigue fracture. This appearance is a characteristic of ductile cast iron under a combination of fatigue flexural and torsional stresses in the crack propagation zone. The secondary cracking and void elongation at the fracture surface represent the mark of fatigue failure of ferritic-pearlytic ductile cast iron [15].

Based on the SEM analysis, porosity was also identified, with the load fluctuations on the crankshaft causing crack propagation through porosity. The causes of porosity include the use of high temperature, imperfect control of gas absorption by the alloy, gas trapping from metal during smelting and casting processes.

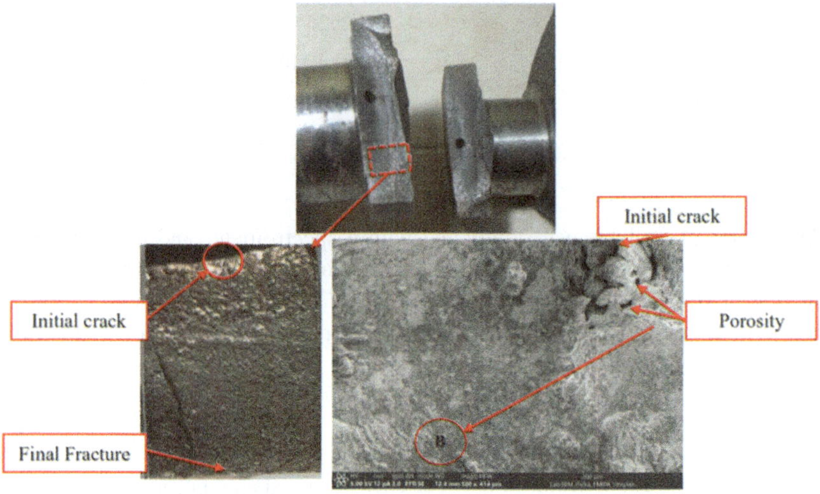

Fig. 6. SEM Observations on the fracture surface and crack initiation. The direction of crack propagation indicated by an arrow (B)

Crankshaft Microstructure. Microstructure testing on the fracture surfaces was conducted with 20x, 50x and 100x enlargements. The results are then used to analyze the microstructure in the crankshaft. Figure 7 provides a representation of the 50x enlargement.

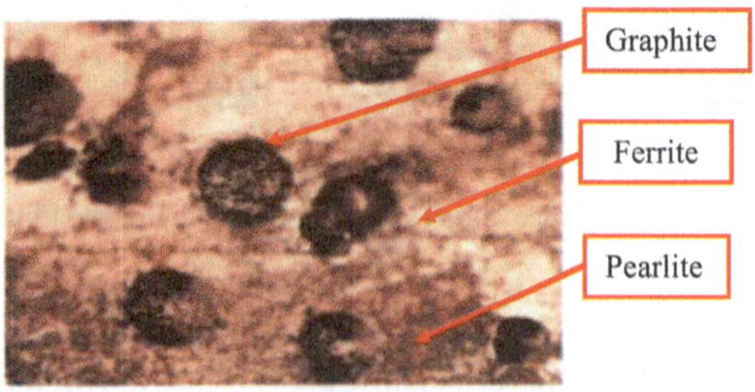

Fig. 7. Microstructure of the Cast Iron with 50x Enlargement

The outcome showed the formation of cyclic graphite by the spherodizing process. Also, a phase ferrite and pearlite on the microstructure were observed. This test describes the microstructure of nodular cast iron with graphite that resembles a perfect circle.

Stress Analysis. Shear stress occurred due to the compressive force of piston, although the opposing effect was responsible for crankshaft rotation. The shear stress obtained a significantly higher estimate of 113.75 MPa, compared to the allowable value at 48.125 MPa. As a consequence, the component tends to eventual fracture.

Dynamic loading also influences crack propagation due to the initial surface defects. This phenomenon also has the probability to accelerate the mechanical crash. Previous study on failure analysis both experimentally and numerically has been conducted by Husaini *et al.* Furthermore, the finite element method (FEM) provides a suitable alternative in obtaining very specific results for the stress intensity factor occurring close to the crack tip [16, 17].

4 Conclusion

Several major factors are known to instigate crankshaft failure. Based on this study, the hardness results showed a minimal value on the surface edges. This lower condition denotes a softer hardness condition at a depth of 4 mm from the crankshaft overlay prone to frictional forces. Furthermore, dynamic loading tends to trigger initial cracks that subsequently propagate until eventual fracture. The shear stress analysis obtained a significant value of 113.75 MPa, compared to the allowable shear stress at 48.125 MPa. As a consequence of this wide variation, the crankshaft ultimately breaks down.

Acknowledgments. The authors would like to express their gratitude to Universitas Syiah Kuala for financial support for research through the Professor's research grant No. 14/UN11.2.1/PT.01.03/ PNBP/2021. The authors are also thankful to the Department of Mechanical and Industrial Engineering, Universitas Syiah Kuala, Banda Aceh, for providing facilities.

References

1. Prasetya, T., Sarifuddin., dan Budi Joko Raharjo: Crank Pin Journal Crankshaft Wear on Diesel Engine Generators in. MV. Kartini Baruna, Jurnal Dinamika Bahari **9**(1) (2018)
2. Shahane, V.C., Pawar, R.S.: Optimization of the crankshaft using finite element analysis approach, automot. Engine Technol. (2018). https://doi.org/10.1007/s41104-016-0014-0
3. Huang, Z.-M., Hameda, S.: Failure Analysis, Book of Failure Analysis, United Kingdom, London (2019)
4. Husaini, Putra, T.E., Zulfikar: Analysis of cracks on a fractured surface of the vehicle crankshaft using the finite element method. Int. J. Eng. Technol. **7**(4.36), 1564–1568 (2018)
5. Putra, T.E., Husaini, Ali, N., Husin, H., Zulfikar: Failure analysis of the fracture surface of the crankshaft of a vehicle. In: IOP Conference Series: Materials Science and Engineering, vol. 523(1), p. 012067 (2019)
6. Kakade, P., Pasarkar, M.D.: Analyzing and identifying various approaches for crankshaft failures. State-Art Rev. J. Multidiscip. Eng. Sci. Technol. JMEST **2**, 3159–0040 (2015)
7. Husaini, Ali, N., Bakhtiar, A.: Analysis on the surface of broken screw spring front car suspension. In: Proceedings of Seminar Nasional Teknologi Rekaysa (SNTR), vol. 3 (2016)
8. Myagkov, L.L., Mahkamov, K., Chainov, N.D., Mahkamova, I.: Chapter: 11 Advanced and conventional internal combustion engine materials, Elsivier (2014)

9. Husaini: Fundamental of Fracture Mechanics (in Indonesia), Syiah Kuala University Press (2015)
10. Witek, L.: Experimental crack propagation and failure analysis of the first stage compressor blade subjected to vibration. Eng. Fail. Anal. **16**, 2163–2170 (2009)
11. Khurmi, R.S.: Machine design 14th edition, Eurasia Publishing House (Pvt). Ltd. New Delhi (2005)
12. Chidambaram, S., Kamaraj, A.: Failure investigation of an industrial crankshaft made of ductile iron, Advances in Natural and Applied Sciences; ISSN: 1995–0772 (2017)
13. Information on DuraBar 2017, ASTM A536 grade 80–55–06 (2017). https://www.dura-bar.com/products/ductile-iron/80-55-06.cfm
14. Asi, O.: Failure analysis of a crankshaft made from ductile cast iron, Elsivier (2006)
15. Bandanadjaja, B.: The improvement of crankshaft mechanical properties by homogenous bainitic structure formation through silicon alloying. In: AIP Conference Proceedings (2016)
16. Husaini, Kishimoto, K., Notomi, M.: Finite element analysis of rubber particles size distribution on fracture toughness of rubber-modified polymer. Key Engineering Materials, vol. 261. Trans Tech Publications Ltd. (2004)
17. Husaini, K.K., Hanji, M., Notomi, M.: Investigations of the mixed mode crack growth behavior of an aluminum alloy. J. Eng. Appl. Sci. **11**(2), 885–890 (2016)

Analysis of the Application of the Warehouse Receipt System (WRS) for Patchouli Oil Commodity

Munziri Abdullah(✉), Hasan Yudie Sastra, and Iskandar Hasanuddin

Magister of Industrial Engineering, Faculty of Engineering, Syiah Kuala University, Banda Aceh, Indonesia
moenziri@gmail.com

Abstract. Patchouli oil has been one of the industrial raw materials for ages. Several industries such as perfumery, pharmaceutical, and herbal industries use this commodity as the main raw materials. In the 90s, this commodity became well known among farmers, especially in the Aceh Jaya district because it has a fairly high selling power which makes most farmers in the area focus on developing patchouli plantations. With this lucrative potential, many people have contributed and taken advantage of the patchouli oil industry opportunities. BSN has issued a quality standard of patchouli oil in the SNI document 06-2385-2006 with the purpose that every industry can refer to a resembling standard. However, along with its development, the phenomenon of a decline in the selling price of patchouli oil began to emerge, which has become one of the main problems faced by stakeholders in the patchouli oil trading network, especially for farmers. It is difficult for farmers to delay their sales when the price drops, which leads to a decrease in their income. Instead of developing patchouli plantations, farmers are now starting to gradually convert the land to other commodities to maintain their economy. The identification of this problem starts from all entities that contribute to the supply chain system to the consideration of the contribution from each entity to the patchouli oil industry in Aceh Jaya Regency. This study examines the benefits of implementing a Patchouli Warehouse Receipt System (WRS) or *Sistem Resi Gudang* (SRG) which aims to provide alternative solutions in stabilizing selling prices, especially to farmers' income. With this price stability, farmers can increase the productivity of patchouli oil in hope that they can optimally supply the world's patchouli oil needs.

1 Background

The scope of the patchouli plantations development in Aceh is over several regions. The district of Aceh Jaya is one of the focus areas for the development of patchouli (*Pogostemon cablin*) in Aceh. Patchouli plantations in Aceh Jaya Regency cover several sub-districts with a total land area of 158 Ha and production of ±36 tons/year. The distribution of land today is not as vast as it was in the 90s for-which made the patchouli farming profession heretofore very promising because of the relatively high bargaining value of patchouli oil. The product is reasonably high on demand, but the quantity of

Akhyar et al. (Eds.): ICECME 2021, LNME, pp. 349–361, 2023.
https://doi.org/10.1007/978-981-19-3629-6_36

patchouli oil supply is still utterly low. One of the factors that influence the low amount of patchouli oil production today is the low selling value of patchouli oil received by farmers, causing a decrease in the tendency of farmers to work on and develop patchouli plantations.

2 Warehouse Receipt System (WRS)/*Sistem Resi Gudang* (SRG)

The Warehouse Receipt System (WRS) program in Indonesia novel carried out in 2008. It is applied for all activities related to the issuance, transfer, guarantee, and settlement of warehouse receipt transactions. Warehouse Receipt is a valuable document that explains the ownership status of goods or commodities stored in the warehouse issued by the warehouse manager.

Some of the benefits of the Warehouse Receipt System (WRS) for agricultural commodities are: (Ministry of Trade of the Republic of Indonesia, 2014)

1. Extending the sales period of agricultural products. This condition is assuredly salutary for farmers. They do not need to rush to sell their agricultural products when commodity prices fall, lieu, they can delay sales until commodity prices rise, thus provides maximum sales results.
2. Warehouse Receipt Documents can be used as collateral to obtain funding by Financial Institutions (Banks/Non-Banks) that have registered as distributors to such an extent that farmers can immediately return to farming in the next season by utilizing existing funding.
3. Realizing a more competitive physical market and futures market. The implementation of the Warehouse Receipt System (WRS) is able to provide information to sellers (goods owners) and buyers (traders, industry, and speculators) in conducting broad commodity trading transactions. Buyers (traders, industry, and speculators) can find out information about the types and availability of commodities available in the warehouse so that buyers (traders, industry, and speculators) can make transactions according to their needs.
4. Farmers can store their agricultural products and obtain Warehouse Receipt documents to be guaranteed at the financial institutions. It allows them to earn fundings to be able to grow crops in the next season. Farmers also continue to monitor if commodity prices are high or stable, then farmers can sell their products.
5. Improving the welfare of society with social justice based on Pancasila and the UUD 1945.
6. Ensuring the stability and sustainability of the production and distribution of the commodities included in the Warehouse Receipt System (WRS). This impacted the price stability in the market.

The implementation of the Warehouse Receipt System (WRS) as a government program involves many cross-sectoral institutions. The thing indicated that it is vital to be regulated by laws and regulations. The law and legal basis regarding its implementation were formulated to provide legal assurance to stakeholders involved.

The practice of the Warehouse Receipt System (WRS), which has been implemented in Indonesia so far, includes several stages until the issuance of warehouse receipts for a commodity. It is supervised by a supervisory body from the Ministry of Trade of the Republic of Indonesia at each of the stages. This results in monitored occurred activities and ensured that they are carried out following the established regulations (Fig. 1).

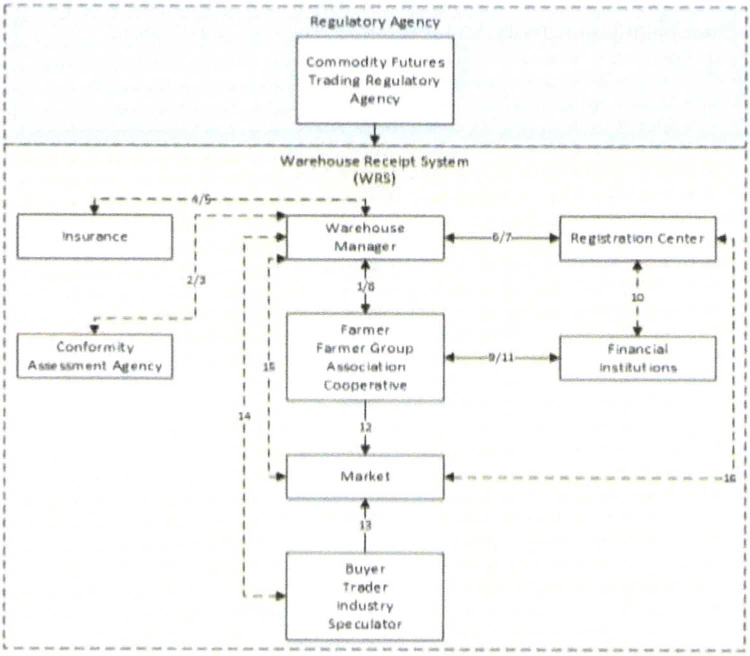

Fig. 1. Schematic of the WRS implementation process (Source: Muflihah, 2014)

Prioritized goods that can be stored in the warehouse with the implementation of the Warehouse Receipt System (WRS) are strategic, superior commodities, export destinations, and/or food security. Technically, some of the requirements related to the prioritized goods must meet at least 3 (three) requirements. Those include having a minimum shelf life of 3 (three) months, meeting defined quality standards or those determined, and meeting the number of allocations of goods specified. By fulfilling all these minimum requirements, the goods abovementioned can be stored in the warehouse of the Warehouse Receipt System (WRS) operator for further issuance of Warehouse Receipt documents.

Based on the Regulation of the Minister of Trade of the Republic of Indonesia No. 33 of 2020 concerning goods and requirements for goods that can be stored in the Warehouse Receipt System (WRS), the types of goods that can be stored in the Warehouse Receipt System (WRS) program are Grain, Rice, Corn, Coffee, Cocoa, Pepper, Rubber, Seaweed, Rattan, Salt, Gambir, Tea, Copra, Tin, Shallots, Fish, Nutmeg, and Carcass Frozen Chicken.

3 Research Method

In carrying out data processing and analysis, there are several stages of identification and analysis that need to be carried out, including identification of existing trading schemes, identification of the concept of Warehouse Receipt System (WRS), commodity analysis and patchouli plant productivity, and feasibility analysis of Warehouse Receipt System (WRS) (Fig. 2).

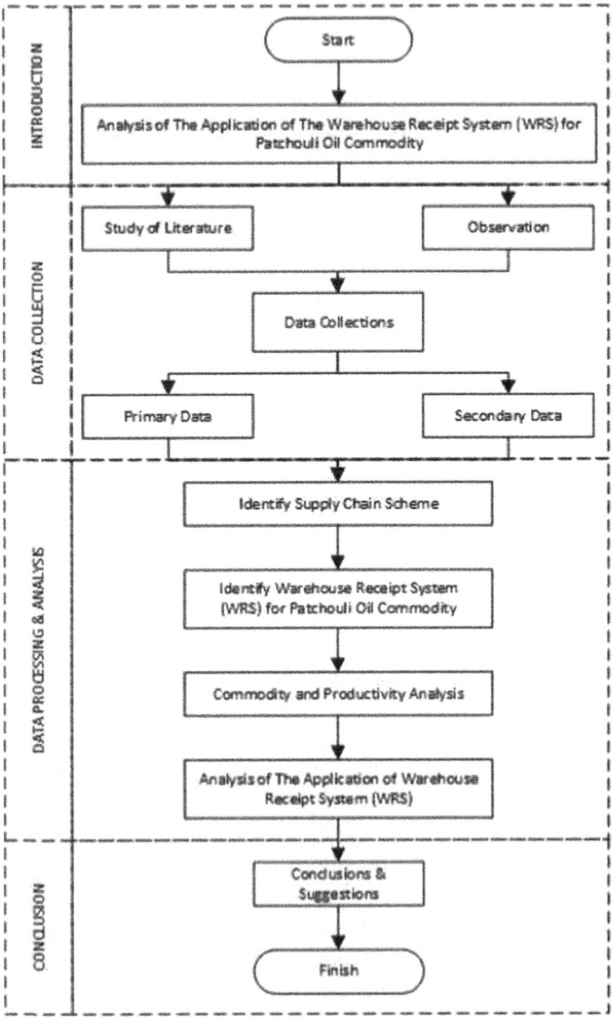

Fig. 2. Research flow chart

4 Patchouli Oil Commodity Profile as Warehouse Receipt System (WRS) Subject

Based on the Regulation of the Minister of Trade of the Republic of Indonesia Number 33 of 2020, only 18 certain commodities were described previously. However, if deemed necessary, a commodity can also be proposed as a WRS commodity. In the procedure for submitting the WRS commodity, of course, there are things that need to be considered and fulfilled so that the commodity is worthy and accepted as a WRS commodity. Based on the characteristics of the type of goods, Patchouli Oil complies with the minimum requirements regarding the types of goods that can be included in the Warehouse Receipt System (WRS). Furthermore, there are several minimum requirements to be able to enter a commodity into the WRS as previously described, so that two points of these requirements have been met (Table 1).

Table 1. The suitability of patchouli oil commodity to WRS

No.	Requirements (*Regulation of the Minister of Trade of the Republic of Indonesia Number 33 of 2020*)	Description	
		Corresponds	Differing
1	Has a shelf life of at least 3 months	√	
2	Meet certain quality standards	√	
3	Minimum number of items stored		√
4	A strategic item	√	
5	Featured commodity	√	
6	Export destination	√	

5 Patchouli Oil Productivity

Patchouli oil plantations located in Aceh Jaya Regency are entirely owned by the people of the Regency. It outspreads over nine sub-districts in the Aceh Jaya Regency area. There is no plantation area specifically developed by the local district government to increase patchouli oil productivity. The data on the potential profile of patchouli oil in Aceh Jaya Regency is presented in Table 2 below (Fig. 3):

The trend of patchouli oil prices in the local market tends to decline from year to year. Many things have contributed to the decline in prices. Of course, this causes a decrease in the interest of the farmers towards the patchouli commodity. This condition must be addressed immediately so that the patchouli commodity, which has become local wisdom in Aceh Jaya Regency, does not come to an end considering the great potential of this commodity, especially in several foreign industries that require patchouli oil as industrial raw material. Below presented the data on patchouli oil sales prices from 2017 to 2020 obtained from field observations (Table 3).

Table 2. Patchouli oil potential in Aceh Jaya Regency

No.	Subdistrict	Land Area (Ha)		Number of farmers	Production (Ton)	Farmers potential	Production potential (tons)
		Unproductive plant	Productive plant				
1	Teunom	18	2	35	1	65	5,80
2	Pasie Raya	13	9	39	3	92	6,38
3	Panga	22	15	67	4	95	10,73
4	Krueng Sabee	20	19	65	6	154	11,31
5	Setia Bakti	9	15	25	4	85	6,96
6	Sampoiniet	14	15	20	4	75	8,41
7	Darul Hikmah	23	16	52	5	183	11,31
8	Jaya	25	21	55	6	47	13,34
9	Indra Jaya	14	12	25	3	65	7,54
	Total	158	124	383	36	861	81,78

Source: Researcher, 2020

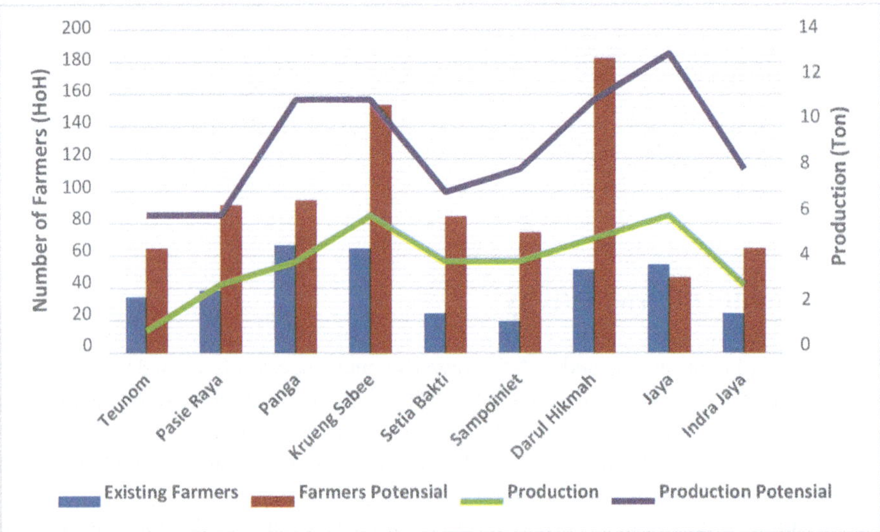

Fig. 3 Farmers' potential and patchouli productivity (Source: Researcher, 2020)

According to the data obtained from the last three years, the lowest patchouli oil price at the farmer's position occurred in 2018 which was Rp. 420,000,-. The decline in prices made farmers helpless because the costs incurred to maintain patchouli plants to the production process into patchouli oil exceeds the profit. Not to mention the pressure

Table 3. Data on the local selling price of patchouli oil in Aceh Jaya Regency

Year	Price on - (Rp)	
	Farmers	Collectors
2017	750.000	770.000
2018	420.000	440.000
2019	580.000	600.000
2020	680.000	700.000

Source: Researcher, 2020

from the daily life needs which forces the farmers unable to delay their sales, making them sell patchouli oil at low prices.

6 Existing Supply Chain

Based on the information and data obtained referring from the field and the literature, the entities participating in the patchouli oil supply chain scheme are farmers, refiners, local collectors, exporters, and consumers. Two supply chain models have been applied in patchouli oil trading in Aceh Jaya District. The difference between the two models lies in the role and function of farmers.

In the first model, farmers only play a role in managing agricultural land up to the pre-refining process. While in the other model, farmers play a role starting from the management of agricultural land, the production process, to distributing products to consumers. In general, farmers are the main actors in the patchouli oil production system in Aceh Jaya Regency (Fig. 4).

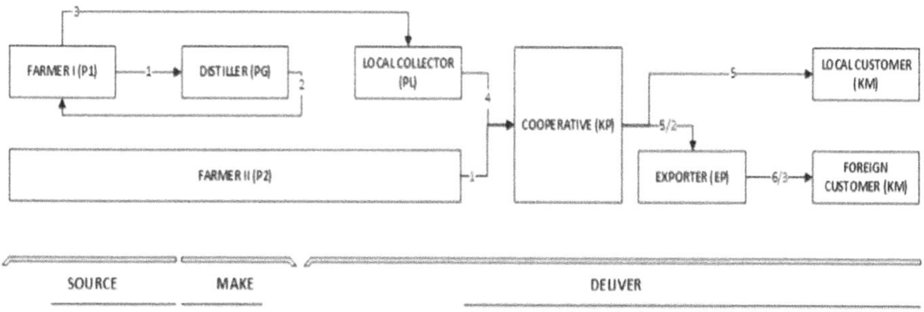

Fig. 4. Patchouli oil supply chain mapping results

After conducting an assessment of the existing supply chain system which is still being implemented by all stakeholders, it is known that most of the activity processes are carried out by farmers. This shows that farmers play the largest role in the patchouli oil supply chain system in Aceh Jaya District.

Each entity in the supply chain system must carry out their respective efforts in increasing the added value to the product. Every effort made by each of these entities forms a value chain system that will affect the selling value of the product to the final consumer. The following presents the results of mapping the patchouli oil value chain in the Aceh Jaya District (Fig. 5).

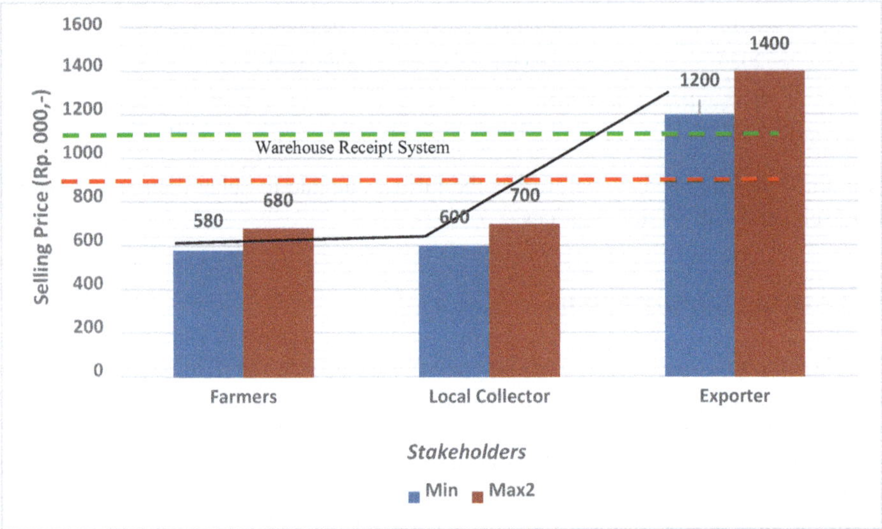

Fig. 5. Price increase for each stakeholder (Source: Researcher, 2020)

Conditions in the field show that farmers play a large role in the patchouli oil industry in Aceh Jaya Regency. Farmers carry out the process of industrial activities including handling pre-planting, planting, harvesting, post-harvesting, pre-distilling, refining, post-refining, and distribution. From the whole process of activities carried out by farmers to become products, the selling value of products that can be marketed to local collectors, cooperatives, and or directly to local consumers is IDR 680,000 per kilogram. Meanwhile, for sales to consumers abroad, farmers still rely on cooperatives that already have an international marketing network. It is known that the selling value of products marketed by cooperatives to consumers in foreign countries ranges from Rp. 1,200,000, - to Rp. 1,400,000, - per kilogram.

So far, there has been no intervention from the government to stabilize the selling price of patchouli oil, especially in the farmer chain. The enthusiasm of the community or farmers is in the certainty and stability of patchouli oil prices, thus farmers can develop plantation land according to the existing potential optimally (Fig. 6).

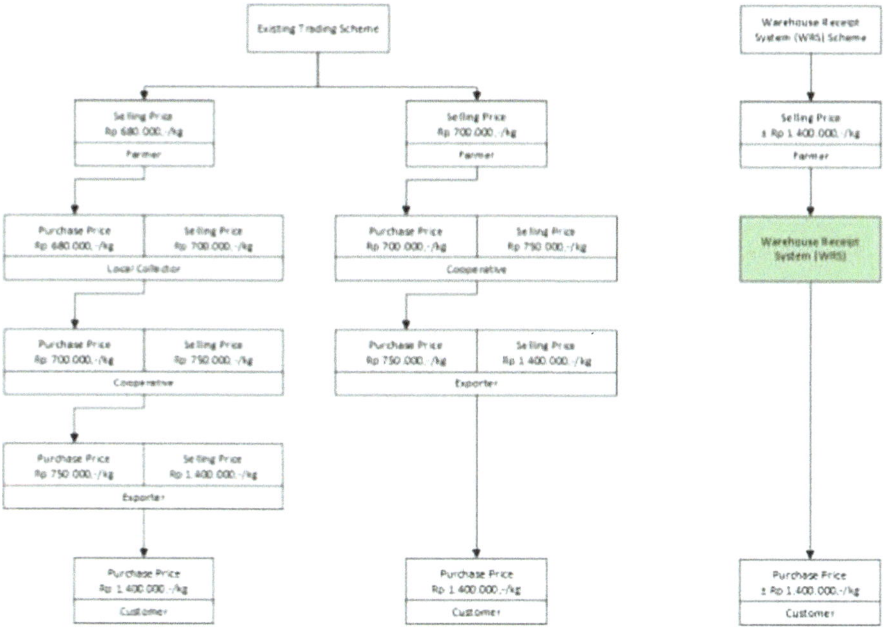

Fig. 6. Patchouli oil commodity trade flow in Aceh Jaya District (Source: Researcher, 2020)

7 Warehouse Receipt System (WRS) Identification

Warehouse managers in the implementation of the Warehouse Receipt System (WRS) can be roled by private parties with legal entities, cooperatives, Regional Owned Enterprises, even the Farmers Group Association/Gabungan Kelompok Tani (Gapoktan), which of course already has a representative storage warehouse to be able to implement the Warehouse Receipt System (WRS). In implementing the

Warehouse Receipt System (WRS) until the issuance of warehouse receipts, the warehouse manager coordinates with various parties involved in the WRS, namely farmers/farmer groups/Gapoktan, Conformity Assessment Institute, Guarantee Institution, and Registration Center as described previously.

The communication established by the warehouse manager and the registration center is supported by the Registration Center Warehouse Receipt Information System. In the communication process, all information on warehouse receipt transaction activities is recorded. The activities among others are transfer, imposition of collateral rights, and transaction settlement. The following is a flowchart of the activity process or procedure in the Warehouse Receipt System (WRS) until the issuance of a warehouse receipt (Fig. 7).

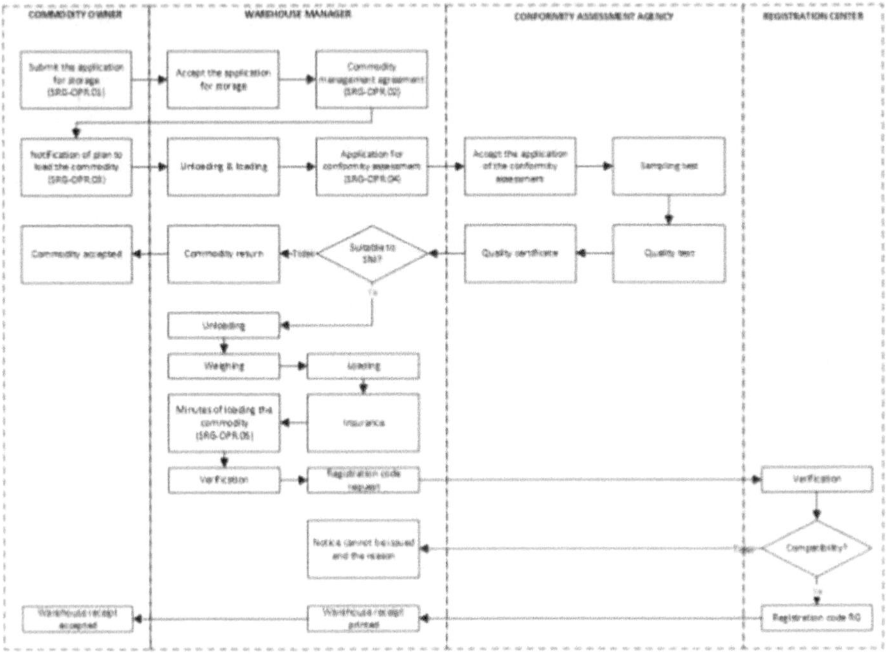

Fig. 7. Warehouse receipt issuance process flow

The implementation of the Patchouli Warehouse Receipt System (WRS) in Aceh Jaya Regency needs to be considered as the patchouli oil has enormous potential. It needs the support of all relevant institutional elements to make it happen. In general, the readiness of institutions and facilities and infrastructure in the implementation of the Warehouse Receipt System (WRS) cannot be ascertained, considering that there is no Warehouse Receipt System (WRS) that has been implemented for any commodity in the area.

8 Identification of the Effects of WRS Implementation

The implementation of the SRG for patchouli oil is projected to have a wide influence, ranging from the superiority of the commodity itself to the prospects and development of the patchouli oil industry in Aceh. The projected impact mapping of patchouli oil WRS implementation includes:

1. The implementation of WRS is advantageous for farmers, both in terms of capital and efforts to stabilize commodity prices. It will be a strong reason for farmers to develop and optimize patchouli plantations.
2. Farmers can take action to delay selling their products so that farmers have the option to hold off selling when commodity prices tend to be low and then sell them when commodity prices are high.

3. WRS assures product quality. The owner of the goods must adjust the quality standards that have been set for the goods to be stored in such a way that prospective buyers have certainty of the quality of the goods to be purchased (Fig. 8).

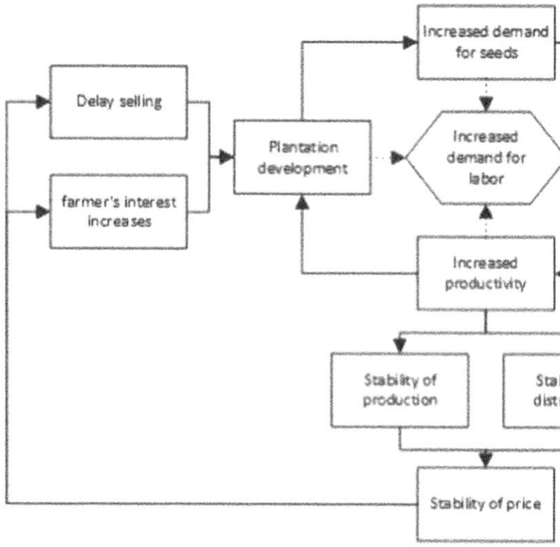

Fig. 8 Scheme of projected impact of WRS implementation

9 Success Factors for WRS Implementation

In the patchouli oil distribution scheme that has been carried out by industry participants in Aceh Jaya, some obstacles tend to harm one party, namely the farmer. The obstacle is the decline in prices on the income side of farmers which causes the low interest of people who work as patchouli farmers to develop and re-cultivate patchouli plants, so that it has an impact on patchouli oil production. The success of the implementation of the Warehouse Receipt System (WRS) generally depends on all elements in the system. It starts from the owner of the commodity and various institutional elements to the market system that is developed as well as the interest of commodity buyers. All components involved have their respective roles which then form a productive system for each stakeholders.

Several aspects that are expected to contribute and have a considerable influence in the effort to succeed in the implementation of the Warehouse Receipt System (WRS) include:

1. Local Government Support

In this case, the local government's support for the Warehouse Receipt System (WRS) is manifested in the local Department of Trade and Agriculture. The Department of Trade plays a role in bringing up warehouse managers and helping to oversee the smooth running of transactions. Meanwhile, the general function of the Department of Agriculture in supporting the Warehouse Receipt System (WRS) is to make efforts to increase patchouli production so that farmers can increase the productivity of the patchouli oil industry in Aceh Jaya Regency.

2. Existence of Commodity Buyers/Auction Market

The warehouse manager must be able to mediate the owner of the goods with prospective buyers/buyers. Warehouse managers also provide access to prospective buyers/buyers to be able to find out the amount of stock of goods/commodities in the warehouse so that the transparency of buying and selling is guaranteed through the Warehouse Receipt System (WRS). Owners of goods/commodities are also very satisfied because they can directly sell their goods/commodities directly to buyers so that the profits received are greater because there are no intermediaries in the trade.

3. Education and Socialization to Farmers

Aceh Jaya Regency has never previously implemented a Warehouse Receipt System (WRS) for any commodity, of course, this is something new for the local community. The community in general and farmers, in particular, must receive proper education about the Warehouse Receipt System (WRS). This is a challenge to be able to successfully implement the Warehouse Receipt System (WRS) in accordance with applicable regulations. Education and socialization must continue to be carried out on an ongoing basis so that farmers and the community understand the intent and purpose of the establishment of the Warehouse Receipt System (WRS) which is none other than to improve the welfare of farmers.

4. Production Increase

For the implementation of the Warehouse Receipt System (WRS) for patchouli oil to operate successfully, farmers must strive to increase their productivity with the intention that the supply of patchouli oil is maintained and available. Thus the warehouse manager can carry out his duties properly in creating a commodity market or auction market that will invite potential buyers of the commodity. If the productivity is not increased, likely, the Warehouse Receipt System (WRS) will not run as effectively as the intent and purpose of its implementation.

10 Conclusion

Based on the research that has been done related to the analysis of the implementation of the Warehouse Receipt System (WRS) in Aceh Jaya Regency on patchouli commodities, several things that can be concluded are as follows:

1. The potential distribution and area for patchouli plantations are vast, but the farmers have not utilized optimally due to the uncertainty of patchouli oil prices which are increasingly declining. This causes many farmers to gradually shift the function of the land by planting other crops that are considered to have more economic value.
2. The quality of patchouli oil produced has met the quality standard of patchouli oil by SNI 06-2385-2006, but due to the non-optimal potential of the land and farmers, the current production capacity is still minimal. This is directly proportional to the

condition of patchouli oil prices on the side of farmers which continues to decline, thereby reducing the interest of farmers to develop their plantations.
3. Patchouli oil can be considered as a commodity in the Warehouse Receipt System (WRS) because patchouli oil is a strategic item, a leading commodity, and an export-oriented item. Besides that, patchouli oil meets the minimum requirements in its implementation as it has a relatively long shelf life and quality standards.
4. The expected impact in implementing the Warehouse Receipt System (WRS) on patchouli commodities is the price stability which can be a stimulus for farmers to be able to re-develop patchouli plantations. This allows an increase in productivity which leads to increased income for farmers in particular and the community on at a large scale.
5. At the beginning of the implementation of the WRS, it is necessary to have support from both the Regional Government and the Central Government to facilitate efforts to increase the stability of patchouli oil production and distribution so that the implementation of the Warehouse Receipt System can be carried out successfully.

References

1. Maizi, H., Hasan Yudie Sastra, A.: Mapping upstream and downstream process in the Patchouli oil industry using supply chain operations reference model version 12.0 (SCOR 12.0). In: IOP Conference Series: Material Science and Engineering, vol. 931-012008 (2020)
2. Fachruddin, A., Rahayu, L.: Evaluasi Prasyarat Keberhasilan Sistem Resi Gudang di Kabupaten Bantul. Penerbit: AGRARIS. J. Agribus. Rural Dev. Res. **3**(2) (2017)
3. Nugroho, S.P.: Kelayakan Usaha Padidengan Menggunakan Skema Resi Gudang untuk Meningkatkan Ketahanan Pangan. Penerbit: University Research Coloquium (ISSN 2407-9189) (2016)
4. Anugrah, I.S., Erwidodo, dan E.S.: Sistem Resi Gudang dalam Perspektif Kelembagaan Pengelola dan Pengguna di Kabupaten Subang: Studi Kasus KSU Annisa. Penerbit: Analisis Kebijakan Pertanian, vol. 13, no. 1, pp. 55–73, Juni 2015
5. Maizi, H.: Perancangan Strategi Mitigasi Risiko Rantai Pasok Industri Minyak Nilam Menggunakan Metode House of Risk (HOR) dan Supply Chain Operations Reference Model (SCOR 12.0): Studi Kasus Industri Minyak Nilam Aceh Jaya. Universitas Syiah Kuala, Penerbit (2019)
6. Bank Indonesia: Kajian Peningkatan Pemanfaatan Sistem Resi Gudang Pilot Project di Kabupaten Kuningan-Jawa Barat (Komoditas Gabah) dan Konawe Selatan-Sulawesi Tenggara (Komoditas Kakao). Bank Indonesia, Penerbit (2017)

Dynamic Stability of a Sieve Shaker Used in a Pachouli Oil Refining Process

Rudi Kurniawan[1], Zahrul Fuadi[1(✉)], Fathurrahman Abubakar[2], and Afdhal Afdhal[3]

[1] Department of Mechanical and Industrial Engineering, Universitas Syiah Kuala, Banda Aceh, Indonesia
{kurniawan,zahrul.fuadi}@unsyiah.ac.id
[2] Undergraduate Study Program, Department of Mechanical Engineering, Universitas Syiah Kuala, Banda Aceh, Indonesia
fathurrahman.me14@gmail.com
[3] Department of Electrical and Computer Engineering, Universitas Syiah Kuala, Banda Aceh, Indonesia
afdhal@unsyiah.ac.id

Abstract. One of the problems causing low quality of essential oils is the high content of solvent resulting from traditional extraction process employing low purity metallic components. This extracted oil therefore requires further process to improve the patchouli alcohol content of the oil by removing unwanted components. One of the process is by mixing the oil with bentonite and zeolite. In this study, a sieve shaker is designed to be used at an integrated equipment for patchouli oil refining process. In order to have a dynamically stable shaker, it is necessary to investigate the effect of various components for the stability, particularly the spring stiffness. The simulation was conducted using Autodesk Inventor using a one degree of freedom model. It is shown from the result that, for the defined condition, the sieve shaker shown a stable condition for the spring stiffness value at least 10 kN/m. Spring stiffness lower than the value could cause detachment of the roll follower from the cam, resulting in impact between the components.

Keywords: Screen vibrator · Shaker · Stability · Spring stiffness · Micro particles

1 Introduction

Patchouli (Pogostemon Cablin Benth) is one of aromatic oil source plants native to South East Asia highly valued for fragrance industries. It has a unique woody smell and is a difficult-to-replace ingredient in many general perfumes and cosmetics [1]. Indonesia is one of the main producers of patchouli oil in the world with annual production of 1,100–1,200MT, comprising 80% of world demand [2]. However, most of the patchouli oil are farmers products which quality is relatively low due to contamination during distillation process. Fe is one of the most common contaminants found in raw patchouli oil produced by farmers, presumably due to oxidized metallic components. Therefore, additional process is necessary in order to remove the contaminants. One of the methods

Akhyar et al. (Eds.): ICECME 2021, LNME, pp. 362–371, 2023.
https://doi.org/10.1007/978-981-19-3629-6_37

to improve the quality of the oil without incurring high cost is by applying absorption process using surfactant materials such as zeolite. The process combines chemical and mechanical procedures and can result in the increase of patchouli alcohol content up to 32.79% [3]. The process consists of preparation of zeolite/surfactant, exposure of the surfactant to the patchouli oil, and extraction of the oil by filtration. A mechanical system containing components for continuous refinement of patchouli oil can be designed by combining surfactant particle screen mechanism, surfactant and patchouli oil mixing chamber, and filtration system.

One of the key steps of such process is obtaining similar size of surfactant particles to be mixed with the patchouli oil. This can be achieved by using vibratory screen such as sieve shaker. Sieve shakers have been used in classifying the particle size in various applications. The sieve is vibrated by its axis to achieve rotary, vertical, or combination of both rotary and vertical motion. Sieve shaker can be operated using electric motor, combination of magnetic and electric, or pulley system [4–6]. In this study, a vibratory sieve shaker for surfactant particle classification of a patchouli oil refining mechanical system is designed and investigated.

Vibratory sieving method involved upward movement of the screen driven by a pair of cam and roll follower. The sample/particle is thrown upwards and fall back down by gravitational force [5]. Here, the performance of the sieve depends on the stability of the vibratory motion of the sieve bed, which is driven to upward and downward movement by a pair of cam and roll follower. Among the components involved in this movement are a cam, a roll follower, and springs.

In this study, the stability of the sieve shaker is investigated to determine necessary parameter for its performance, particularly the spring stiffness. Therefore, the objective of this study is to determine the suitable value of spring stiffness for a stable sieve shaker operation. The analysis was conducted by numerical simulation using Autodesk Inventor Software package.

2 Methodology

2.1 Components and Specifications

The sieve shaker (Fig. 1) is designed to have three stacking screens each having different mesh dimension. In application, the sieve shaker is filled with zeolite or bentonite particles where screening take place. The particles with smaller size will go through to the mixing chamber while those having larger size will be removed from the outlet.

The sieve shaker has six main components as indicated by numbers in Fig. 1 i.e. 1) sieve, 2) frame, 3) base, 4) coil spring, 5) electric motor, and 6) cam and roll follower. The detail specification of each components is given as follows.

1. *Sieve.* The detail components of the sieve shaker are given in Fig. 2. The sieve has a dimension of 480 mm of length, 230 mm of width and 290 cm of height. It consists of three stacking screen meshes with dimension of 60 mesh, 80 mesh and 100 mesh screens. The sieve is installed on a frame structure.

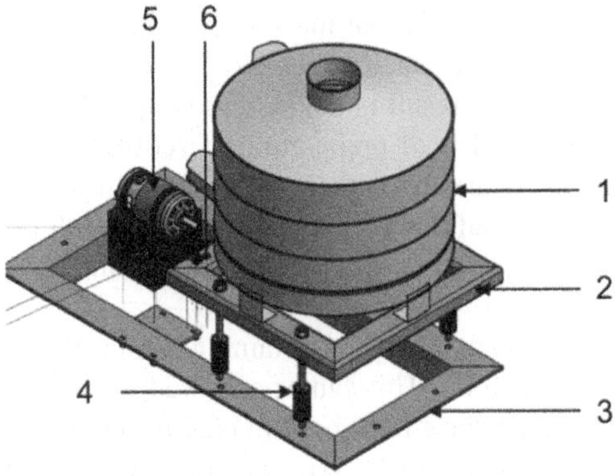

Fig.1. Sieve shaker design used for stability analysis

2. Frame. The detail component of the frame is given in Fig. 3. The frame has a dimension of 230 mm × 230 mm (length x width) and thickness of 30 mm. The frame has a motor base at which the cam and follower are installed. The frame is connected to the main base with four coil springs.

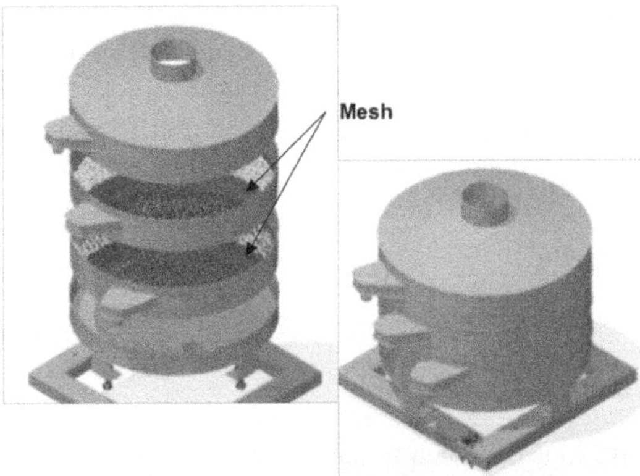

Fig. 2. Sieve with two stacking screen meshes

3. Coil springs. Four coil springs were designed to connect the frame to the base. The length of the coil spring is designated as 35 mm. As indicated in Fig. 4, the spring is mounted to the frame and the base using a pair of bolt and nut. The spring is one of the main components that determines the stability of the sieve shaker. In this analysis, the simulation was conducted to find out the effect of the spring stiffness on the stability of the shaker. The position of each coil spring is indicated in Fig. 1. An optimum spring stiffness is determined based on the simulation result (Fig. 12).

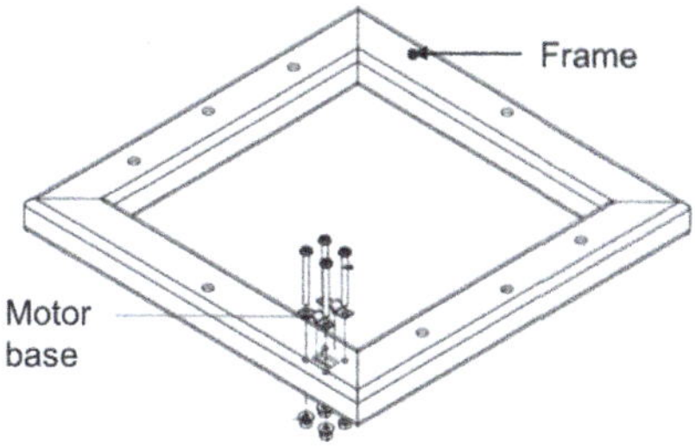

Fig. 3. The frame for sieve base with motor base support

4. 4. Base. The base is made of steel frame to connect the shaker part to the main frame where components for oil refining process are installed. Four coil springs and a motor base were installed on the base, as indicated in Fig. 5.

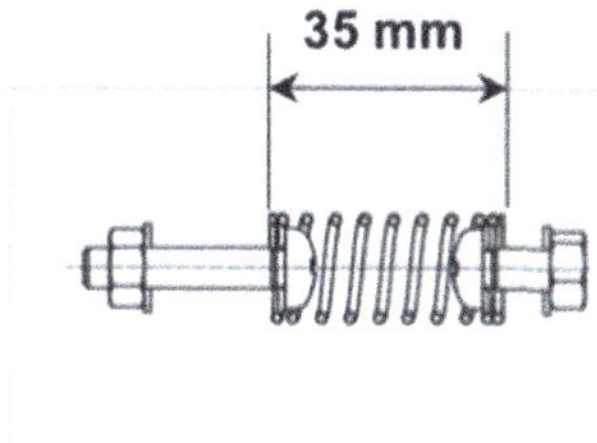

Fig. 4. Coil spring used for sieve shaker

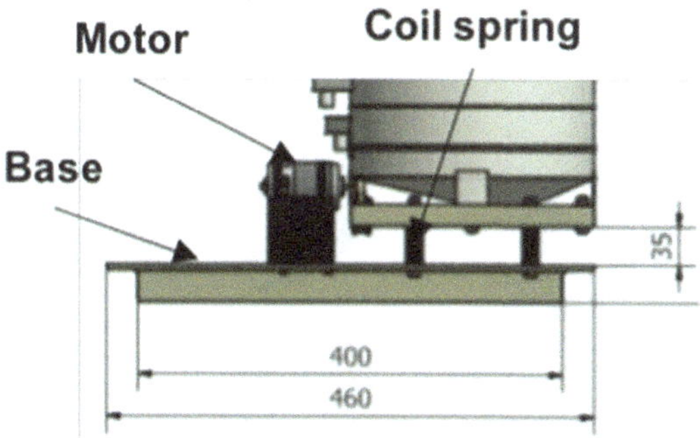

Fig. 5. Base with shaker and motor installed, units in millimetres

5. Electric motor, cam, and roll follower. An electric motor is chosen to rotate the cam to push the roll follower and drive the sieve shaker into vibratory motion, as illustrated in Fig. 6. The motor is installed on the base using C-clamp. A cam is installed on the motor to generate a displacement of 2 mm to the shaker in normal direction. The roll follower is mounted on the frame. Besides spring stiffness, the shape of the cam also become an important factor affecting the stability of the shaker. Here, a displacement of 2 cm is about 5.7% of the total length of the spring.

2.2 Stability Criterion

The shaker is designed to have a displacement of 2 mm, as shown in Fig. 7. Therefore, the shaker is considered stable if the spring has a displacement 2 mm. A 1% tolerance is considered so that the stability criterion is defined as the spring displacement of 2 + 0.02 mm. Using this stability criterion, a suitable stiffness value of the coil spring can be determined.

2.3 Simulation Parameters

The following conditions are assumed in the simulation:

a. Solver package: Autodesk Inventor Software Package
b. Simulation type: Dynamic
c. Numbers of DOF: one
d. d. Damping coefficient: 0.11 N.s/mm e. Simulation time: 0.5 s

Fig. 6. An electric motor with cam installed

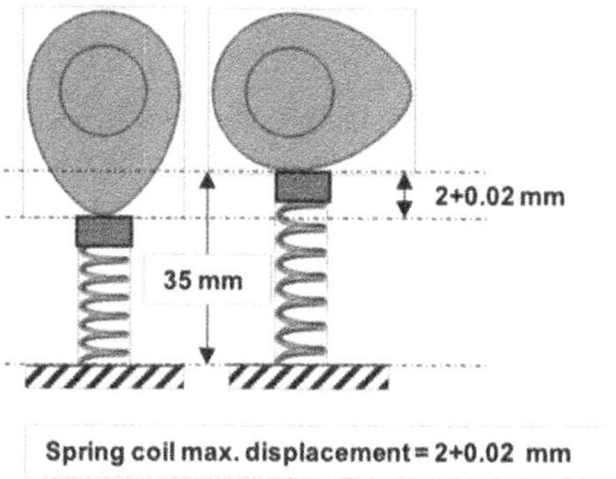

Fig. 7. The stability criterion for the sieve shaker; coil spring displacement of 2 + 0.02 mm

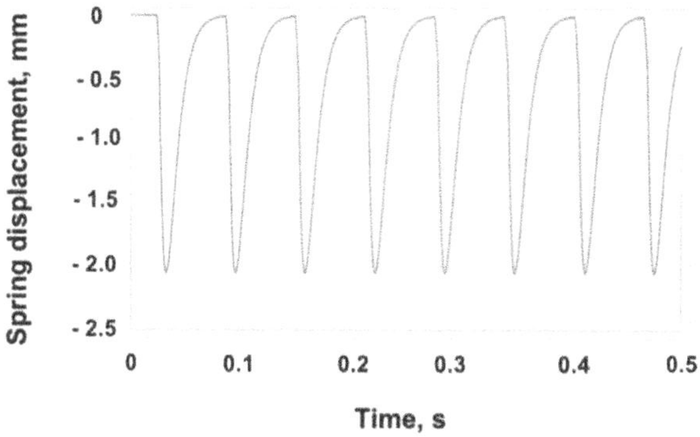

Fig. 8. Simulation results showing coil spring displacement for spring stiffness of 7.5 kN/m

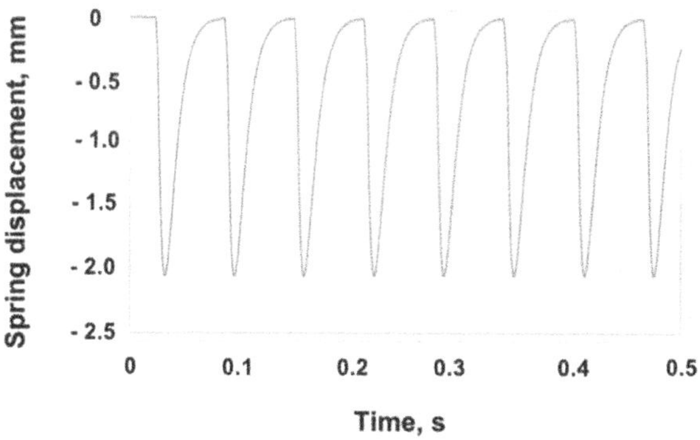

Fig. 9. Simulation results showing coil spring displacement for spring stiffness of 7.5 kN/m

3 Results and Discussion

Figure 8 shows the spring displacement for a spring stiffness of 5 kN/m. At 0 s, the spring is at initial position with 0 mm displacement. As the cam pushes the roll follower, the spring moves to the minimum position. In this case, the spring has a displacement of 2.12 mm. This means that at this position, the spring has been compressed to a length of 32.88 mm.

At this condition, the spring undergo a displacement more that the maximum allowable by the design criterion, i.e. $2 + 0.02$ mm. A displacement more that 2 mm means that the cam has been detached from the roll follower. Furthermore, the spring returns back to its initial position. However, before it reached the initial position, the roll touches the cam at a position of 0.1 mm from 0 point. The spring is then pushed back downward.

Since the roll follower has been detached from the cam, impact occurs as they reconnect. The impact occurs at a position before the spring returned to 0 position. In practice, this impact could cause instability to the system resulting in components unexpected failure. Therefore, at the spring stiffness of 0.5 kN/m, the designed sieve shaker is not stable.

Figure 9 shows the spring displacement curve for a spring stiffness of 7.5 kN/m. In this case, the cam pushed the spring to a position of 2.05 mm from the initial point. This value of displacement also causes detachment of the roll follower from the cam, similar to previous condition at spring stiffness of 5 kN/m. Because the roll follower is detached from the cam, impact could also take place. Since the displacement occurred is larger than $2 + 0.01$ mm, the sieve shaker at this stiffness value also fall into unstable condition according to the criterion given in Fig. 7.

Figure 10 gives the result of simulation showing the spring displacement at the stiffness of 10 kN/m. In this case, the cam pushes the roll follower so that the spring has a displacement of 2.02 mm. The spring returns back quickly to its initial position, indicated by the flat curve of the spring displacement at initial position. This means that the roll follower does not detached from the cam. Since the maximum displacement of the spring is 2.02 mm, the sieve shaker is in stable condition according to the defined criterion in Fig. 7.

The comparison of the spring displacement for the three different value of spring stiffness is given in Fig. 11. Here, it can be seen that detachment of the roll follower from the cam is expected for the spring stiffness lower than 10 kN/m. The figure also indicates the rate of spring recovery from its minimum position. Using this result, it can be projected that a lower stiffness of the spring will cause larger displacement and lower recovery rate. On the other hand, a higher value of stiffness will result in faster recovery rate and avoidance of detachment of the roll follower from the cam.

Based on the results, a stability condition for the designed sieve shaker based on the spring stiffness can be proposed, as shown in Fig. 12. The stability condition is defined by a displacement line of the spring at 2.0 mm. Stable condition can be achieved if the displacement of the spring does not exceed the value of 2.02 mm. Here, the maximum designed displacement is 2 mm (decided by cam dimension). However, displacement less that 2 mm can occur due to wear of the cam or the roll follower.

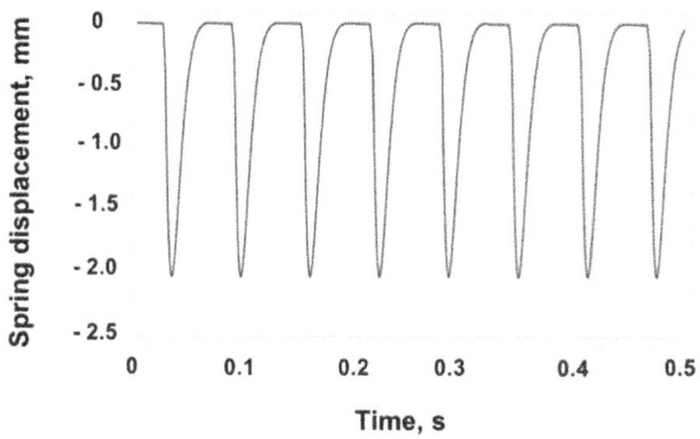

Fig. 10. Simulation results showing coil spring displacement for spring stiffness of 10 kN/m

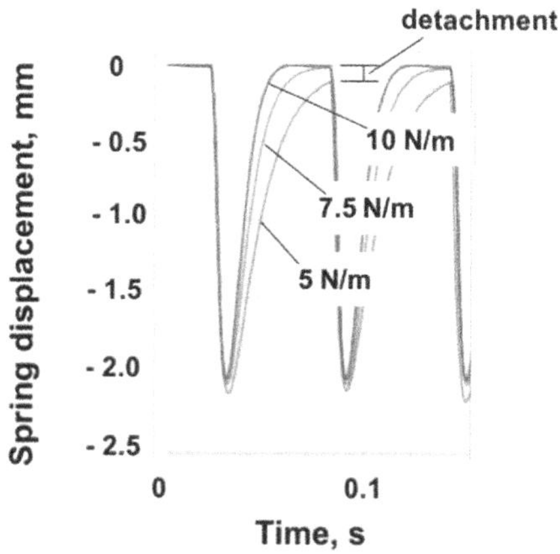

Fig. 11. Simulation results showing the effect of spring stiffness on the spring displacement

Figure 12 indicates that the minimum spring stiffness required to achieve stable condition is 10 kN/m. The spring stiffness lower than 10 kN/m causes dynamic instability to the system, triggered by impact of the cam and roll follower. A spring stiffness more that 10 kN/m can provide displacement of exactly 2.00 mm but excessively high spring stiffness will require higher motor power. Therefore, in this condition, a spring stiffness of 10 kN/m is considered optimum.

The dynamic stability simulation is conducted by taking the spring stiffness as parameter. However, the dynamic instability of the system can be affected by various other parameters such as mass, damping ratio, and cam rotation speed. Therefore, further analysis are necessary to investigate the effect of other parameters for the sieve stability.

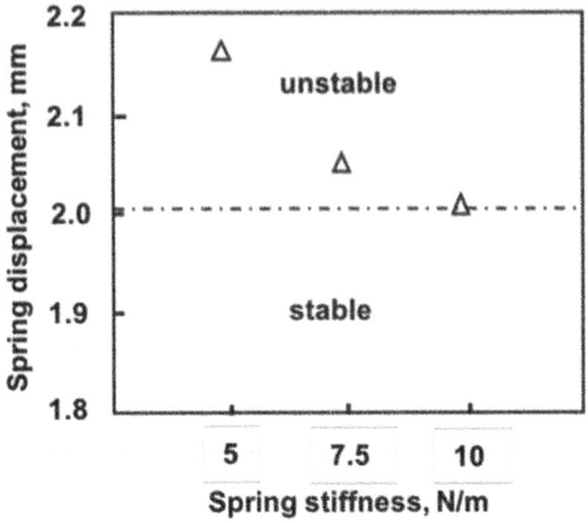

Fig. 12. Stability condition of the sieve shaker as a function of spring stiffness

4 Conclusion

The stability analysis of a sieve shaker used for a patchouli oil refining process was conducted. The stability simulation was conducted to determine the suitable spring stiffness value for the shaker. The results can be concluded as follows:

1. The shaker is stable for coil stiffness value more than 10 kN/m. At this value, the spring displacement does not exceed the stability criteria, i.e. spring displacement more than 2 + 0.2 mm.
2. The stability criteria is true for the parameter condition assumed, i.e. damping coefficient of 0.11 N.s/mm. Stability criteria for other value of damping coefficient has not been considered.
3. The stability analysis was conducted for one degree-of- freedom model. In order to have a more realistic result, further analysis should be conducted by considering three degree-of-freedom model.

Acknowledgment. This study is funded by Kementerian Pendidikan dan Kebudayaan Republik Indonesia through research scheme Penelitian Terapan Unggulan Perguruan Tinggi (PTUPT) 2021 with reference number: 301/SP2H/LT/DRPM/2021, which is highly acknowledged.

References

1. van Beek, T.A., Joulain, D.: The essential oil of patchouli, Pogostemon cablin: A review. Flavour Fragr. J. **33**(1), 6–51 (2018). https://doi.org/10.1002/ffj.3418
2. Ultra International BV. Patchouli Oil (Indonesia). http://ultranl.com/products/patchouli-oil-indonesia/. Accessed 2 Aug 2021

3. Kuntari, T.E., Purbaningtias, B., Wiyantoko, P., Kurniawati, D., Prasetyoko, Suprapto: Improving the quality of patchouli oil by adsorption process using surfactant modified of natural zeolite. In: AIP Conference Proceedings, vol. 1823 (2017), https://doi.org/10.1063/1.4978179
4. Liu, K.S.: Some factors affecting sieving performance and efficiency. Powder Technol. **193**(2) (2009). https://doi.org/10.1016/j.powtec.2009.03.027
5. Oladeji, A.O.: Design, construction and testing of a dry sand sieving machine. J. Appl. Sci. Environ. Manag. **16**(3), 241–243 (2012). https://doi.org/10.4314/jasem.v16i3
6. Simolowo, O.E., Adeniji, O.O.: Prototype design and performance analysis of a sieving machine for improved food-slurry processing in Nigeria. Indian J. Sci. Technol. **4**(12) (2011). https://doi.org/10.17485/ijst/2011/v4i12/30299

Author Index

Akhyar et al. (Eds.): ICECME 2021, LNME, pp. 373–374, 2023.
https://doi.org/10.1007/978-981-19-3629-6

The manufacturer's authorised representative in the EU is Springer Nature Customer Service Centre GmbH, Europaplatz 3, 69115 Heidelberg, Germany. If you have any concerns regarding our products, please contact ProductSafety@springernature.com

Printed and bound by CPI Group (UK) Ltd, Croydon, CR0 4YY
30/09/2026
02225238-0010